Mark Last, Piotr S. Szczepaniak, Zeev Volkovich, Abraham Kandel (Eds.)

Advances in Web Intelligence and Data Mining

Studies in Computational Intelligence, Volume 23

Editor-in-chief
Prof. Janusz Kacprzyk
Systems Research Institute
Polish Academy of Sciences
ul. Newelska 6
01-447 Warsaw
Poland
E-mail: kacprzyk@ibspan.waw.pl

Further volumes of this series can be found on our homepage: springer.com

Vol. 7. Bruno Apolloni, Ashish Ghosh, Ferda Alpaslan, Lakhmi C. Jain, Srikanta Patnaik (Eds.)
Machine Learning and Robot Perception, 2005
ISBN 3-540-26549-X

Vol. 8. Srikanta Patnaik, Lakhmi C. Jain, Spyros G. Tzafestas, Germano Resconi, Amit Konar (Eds.)
Innovations in Robot Mobility and Control, 2005
ISBN 3-540-26892-8

Vol. 9. Tsau Young Lin, Setsuo Ohsuga, Churn-Jung Liau, Xiaohua Hu (Eds.)
Foundations and Novel Approaches in Data Mining, 2005
ISBN 3-540-28315-3

Vol. 10. Andrzej P. Wierzbicki, Yoshiteru Nakamori
Creative Space, 2005
ISBN 3-540-28458-3

Vol. 11. Antoni Ligęza
Logical Foundations for Rule-Based Systems, 2006
ISBN 3-540-29117-2

Vol. 12. Jonathan Lawry
Modelling and Reasoning with Vague Concepts, 2006
ISBN 0-387-29056-7

Vol. 13. Nadia Nedjah, Ajith Abraham, Luiza de Macedo Mourelle (Eds.)
Genetic Systems Programming, 2006
ISBN 3-540-29849-5

Vol. 14. Spiros Sirmakessis (Ed.)
Adaptive and Personalized Semantic Web, 2006
ISBN 3-540-30605-6

Vol. 15. Lei Zhi Chen, Sing Kiong Nguang, Xiao Dong Chen
Modelling and Optimization of Biotechnological Processes, 2006
ISBN 3-540-30634-X

Vol. 16. Yaochu Jin (Ed.)
Multi-Objective Machine Learning, 2006
ISBN 3-540-30676-5

Vol. 17. Te-Ming Huang, Vojislav Kecman, Ivica Kopriva
Kernel Based Algorithms for Mining Huge Data Sets, 2006
ISBN 3-540-31681-7

Vol. 18. Chang Wook Ahn
Advances in Evolutionary Algorithms, 2006
ISBN 3-540-31758-9

Vol. 19. N. Ichalkaranje, A. Ichalkaranje, L.C. Jain (Eds.)
Intelligent Paradigms for Assistive and Preventive Healthcare, 2006
ISBN 3-540-31762-7

Vol. 20. Wojciech Penczek, Agata Półrola
Advances in Verification of Time Petri Nets and Timed Automata, 2006
ISBN 3-540-32869-6

Vol. 21 Cândida Ferreira
Gene Expression Programming: Mathematical Modeling by an Artificial Intelligence, 2006
ISBN 3-540-32796-7

Vol. 22. N. Nedjah, E. Alba, L. de Macedo Mourelle (Eds.)
Parallel Evolutionary Computations, 2006
ISBN 3-540-32837-8

Vol. 23. M.Last, P.S. Szczepaniak, Z. Volkovich, A. Kandel (Eds.)
Advances in Web Intelligence and Data Mining, 2006
ISBN 3-540-33879-9

Mark Last
Piotr S. Szczepaniak
Zeev Volkovich
Abraham Kandel
(Eds.)

Advances in Web Intelligence and Data Mining

Springer

Mark Last
Department of Information
Systems Engineering
Ben-Gurion University of the
Negev
Beer-Sheva 84105, Israel
E-mail: mlast@bgu.ac.il

Zeev Volkovich
Department of Software
Engineering
ORT Braude College
POB. 78,
21982 Karmiel, Israel
E-mail: vlvolkov@ort.org.il

Abraham Kandel
Department of Computer
Science and Engineering
University of South Florida
4202 E. Fowler Ave., ENB 118
Tampa, FL 33620, USA
E-mail: kandel@csee.usf.edu

Piotr S. Szczepaniak
Institute of Computer Science
Technical University of Lodz
ul.Wolczanska 215
93-005 Lodz, Poland
and
Systems Research Institute
Polish Academy of Sciences
ul. Newelska 6
01-447 Warsaw, Poland

ISSN print edition: 1860-949X
ISSN electronic edition: 1860-9503
ISBN 978-3-642-07045-7 e-ISBN 978-3-540-36810-6

Springer is a part of Springer Science+Business Media
springer.com

Printed in The Netherlands

Cover design: deblik, Berlin

Preface

Today, in the middle of the first decade of the 21st century, the Internet has become a major communication medium, where virtually any kind of content can be transferred instantly and reliably between individual users and entire organizations located in any part of the globe. The World Wide Web (WWW) has a tremendous effect on our daily activities at work and at home. Consequently, more effective and efficient methods and technologies are needed to make the most of the Web's nearly unlimited potential. The new Web-related research directions include intelligent methods usually associated with the fields of computational intelligence, soft computing, and data mining.

AWIC, the "Atlantic Web Intelligence Conferences" continue to be a forum for exchange of new ideas and novel practical solutions in this new and exciting area. The conference was born as an initiative of the WIC-Poland and the WIC-Spain Research Centers, both belonging to the Web Intelligence Consortium – WIC (http://wi-consortium.org/). Prior to this year, three AWIC conferences have been held: in Madrid, Spain (2003), in Cancun, Mexico (2004), and in Łódź, Poland (2005). AWIC 2006 took place in Beer-Sheva, Israel during June 5–7, 2006, organized locally by Ben-Gurion University of the Negev.

The book presents state-of-the-art developments in the field of computationally intelligent methods applied to various aspects and ways of Web exploration. Contributions cover such diverse Web applications as adaptive Web, conversational systems, electronic commerce, information retrieval, information security, recommender systems, user profiling/clustering, and Web design. The papers presented at the Second Workshop on Algorithmic Techniques for Data Mining (ATDM 2006), which was co-located with AWIC 2006, describe novel data mining algorithms for such popular data mining tasks as clustering, classification, and feature selection. The proposed data mining techniques can lead to more effective and intelligent Web-based systems.

All conference and workshop papers were selected after a peer-review process. The material published in the book is divided into two main parts: contributions of AWIC 2006 participants and ATDM 2006 Workshop papers. AWIC contributions and workshop papers are arranged in alphabetical order according to the name of the first author.

We deeply appreciate the effort of our plenary speaker, Prof. Janusz Kacprzyk (Systems Research Institute, Polish Academy of Sciences, Poland), and thank him for his presentation. We are indebted to the reviewers for their reliability and hard work done in a short time. We also highly appreciate the remarkable effort made by Ifat Zoltan, the Scientific Secretary of the conference, and the Public Relations Department of Ben-Gurion University. True thanks are also given to the series editor and to the Springer team for their friendly help. The technical cooperation of Ort Braude Academic College (Israel), National Institute for Applied Computational Intelligence (USA), and Technical University of Łódź (Poland) is highly appreciated.

Our hope is that the readers will find many inspiring ideas in this volume.

March 2006

Mark Last
Piotr S. Szczepaniak
Zeev Volkovich
Abraham Kandel

Organization

Organizers
Ben-Gurion University of the Negev, Israel
Ort Braude Academic College, Israel
National Institute for Applied Computational Intelligence, USA
Technical University of Lodz, Poland

Conference Chairs
Mark Last, Ben-Gurion University of the Negev, Beer-Sheva, Israel
Piotr S. Szczepaniak, Technical University of Lodz, Poland, and
Systems Research Institute, Polish Academy of Sciences, Warsaw, Poland

Steering Committee
Jesus Favela, CICESE, Mexico
Janusz Kacprzyk, Polish Academy of Sciences, Poland
Jiming Liu, Hong Kong Baptist University
Masoud Nikravesh, University of California, Berkeley, USA
Javier Segovia, UPM, Spain
Ning Zhong, Maebashi Institute of Technology, Japan

Advisory Committee
Abraham Kandel, University of South Florida, USA
Witold Pedrycz, University of Alberta, Canada
Zeev Volkovich, ORT Braude College, Israel

Scientific Secretary
Ifat Zoltan, Ben-Gurion University of the Negev, Israel

Program Committee

Ricardo Baeza-Yates
University of Chile, Chile

Patrick Brezillon
Universite Paris 6, France

Alex Buchner
University of Ulster, Northern Ireland, UK

Edgar Chavez
Universidad Michoacana, Mexico

Pedro A. da Costa Sousa
Uninova, Portugal

Alfredo Cuzzocrea
University of Calabria, Italy

Lipika Dey
Indian Institute of Technology, India

Santiago Eibe
UPM, Spain

Jesus Favela
CICESE, Mexico

Michael Hadjimichael
Naval Resarch Laboratory, USA

Enrique Herrera Viedma
Universidad de Granada, Spain

Pilar Herrero
UPM, Spain

Esther Hochsztain
ORT, Uruguay

Andreas Hotho
University of Karlsruhe, Germany

Janusz Kacprzyk
Systems Research Institute, Polish Academy of Sciences, Poland

Abraham Kandel
University of South Florida, USA

Samuel Kaski
Helsinki University of Technology, Finland

Jozef Korbicz
Technical University of Zielona Gora, Poland

Jacek Koronacki
Institute of Computer Science, Polish Academy of Sciences, Poland

Rudolf Kruse
Otto-von-Guericke University of Magdeburg, Germany

Mark Last
Ben-Gurion University of the Negev, Israel

Jiming Liu
Hong Kong Baptist University, Hong Kong

Vincenzo Loia
Univesrsity of Salerno, Italy

Aurelio Lopez
INAOE, Mexico

Oscar Marban
UPM, Spain

Oscar Mayora
ITESM-Morelos, Mexico

Ernestina Menasalvas
Technical University of Madrid, Spain

Bamshad Mobasher
DePaul University, USA

Manuel Montes-y-Gomez
INAOE, Mexico

Alex Nanolopoulos
Aristotle University, Greece

Marian Niedzwiedzinski
University of Lodz, Poland

Masoud Nikravesh
University of California, Berkeley, USA

Witold Pedrycz
University of Alberta, Canada

Maria Perez
UPM, Spain

Mario Piattini
Universidad Castilla-La Mancha, Spain

Paulo Quaresma
Universidade de Evora, Portugal

Victor Robles
UPM, Spain

Danuta Rutkowska
Technical University of Czestochowa, Poland

Leszek Rutkowski
Technical University of Czestochowa, Poland

Eugenio Santos
UPM, Spain

Javier Segovia
UPM, Spain

Andrzej Skowron
University of Warsaw, Poland

Roman Slowinski
Poznan University of Technology, Poland

Myra Spiliopoulou
University of Magdeburg, Germany

Ryszard Tadeusiewicz
AGH University of Science and Technology, Poland

Andromaca Tasistro
Universidad de la Republica, Uruguay

Kathryn Thornton
University of Durham, UK

Maria Amparo Vila
University of Granada, Spain

Zeev Volkovich
Ort Braude Academic College, Israel

Anita Wasilewska
Stony Brook New York University, USA

Jan Weglarz
Poznan University of Technology, Poland

Katarzyna Wegrzyn-Wolska
ESIGETEL, Avon-Fontainebleau, France

Ronald Yager
Iona College, USA

Yiyu Yao
University of Regina, Canada

Slawomir Zadrozny
Systems Research Institute, Polish Academy of Sciences, Poland

Ning Zhong
Maebashi Institute of Technology, Japan

Wojciech Ziarko
University of Regina, Canada

Plenary Speaker
Janusz Kacprzyk, Polish Academy of Sciences, Warsaw, Poland

Algorithmic Techniques for Data Mining (ATDM 2006)

Workshop Chairs
Mark Last, Ben-Gurion University of the Negev, Beer-Sheva, Israel
Zeev Volkovich, ORT Braude Academic College, Israel

ATDM 2006 Advisory Committee
Abraham Kandel, University of South Florida, USA
Zeev Barzily, ORT Braude Academic College, Israel
Leonid Morozensky, ORT Braude Academic College, Israel

ATDM 2006 Organizing Committee
Mark Last, Ben-Gurion University of the Negev, Israel
Zeev Volkovich, ORT Braude Academic College, Israel
Zeev Barzily, ORT Braude Academic College, Israel
Leonid Morozensky, ORT Braude Academic College, Israel

ATDM 2006 Program Committee

Zeev Barzily, ORT Braude Academic College, Israel
Efstratios Gallopoulos, University of Patras, Greece
Ehud Gudes, Ben-Gurion University of the Negev, Israel
Abraham Kandel, University of South Florida, USA
Jacob Kogan, Univ. of Maryland, Baltimore County, USA
Mark Last, Ben-Gurion University of the Negev, Israel
Oded Maimon, Tel Aviv University, Israel
Leonid Morozensky, ORT Braude Academic College, Israel
Charles Nicholas, Univ. of Maryland, Baltimore County, USA
Lior Rokach, Ben-Gurion University of the Negev, Israel
Volker Roth, ETH Zurich, Switzerland
Assaf Schuster, TECHNION - Israel Institute of Technology, Israel
Eyal Shimony, Ben-Gurion University of the Negev, Israel
Armin Shmilovich, Ben-Gurion University of the Negev, Israel
Marc Teboulle, Tel-Aviv University, Israel
Zeev Volkovich, ORT Braude Academic College, Israel
Gerhard-Wilhelm Weber, Middle East Technical University, Turkey

Table of Contents

Part 1

Part 2

DataRover: An Automated System for Extracting Product Information From Online Catalogs

Syed Toufeeq Ahmed, Srinivas Vadrevu, and Hasan Davulcu

Department of Computer Science and Engineering,
Arizona State University,
Tempe, AZ, 85287, USA
{toufeeq, svadrevu, hdavulcu}@asu.edu

Abstract. The increasing number of e-commerce Web sites on the Web introduces numerous challenges in organizing and searching the product information across multiple Web sites. This problem is further exacerbated by various presentation templates that different Web sites use in presenting their product information, and different ways of product information they store in their catalogs. This paper describes the DataRover system, which can automatically crawl and extract all products from online catalogs. DataRover is based on pattern mining algorithms and domain specific heuristics which utilize the navigational and presentation regularities to identify taxonomy, list-of-product and single-product segments within an online catalog. Next, it uses the inferred patterns to extract data from all such data segments and to automatically transform an online catalog into a database of categorized products. We also provide experimental results to demonstrate the efficacy of the DataRover.

1 Introduction

The advent of e-commerce has created a trend that has brought thousands of product catalogs online. Most of data-intensive shopping Web sites are made up of a combination of static and dynamic content, which is generated from an underlying database. Each of these data-intensive Web sites present their product information in different presentation templates with different schema. In order to effectively make use of this information, we need to organize it and make it searchable for effective mediation over the Web.

Information extraction from Web is a well-studied problem and related work can be categorized as wrapper development tools, semi-automated wrapper learning, ontology based approaches and template based automated algorithms. Wrappers [1, 2] are scripts that are created either manually or semi-automatically after analyzing the location of the data in the HTML pages. Wrappers tend to be brittle against variations and require maintenance and human intervention when the underlying Web sites change. Wrapper induction systems [3] generate extraction rules from semi-structured Web pages. These extraction rules can be

Mark Last et al. (Eds.): Advances in Web Intelligence and Data Mining (SCI) **23**, 1-10 (2006)
www.springerlink.com

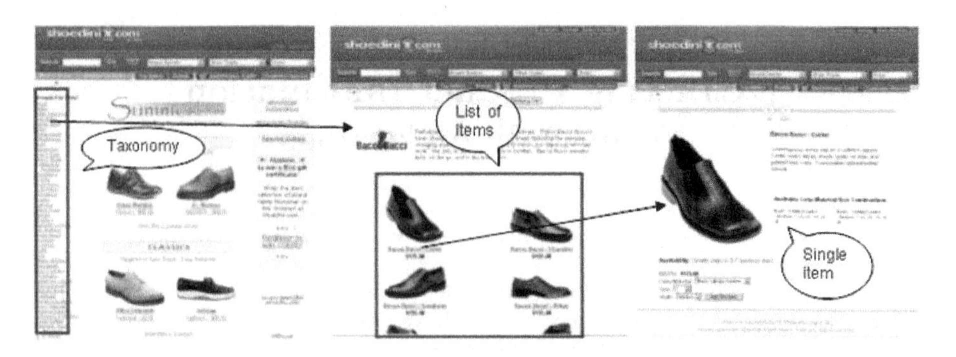

Fig. 1. An Example for a taxonomy-based data-intensive online catalog

applied on other new pages to extract the data. These systems utilize the natural language processing techniques and html tags to infer extraction patterns. But these wrapper induction systems require labeled training examples. The template based systems employ a strong bias on the expected presentation of items within a list of products segment, such as product descriptions should reside on a single line [4] and they may not have missing or repeating attributes [4].

Many of the shopping Web sites organize their content in a taxonomy of categories and present the instances of each category in a regular fashion. A "taxonomy-directed" Web site organizes its contents into a sequence of taxonomy segments which leads to either list of items pages or single item pages. Similarly, a list of items page might lead to a set of single item pages. DataRover is based on pattern mining algorithms and domain specific heuristics which utilize these navigational and presentation regularities to identify taxonomy, list-of-products segments and single-product pages within an online catalog. Next, it uses the inferred patterns to extract data from all such segments and to automatically turn a taxonomy-directed data-intensive catalog into a database of categorized products.

For attribute extraction from single-item pages, template learning algorithms [5, 6] were developed to separate the dynamic plug in values from the static template contents. The major difference between these template learning algorithms and the DataRover's algorithm is that: DataRover learns the path expressions of the plug in values from within the DOM trees by comparing and aligning item segments, where as RoadRunner [5] tokenizes, compares and aligns the HTML token sequences tag by tag. Our procedure of comparing segment by segment and then learning of requisite data paths requires only two examples and hence the learning is faster and also, the learned path expressions are more resilient against missing, repeated or reordered data values – where as such resilient grammar learning from token sequences requires that their learning algorithms should be presented with examples of all possible variations.

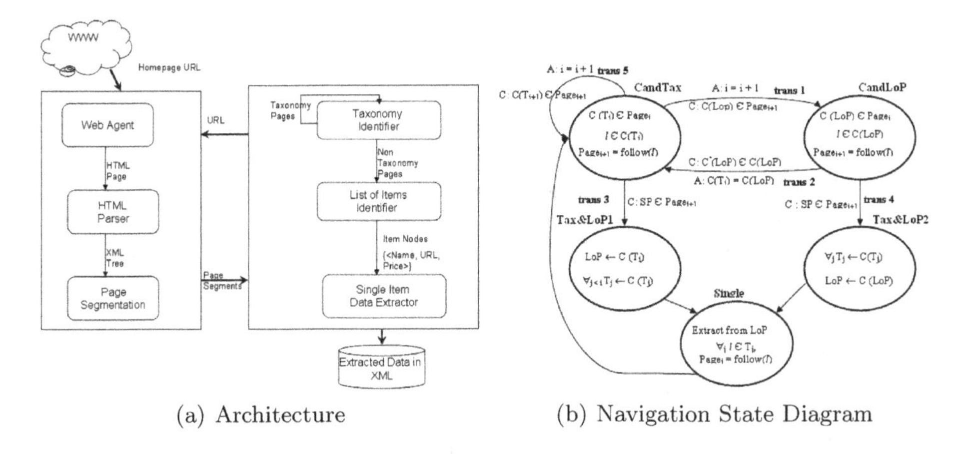

(a) Architecture (b) Navigation State Diagram

Fig. 2. Architecture and Navigation State Diagram of DataRover

For example, the Shoedini[1] Web site is an online catalog that sells shoes. As can be seen in Figure 1, all of its top level product categories are listed in a taxonomy segment in its home page, and the links within this segment leads to list of products pages and which leads to single item pages. This kind of structure is characteristic of many online catalogs.

Our contributions are three-fold as follows:

- A page segmentation algorithm that partitions the HTML page into logical segments based on its DOM representation
- A taxonomy detection and extraction algorithm that gathers the category information of the products
- A pattern mining algorithm to extract product information from single-item pages

Our pattern mining algorithm can find the individual products whenever they are presented regularly and together. It can accommodate some noise in the patterns in the form of optional characters in the regular expressions, allowing it to extract the products correctly even in the present of certain irregularities in the presentation. The rest of the paper is organized as follows. Section 2 presents an overview of the system and describes the navigation process for the DataRover's crawler. Section 3 describes its data extraction algorithms. Section 4 presents experimental results and Section 5 concludes the paper.

2 System Overview

DataRover is a fully automated system that extracts product records from data intensive catalogs, which organize their product categories using a taxonomy. The architecture of the system is as shown in the Figure 2(a). The input to

[1] `http://store.yahoo.com/shoedini`

DataRover is the home page of the catalog of interest and the output is a set of product records. Page segmentation component finds logical segments in a DOM tree of the Web page. It groups similar contiguous substructures in the Web pages into logical segments by detecting a high concentration of neighboring nodes with similar root-to-leaf tag paths. Each page segment is analyzed and depending on the type of the segment one of the following three components is employed: *Taxonomy Identifer* to extract category links from taxonomies, *List-of-Items Identifier* to recognize distinct products within a list of items segment, and *Single Item Data Extractor* to extract the single item information from within a single item segment.

Navigation State Diagram: The navigation state diagram of DataRover is described in Figure 2(b). Each state in the diagram corresponds to the DataRover's assumption about the type of current data segment. A data segment can be a candidate taxonomy (CandTax) or a candidate list of products (CandLoP) or a single item segment (Single). The labels marked with "C" on the edges correspond to the transition conditions about the type of segments found in the next page when a link is followed from the current segment. The labels marked with "A" correspond to state changes that should be made upon transitioning to the next state. State transitions within the diagram always occur as the DataRover crawls from a link within the current data segment to another data segment in the next page.

The navigation state diagram has five states labeled as "CandTax", "CandLop", "Tax&Lop1", "Tax&Lop2" and "Single". First, the state is initialized as "CandTax" which indicates that the current page i contains a candidate taxonomy, C(T). If a link l within a candidate taxonomy segment leads to a page which has a candidate list of products segment, C(Lop), then upon $trans1$, the current state is changed to "CandLop" and i is incremented. Alternatively, if a link from within the current C(Lop) segment leads to a single item segment, then upon $trans4$, the current segment becomes a list of products (Lop) segment and all the segments that were visited earlier become taxonomy segments. Upon finding a single item segment, the product information is extracted from the current list of products segment, as well as from all of the other reachable single item segments from the categories of previous taxonomies.

3 Segment Classification and Data Extraction

In this section, we present algorithms for classifying segments as candidate taxonomy, as candidate list of products, as single item segments and the data extraction algorithms from these types of segments.

The classification and extraction phase involves identifying candidate taxonomy, candidate list of items and single item segments and extracting data from those. Before a page can be processed, it has to be segmented using the following page segmentation algorithm.

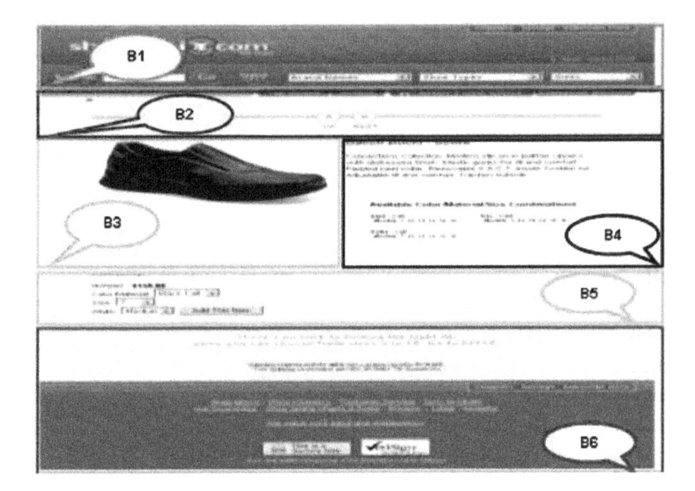

Fig. 3. Snapshot of a web page and its logical segments

3.1 Page Segmentation

Page Segmentation component identifies logical segments of a given web page. One way to achieve this is by grouping similar contiguous substructures within a Web page by detecting a high concentration of neighboring nodes with similar root-to-leaf tag paths. Consider the product page from online Yahoo Store Web site. Various logical segments of this Web page are marked by boxes in Fig 3. We do not discuss the details of the page segmentation component here since various approaches like the flat partitioner algorithm in [7, 8], record boundary discovery using an ontology in [9], and schema discovery heuristics in [10] can be used to perform this function. We have implemented and used the approach discussed in [7] for page segmentation.

3.2 Candidate Taxonomy Identification and Category Extraction

This component finds all taxonomy candidates C(T). The classification process is based on a weighted combination of a number of syntactic properties of the segment under consideration. The three syntactic features that are used are as follows:

- The ratio of total number of URLs to the total number of leaf nodes
- The ratio of maximum number of URL nodes with the same root-to-leaf tag path to the total number of segment URLs
- The ratio of maximum number of content similar URLs to the total number of URLs (two URLs are content similar if they differ only by the values of their dynamic page arguments)

Next, a weighted average of the above three ratios is computed. A segment is classified as a taxonomy candidate segment, only if this ratio is greater than an experimentally determined threshold value which is currently set to 0.65.

Algorithm 1 Pattern Finder Algorithm

Input: String *S* *Ouput: String P marked with patterns in the String S*

```
1: pat_list := φ; pat := φ
2: cur := S.first_element; found := true
3: while cur <> end_of(S) do
4:    c_items[] := get_cand_items(cur, S)
5:    if c_items <> φ then
6:       for ind := 1 to |c_items| do
7:          while ! aligned && comp ε c_items && comp <> c_items[ind] do
8:             cand_sig := PatternFinder(c_items[ind])
9:             comp_sig := PatternFinder(comp)
10:            alignment := align(cand_sig, comp_sig)
11:         end while
12:         if aligned then
13:            pat_list.add(cand_sig); pat_list.add(comp_sig)
14:         else
15:            found := false; break
16:         end if
17:      end for
18:      if found then
19:         pat := append(pat, pat_list.sort)
20:         cur := alignment.last_index()
21:      else
22:         pat := append(pat, S.element(cur));
23:         cur := cur + 1
24:      end if
25:   else
26:      pat := append(pat, S.element(cur));
27:      cur := cur + 1
28:   end if
29: end while
30: return pat
```

3.3 Identifying List of Products - Pattern Finder

A segment is classified as a candidate list of products segment if there is at least two consecutive products in it. The Pattern Finder algorithm in Algorithm 1 identifies all the consecutive product items and their boundaries. The algorithm works by detecting contiguous repeating structures inside a candidate segment. Since, the standard algorithms to learn regular expressions that might capture product sequences require *large* sets of *labeled* examples [11] and the seminal work of Gold [12] showed that the problem of inferring a DFA of minimum size from positive examples is NP-complete we have developed a heuristic algorithm that seems to work well within the list of product segments. In order to detect such repeating structures first we recursively identify and standardize the representation of all the repeating sub-pattern structures. Next, we identify the set of all recurring substructures within a candidate item and then we obtain a standardized repeating pattern signature by appending the repeating patterns in lexicographical order. Once standardized repeating pattern signatures are identified, a simple alignment based test which checks if an item's signature is a subsequence of another item's signature can be used to detect repetition.

Each unique attributed root-to-leaf tag path within the segment is labeled with a unique identifier. Then, these identifiers are concatenated to form a path-sequence. For example, for the "list of products" page from http://www.walmart.com, the products are marked within dotted lines as shown in Figure 4. The constructed "path-sequence" for this example is "ABBCDE-ABBFCCDEABBFFCDEABBFCCDE".

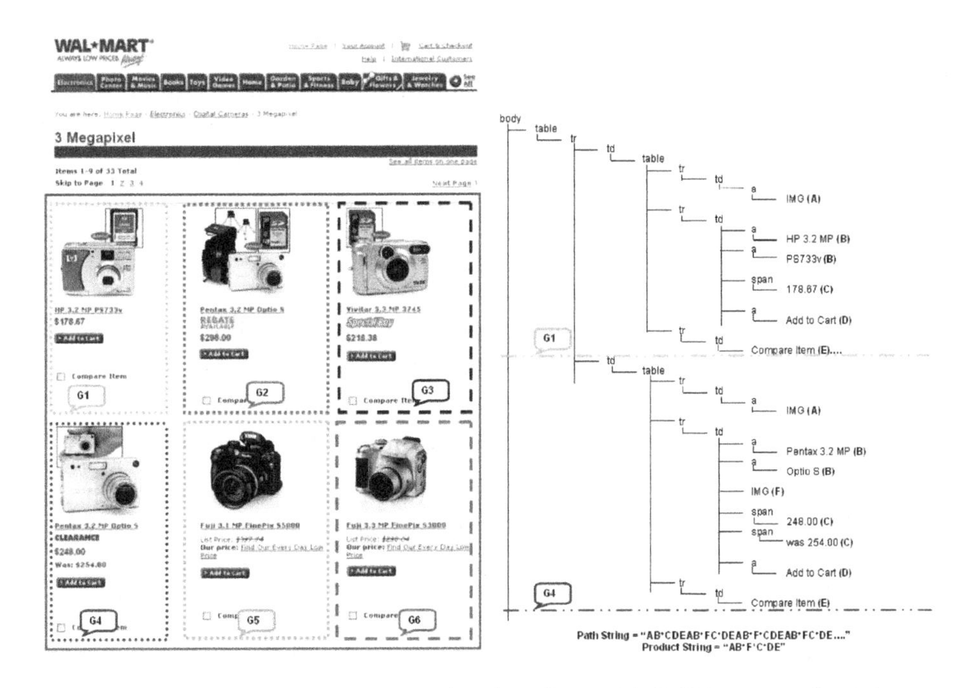

Fig. 4. Snapshot of Walmart.com Product Page with DOM tree on the right

In order to demonstrate the algorithm we will use the path-sequence "ABCBGCBGCDABCBHCBCDABCD" First A is identified as a product header (*cur* in Line 2) and the above sequence is partitioned into candidate product sequences "ABCBGCBGCD", "ABCBHCBCD", "ABCD" (in Line 4). In Lines 5-20, the algorithm obtains the set of recurring patterns within every candidate product sequence as follows: for the candidate product "ABCBHCBCD" the repeating patterns are "BC", "BHC" and hence the product pattern signature is identified as "$A(BC)^*BHCD$". In Line 19, the current set of repeating patterns are lexicographically sorted and appended to the current prefix and the current header is advanced to be next identifier right after the repeating

Domain	Total Number of Web pages	Taxonomy counts in home pages				Total Number of Categories	Extracted Product Precision	Extracted Product Recall
		Tax Count	Tax Found	Category Count	Category Found			
www.backcountrystore.com	46	4	4	64	47	19	51%	81%
www.etronics.com	53	2	2	24	18	14	42%	89%
www.compactappliance.com	150	7	7	34	34	19	69%	81%
www.rugsusa.com	225	1	1	21	20	730	51%	46%
www.cooking.com	297	3	3	49	49	80	95%	75%
www.drugstore.com	361	3	3	71	71	85	100%	86%
www.basspro.com	386	10	10	64	64	80	98%	74%
www.shoes.com	421	4	4	93	93	15	93%	88%
www.overstock.com	426	9	9	73	73	12	97%	90%
www.officedepot.com	473	9	9	87	87	34	85%	83%
www.hammacher.com	518	4	4	42	30	60	72%	80%
www.walmart.com	539	6	6	67	56	70	96%	92%
www.target.com	585	4	4	69	69	68	94%	89%
www.kmart.com	610	5	5	49	49	172	97%	81%
www.stacksandstacks.com	722	4	4	44	44	92	97%	73%
www.homevisions.com	930	1	1	51	51	127	93%	77%
www.boscovs.com	1102	3	3	18	18	212	98%	82%
www.smartbargains.com	1207	4	4	95	93	410	92%	67%
www.zappos.com	1243	3	3	170	170	420	93%	77%
Average							**82%**	**79%**

Table 1. Experimental results for the DataRover system for product extraction on various Web sites.

patterns. At the end, the prefix is returned as the signature. The complexity of the algorithm is $O(n^3)$ where n is the length of the input string.

3.4 Identification and Extractom from Single Item Pages

A candidate list of items segment is classified as a list of items segment if we follow two sample links and we can identify matches for a price with identical root-to-leaf tag paths. Then, target pages also become single item pages.

The data extraction from single item pages is based on the instance extraction algorithm discussed in [7]. Any two single item pages are segmented using the page segmentation algorithm presented in Section 3.1 and the segments of both pages are aligned, based on their content similarity. The dissimilar segments denote the dynamic content areas where as the similar segments denote the template regions (such as header, footers, navigation bars) of the page. Root to leaf HTML tag paths of all leaf nodes within the content areas are used to extract the product details from all the remaining single item pages. Finally the extracted data is labeled based on the schema <name, price, image, URL, description> by choosing them as the closest ones to the product itself.

4 Experimental Results

We used 19 online catalogs (E-commerce Web sites) to test the DataRover system for precision and recall. Online catalogs selected have all different templates, and provide an excellent diversity to test the system. The experimental results are

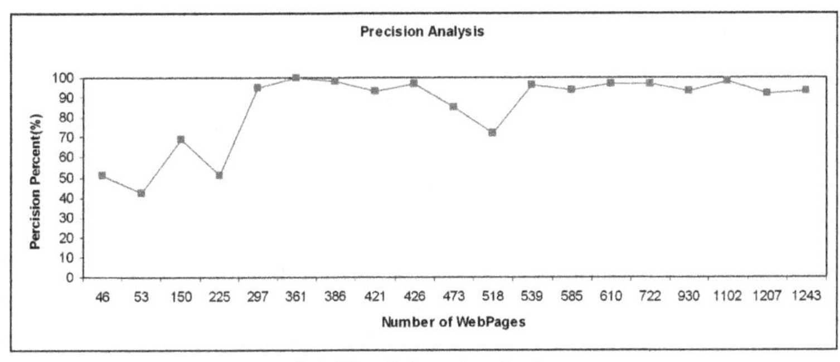

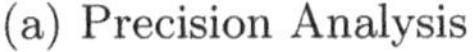
(a) Precision Analysis

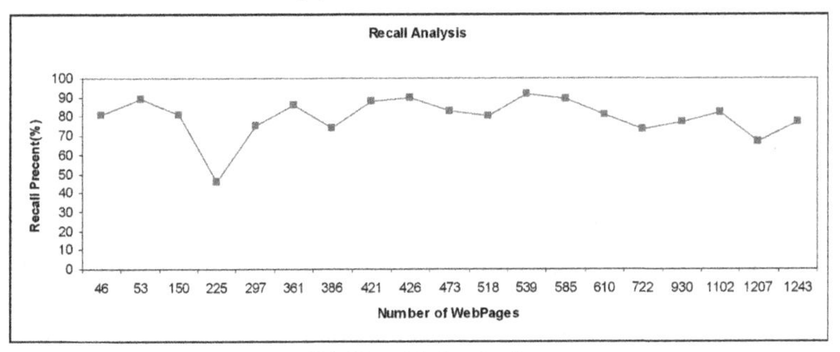

(b) Recall Analysis

Fig. 5. Performance of DataRover system with respect to the number of Web pages in the domain.

presented in Table 1 and Figures 4 (a) and (b). The results strongly show that our algorithms were able to extract the product information accurately except in few cases.

Precision Analysis: Precision of the system is high whenever number of pages in the domain are high. As seen in Figure 4(a), when the number of pages are more than 300, the precision seems to be at median 95%. As most of the E-Commerce Web sites have high number of pages, our algorithm will perform satisfactorily.

Recall Analysis: Recall of the system is around 80-85% for most of the Web sites. As the number of pages increase, there is a slight drop in recall. One of the possible reasons for this is the misclassification of certain segments within the single product pages as a header segment. This was because we select top two product links which happened to have identical product names and prices. Hence, during alignment these segments were identified as part of the header segment in the single item page. This problem can be solved by always selecting products with different prices and items names, if they exist. By omitting the tag-path information from our syntactic features and adding a more robust feature we hope to arrive at a better candidate taxonomy identifier.

5 Conclusions

We presented the DataRover system that can crawl and automatically extract the individual products from taxonomy-directed data-intensive online catalogs. The experimental results with various online shopping Web sites indicate that our system is able to extract the product information with 82% precision and 79% recall. By upgrading to a better parser and with an improved taxonomy identifier we hope improve our extraction accuracy from all catalogs. DataRover utilizes heuristic rules and pattern mining algorithms to discover the structural regularities among taxonomy segments, list of items segments and single item pages and uses these regularities and mined patterns to transform the online catalogs into a database of categorized items without the need for user interaction or the wrapper maintenance burden.

References

1. J. Hammer, H. Garcia-Molina, S. Nestorov, R. Yerneni, M. M. Breunig, and V. Vassalos. Template-based wrappers in the tsimmis system. In *ACM SIGMOD*, 1997.
2. Gustavo O. Arocena and Alberto O. Mendelzon. Weboql: Restructuring documents, databases, and webs. In *ICDE*, pages 24–33, 1998.
3. Nickolas Kushmerick, Daniel S. Weld, and Robert B. Doorenbos. Wrapper induction for information extraction. In *Intl. Joint Conference on Artificial Intelligence (IJCAI)*, pages 729–737, 1997.
4. Robert B. Doorenbos, Oren Etzioni, and Daniel S. Weld. A scalable comparison-shopping agent for the world-wide web. In W. Lewis Johnson and Barbara Hayes-Roth, editors, *Proceedings of the First International Conference on Autonomous Agents (Agents'97)*, pages 39–48, Marina del Rey, CA, USA, 1997. ACM Press.
5. Valter Crescenzi, Giansalvatore Mecca, and Paolo Merialdo. Roadrunner: Towards automatic data extraction from large web sites. In *Intl. Conf. on Very Large Data Bases*, 2001.
6. A. Arasu and H. Garcia-Molina. Extracting structured data from web pages. In *ACM SIGMOD*, 2003.
7. Hasan Davulcu, Srinivas Vadrevu, and Saravanakumar Nagarajan. Ontominer: Bootstrapping and populating ontologies from domain specific web sites. *IEEE Intelligent Systems*, 18(5), September 2003.
8. Hasan Davulcu, Sukumar Koduri, and Saravanakumar Nagarajan. Datarover: A taxonomy based crawler for automated data extraction from data-intensive web sites. In *Proceedings of the ACM International Workshop on Web Information and Data Management*, pages 9–14, 2003.
9. D. W. Embley, Y. Jiang, and Y.-K. Ng. Record-boundary discovery in Web documents. pages 467–478, 1999.
10. Christina Yip Chung, Michael Gertz, and Neel Sundaresan. Reverse engineering for web data: From visual to semantic structures. In *Intl. Conf. on Data Engineering*, 2002.
11. R. C. Berwick and S. Pilato. Learning syntax by automata induction. In *Machine Learning 2*, pages 9–38, 1987.
12. E. Mark Gold. Complexity of automaton identification from given sets. In *Information and Control*, pages 37:302–320, 1978.

A New Path Generalization Algorithm for HTML Wrapper Induction

Costin Bădică[1], Amelia Bădică[2], and Elvira Popescu[1]

[1] University of Craiova, Software Engineering Department
Bvd.Decebal 107, Craiova, RO-200440, Romania
{badica_costin, elvira_popescu}@software.ucv.ro

[2] University of Craiova, Business Information Systems Department
A.I.Cuza 13, Craiova, RO-200585, Romania
ameliabd@yahoo.com

Summary. Recently it was shown that Inductive Logic Programming can be successfully applied to data extraction from HTML. However, the approach suffers from two problems: high computational complexity with respect to the number of nodes of the target document and to the arity of the extracted tuples. In this note we address the first problem by proposing an efficient path generalization algorithm for learning rules to extract single information items. The presentation is supplemented with a description of a sample experiment.

1 Introduction

The Web was originally designed as a major information provider for the human consumer, but the interest has rapidly shifted to make that information available for machine consumption. For example, Web directories and search engines are Web applications that are capable of providing useful information upon request to individual users, businesses or software agents.

However, despite the fact that technologies have been put forward to enable automated processing of information published on the Web (semantic markup, Web services), most of the practices in Web publishing are still being based on the combination of traditional HTML – *lingua franca* for Web publishing, with server-side dynamic content generation from databases. Moreover, many Web pages are using HTML elements that were originally intended for use to structure content (e.g. elements related to tables), for layout and presentation effects, even if this practice is not encouraged in theory. Therefore, automatic information extraction from documents published on the Web has attracted a lot of researches during the last decade and this interest is expected to grow, as the Web is also growing in both size and complexity.

Information extraction is concerned with locating and extracting specific values in documents, and then using them to populate a database or structured document. The information extraction research community has proposed a quite large variety of machine learning techniques for automatic information extraction ([7]).

Mark Last et al. (Eds.): Advances in Web Intelligence and Data Mining (SCI) **23**, 11-20 (2006)
www.springerlink.com

Inductive Logic Programming is one of the success stories in the application area of wrapper induction for information extraction ([1, 2, 4, 7]). However, this approach suffers from two problems: high computational complexity with respect to the number of nodes of the target document and to the arity of the extracted tuples. In this paper we address the first problem by proposing a path generalization algorithm for learning rules to extract single information items (a task similar to [1]). The algorithm produces an XPath ([10]) extraction path from positive examples and is proven to have good computational properties. The presentation is supplemented with a detailed description of a sample experiment that shows how the technique performs in practice on real Web pages.

We proceed as follows. In section 2 we define extraction paths. In section 3 we describe an algorithm for learning extraction paths. In section 4 we show how extraction paths can be translated to XPath. In section 5 we describe an experiment showing our technique at work on a real Web site. In section 6 we present researches connected to our work. Last section concludes and points to future work.

2 Extraction Path

We model well-formed HTML documents as labeled ordered trees. An extraction path takes a labeled ordered tree and returns a subset of extracted nodes. An extracted node can be viewed as a subtree rooted at that node. The node labels of a labeled ordered tree correspond to tags in HTML texts. Let Σ be the set of all node labels of a labeled ordered tree.

For our purposes, it is convenient to abstract labeled ordered trees as sets of nodes on which certain relations and functions are defined. Figure 1 shows a labeled ordered tree with 25 nodes and tags in the set $\Sigma = \{a, b, c\}$.

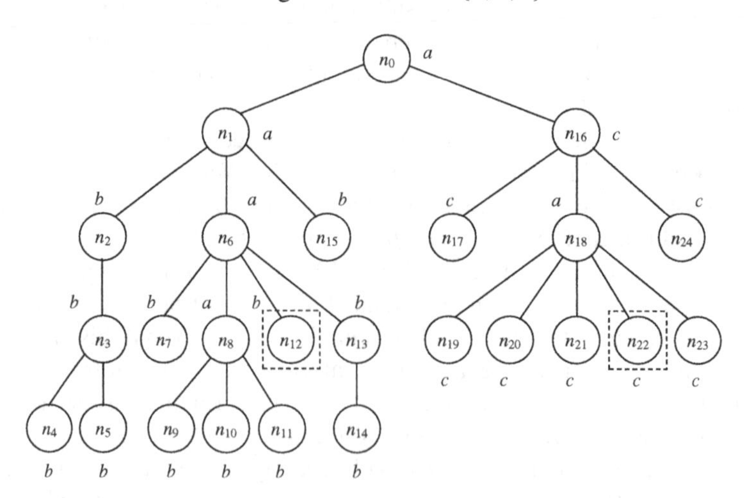

Fig. 1. Document as labeled ordered tree

An *extraction path* is a labeled directed graph. Arc labels denote conditions that specify the tree delimiters of the extracted information, according to parent-child and next-sibling relationships (eg. is there a parent node ?, is there a left sibling ?, a.o). Vertex labels specify conditions on nodes (eg. is the tag label *td* ?, is it the first child ?, a.o). A special vertex of this graph is used for selecting the nodes for extraction.

Intuitively, an arc labeled $'n'$ denotes the "next-sibling" relation while an arc labeled $'c'$ denotes the "parent-child" relation. As concerning vertex labels, label $'f'$ denotes "first child" condition, label $'l'$ denotes "last child" condition and label $\sigma \in \Sigma$ denotes "equality with tag σ" condition.

Note that we use the term'node' when referring to document trees and the term 'vertex' when referring to the graph of an extraction path.

Definition 1. *(Extraction path) An* extraction path *is a labeled directed graph that is described as a list* $[t_0, t_1, \ldots, t_k]$, $k \geq 0$ *with the following properties:*

1. *Each element* t_i, $0 \leq i \leq k$ *is a list* $[v_{-l}, \ldots, v_{-1}, v_0, v_1, \ldots, v_r]$, $l \geq 0$, $r \geq 0$ *such that: i)* v_i, $-l \leq i \leq r$ *are vertices; ii)* (v_i, v_{i+1}), $-l \leq i < r$ *are arcs labeled with* $'n'$, *and iii) for each pair of adjacent lists* t_i, t_{i+1}, $1 \leq i < k$ *in the extraction path,* (v_0^{i+1}, v_0^i) *is an arc labeled with* $'c'$.
2. *Vertex labels are defined as: i) if* $l > 0$ *then* v_{-l} *is labeled with a subset of* $\{'f', \sigma\}$, $\sigma \in \Sigma$; *ii)* v_i, $-l < i < 0$ *is labeled with a subset of* $\{\sigma\}$, $\sigma \in \Sigma$; *iii) if* $r > 0$ *then* v_r *is labeled with a subset of* $\{'l', \sigma\}$, $\sigma \in \Sigma$; *ii)* v_i, $1 < i < r$ *is labeled with a subset of* $\{\sigma\}$, $\sigma \in \Sigma$; *v) If* $l, r > 0$ *then* v_0 *is labeled with a subset of* $\{\sigma\}$, $\sigma \in \Sigma$; *if* $l = 0, r > 0$ *then* v_0 *is labeled with a subset of* $\{'f', \sigma\}$, $\sigma \in \Sigma$; *if* $l > 0, r = 0$ *then* v_0 *is labeled with a subset of* $\{'l', \sigma\}$, $\sigma \in \Sigma$; *if* $l = r = 0$ *then* v_0 *is labeled with a subset of* $\{'f', 'l', \sigma\}$, $\sigma \in \Sigma$.
3. *Vertex* v_0^k *(i.e.* v_0 *in list* t_k*) is matched against the extraction node and consequently is called* extraction vertex.

It is not hard to see that an extraction path models a conjunctive query of unary and binary conditions on a labeled ordered tree ([6]). Actually, an extraction path is a special kind of extraction pattern of arity 1 according to [3].

Figure 4a shows an extraction path. The extraction vertex is marked with a small arrow (vertex C in figure 4a). A node is extracted by this path if it has the following properties: i) it has two preceding left siblings; ii) it has one following right sibling that is the last child of its parent node; iii) it has a parent labeled with a; iv) its parent has a following right sibling that is the last child of its parent node; v) its parent has a preceding left sibling that is the first child of its parent node; vi) it has a grand-parent; vii) it has a grand-grand-parent labeled with a that is the unique child of its parent.

Consider an extraction path $p = [t_0, t_1, \ldots, t_k]$. For a list $t_i = [v_{-l}, \ldots, v_{-1}, v_0, v_1, \ldots, v_r]$ let $left(t_i) = l$ and $right(t_i) = r$. The following definition introduces height, together with left and right widths of an extraction path.

Definition 2. *(Height and widths of an extraction path) Let* $p = [t_0, t_1, \ldots, t_k]$ *be an extraction path.*

1. *The value* $height(p) = k$ *is called the* height *of* p.

2. *The value* $left(p) = \max_{i=0}^{k} left(t_i)$ *is called the* left width *of p. The value* $right(p) = \max_{i=0}^{k} right(t_i)$ *is called the* right width *of p.*

In practice it is useful to limit the height and the widths of an extraction path, yielding a *bounded extraction path.*

Definition 3. *(Bounded extraction paths) Let* H, L, R *be three positive integers. An extraction path* $p = [t_0, t_1, \ldots, t_k]$ *is called* (H, L, R)-bounded *if* $height(p) \leq H$, $left(p) \leq L$ *and* $right(p) \leq R$.

Note that the extraction path shown in figure 4a is a $(3, 2, 1)$-bounded extraction path. Moreover, if we restrict $H = 2$ and $L = 1$, then nodes I and A will be pruned resulting a $(2, 1, 1)$-bounded extraction path, that obviously, is less constrained than the initial path.

3 A Path Generalization Algorithm

The practice of Web publishing assumes dynamically filling-in HTML templates with structured data taken from relational databases. Thus, we can safely assume that a lot of Web data is contained in sets of documents that share similar structures. Examples of such documents are: search engines result pages, product catalogues, news sites, product information sheets, travel resources, etc.

We consider a Web data extraction scenario which assumes the manual execution of a few extraction tasks by the human user. An inductive learning engine could then use the extracted examples to learn a general extraction rule that can be further applied to the current or other similar Web pages.

Usually the extraction task is focused on extracting similar items (like book titles in a library catalogue or product features in a product information sheet). One approach to generate an extraction rule from a set of examples is to discover a common pattern of their neighboring nodes in the tree of the target document.

In what follows we discuss an algorithm that takes: i) an XML document (possibly assembled from more Web pages, previously converted to XHTML) modeled as a labeled ordered tree t; ii) a set of example nodes $\{e_1, e_2, \ldots, e_n\}$; iii) three positive integers H, L, R and produces an (H, L, R)-bounded extraction path p that generalizes the set of input examples. Intuitively, this technique is guaranteed to work if we assume that semantically similar items will exhibit structural similarities in the target Web document. This is a feasible assumption for the case of Web documents that are generated on-the-fly by filling-in HTML templates with data taken from databases. Moreover, based on experimental results recorded in previous work ([1, 2]), we have noticed that in practice an extraction rule only needs to check the proximity of nodes. This explains why we focused on the task of learning bounded extraction paths.

The basic operation of the learning algorithm is the generalization operator of two extraction paths. This operator takes two extraction paths p_1 and p_2 and produces an appropriate extraction path p that generalizes p_1 and p_2.

The idea of the learning algorithm is as follows. For each example node we generate a bounded extraction path (of given input parameters H, L, R) by following sibling and parent links in the document tree. We initialize the output path with the first extraction path and then we proceed by iterative application of the generalization operator to the current output path and the next example extraction path, yielding a new output path. The result is a bounded extraction path that represents an appropriate generalization of the input examples.

The generalization of two paths assumes the generalization of their elements, starting with the elements containing the extraction vertices and moving upper level by level. The generalization of two levels assumes the generalization of each pair of corresponding vertices, starting with vertices with index 0 and moving to left and respectively to right in the lists of vertices. Generalization of two vertices is as simple as taking the intersection of their labels. The algorithm is shown in figure 2.

```
LEARN(p1,...,pn,n)
1. p ← p1
2. for i = 2,n do
3.     p ← GEN-PATH(p,pi)
4. return p

GEN-PATH(p1,p2)
1. let p1 = [t^1_0,...,t^1_k1]
2. let p2 = [t^2_0,...,t^2_k2]
3. k ← min(k1,k2)
4. i ← k, i1 ← k1, i2 ← k2
5. while i ≥ 0 do
6.     ti ← GEN-LEVEL(t^1_i1, t^2_i2)
7.     i ← i − 1, i1 ← i1 − 1, i2 ← i2 − 1
8. return p = [t0,...,tk]

GEN-LEVEL(t1,t2)
1. let t1 = [v^1_-l1,...,v^1_-1,v^1_0,...,v^1_r1]
2. let t2 = [v^2_-l2,...,v^2_-1,v^2_0,...,v^2_r2]
3. l ← min(l1,l2)
4. r ← min(r1,r2)
5. for i = 0,r do
6.     vi ← GEN-VERTEX(v^1_i, v^2_i)
7. for i = 1,l do
8.     v-i ← GEN-VERTEX(v^1_-i, v^2_-i)
9. return t = [v-l,...,v-1,v0,...,vr]

GEN-VERTEX(v1,v2)
1. let λ1 be the label of v1
2. let λ2 be the label of v2
3. λ ← λ1 ∩ λ2
4. return node v with label λ
```

Fig. 2. Path generalization algorithm

Function LEARN generalizes the extraction paths of the example nodes. We assume that paths $p_1, \ldots, p_n$ are generated as bounded extraction paths before function LEARN is called. Function GEN-PATH takes two extraction paths p_1, p_2 and computes the generalized path p. Function GEN-LEVEL takes two lists of vertices t_1 and t_2 that are members of the extractions paths and computes a generalized list t that is member of the generalized path. Function GEN-VERTEX takes two vertices v_1, v_2 and computes a generalized vertex v.

It is easy to see that the execution of algorithm LEARN takes time $O(n \times H \times (L+R))$ because GEN-VERTEX takes time $O(1)$, GEN-LEVEL takes time $O(L+R)$ and GEN-PATH takes time $O(H \times (L+R))$. Note also that if we set $H = L = R = \infty$ then the complexity of the algorithm is $O(n \times H^* \times W^*)$ where H^* and W^* are the height and the width of the target document tree.

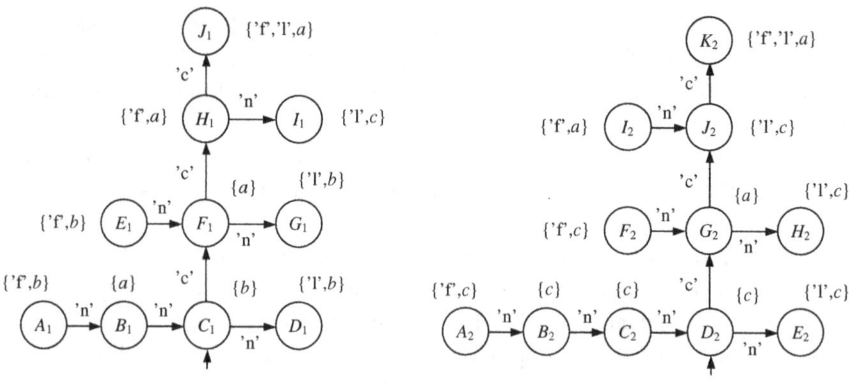

a.Extraction path for example node n_{12} b.Extraction path for example node n_{22}

Fig. 3. Extraction paths for example nodes from figure 1

Consider again the labeled ordered tree shown in figure 1 and the example nodes marked with dashed rectangles (n_{12} and n_{22}). The extraction paths corresponding to these nodes are shown in figure 3. The result of applying the generalization algorithm on those paths is shown in figure 4a.

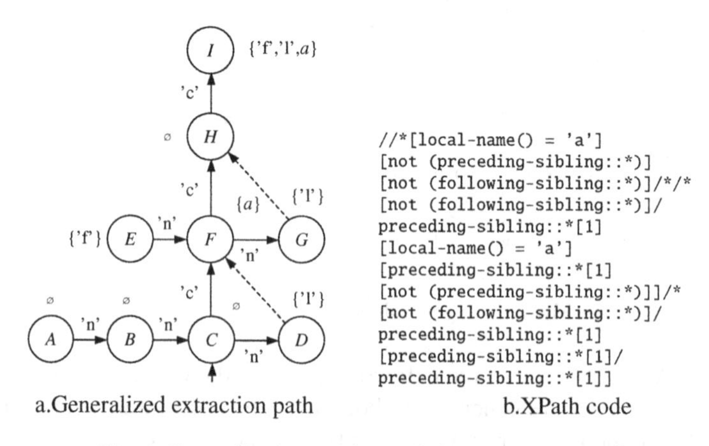
//*[local-name() = 'a']
[not (preceding-sibling::*)]
[not (following-sibling::*)]/*/*
[not (following-sibling::*)]/
preceding-sibling::*[1]
[local-name() = 'a']
[preceding-sibling::*[1]
[not (preceding-sibling::*)]]/*
[not (following-sibling::*)]/
preceding-sibling::*[1]
[preceding-sibling::*[1]/
preceding-sibling::*[1]]

a.Generalized extraction path b.XPath code

Fig. 4. Generalized extraction path for example from figure 1

4 Translating Extraction Paths to XSLT

An extraction path can be translated to an XPath query. The XPath query can be embedded into an XSLT stylesheet ([5]) to finally extract the information and store it into a database or another structured document ([3]).

Figure 5 shows an algorithm for translating an extraction path into an XPath query. The translation algorithm takes an extraction path $p = [t_0, \ldots, t_k]$ and explores it starting with t_0 and moving to t_k. For each element t_i, $0 \leq i \leq k$, the algorithm maps t_i to a piece of the output XPath query. Actually the algorithm takes the following route of vertices: $v^0_{r_0} \to \ldots \to v^0_0 \to v^1_{r_1} \to \ldots \to v^1_0 \to \ldots v^k_{r_k} \to \ldots \to v^k_0$. Note that when moving from element i to element $i+1$, $0 \leq i < k$, the algorithm takes the route $v^i_0 \to v^{i+1}_{r_{i+1}}$ (opposite direction of dotted arrows in figure 4a) rather than the route $v^i_0 \to v^{i+1}_0$. For each vertex v^i_0, $0 \leq i \leq k$, the algorithm also generates a condition that accounts for their left siblings by taking the route $v^i_0 \to v^i_{-1} \to \ldots \to v^i_{-l_i}$. It is easy to see that if p is an (H, L, R)-bounded extraction path then the time complexity of the translation algorithm PATH-TO-XPATH is $O(H \times (L + R))$.

Fig 4b shows the result of applying this algorithm to the extraction path from figure 4a. The algorithm will explore the following route of vertices: $I \to H \to G \to F \to D \to C$. For each vertex the algorithm generates a location step comprising an axis specifier, a node test and a sequence of predicates written between '[' and ']'. The node test is always *. The axis specifier is determined by the relation of the current vertex with its preceding vertex on the route explored by the translation algorithm. For example, the axis specifier that is generated for vertex F is `preceding-sibling::`. In this later case, an additional predicate `[1]` that constraints the selection of exactly the preceding node, is added. The algorithm also generates a predicate for each element of the label of a vertex. For example, predicate `[local-name() = 'a']` is generated for vertex F, that checks the node tag, and predicate `[not (following-sibling::*)]` is generated for vertex G, that checks if the matched node is the last child of its parent node. Moreover, for vertices F and C the algorithm generates an additional predicate that accounts for their left siblings E (of F) and respectively $B \to A$ (of C). For example, additional predicate `[preceding-sibling::*[1][not (preceding-sibling::*)]]` is generated for vertex F. This predicate checks if the document node matched by vertex F has a predecessor and if the predecessor is the first child of its parent node.

Note that running the XPath query from figure 4b on the labeled ordered tree from figure 1 produces the following two answers `/a[1]/a[1]/a[1]/b[3]` and `/a[1]/c[1]/a[1]/c[4]` that correspond to nodes n_{12} and n_{22}.

```
PATH-TO-XPATH(p)
1. let p = [t_0, ..., t_k]
2. xp ← "/"
3. for i = 0, k do
4.     let t_i = [v_{-l}, ..., v_{-1}, v_0, ..., v_r]
5.     t ← t_i
6.     xp ← xp+ "/*" + COND(v_r)
7.     for j = r − 1, 0 do
8.         xp ← xp+ "/preceding-sibling::*[1]" +
8.             COND(v_j)
9.     if l > 0 then
10.        xp ← xp+ "[preceding-sibling::*[1]" +
11.            COND(v_{-1})
11.        for j = 2, l do
12.            xp ← xp+ "/preceding-sibling::*[1]" +
13.                COND(v_{-j})
13.        xp ← xp+ "]"
14.return xp
```

```
COND(v)
1. let λ be the label of v
2. xc ← ""
3. if there is σ ∈ λ ∩ Σ then
4.     xc ← xc+
5.         "[local-name()=σ]"
5. if there is 'f' ∈ λ ∩ Σ then
6.     xc ← xc+
7.         "[not (preceding-sibling::*)]"
7. if there is 'l' ∈ λ ∩ Σ then
8.     xc ← xc+
9.         "[not (following-sibling::*)]"
9. return xc
```

Fig. 5. Algorithm for translating an extraction path into XPath

5 Experiment

We performed a simple (but realistic) experiment using data harvested from the Expedia Web site with the task of extracting hotel names. We followed a process consisting of the following stages: page collection, pre-processing, manual information extraction, conversion to the input format of the learning program, learning, wrapper compilation, wrapper execution ([2]).

We selected a sample page set of 50 pages containing all the results we got by searching Expedia for hotels in Paris (see figure 6). This set contained 1248 hotels, 25 hotels per page (excepting last page with only 23 hotels). We converted each page to XHTML using the Tidy program and then we assembled all these files into a single XML file by concatenating them under a new root element. The resulting file had about 29 Mb comprising a total of 191816 nodes. Note that the size of this training document is about two orders of magnitude larger than the one used in [1].

Fig. 6. Sample page resulted by searching hotels on Expedia

We run the learner on this sample file for 5 training examples that were randomly selected from result pages 1, 5, 11, 24 and 46 and parameters values set to $H = 5$ and $L = R = 3$. The resulted extraction path was converted to XPath and then the XPath query was run on the initial document and on other 80 result pages, obtained by searching other hotels on Expedia. As result, excellent values were recorded for both precision and recall – both values were 1. The resulted wrapper expressed in XSLT is shown in the appendix.

6 Related Works

The rapid expansion of the Web attracted a lot of researches in the area of information extraction and wrapper induction. Overviews of related technologies and systems can be found in: [3], [7], and [8]. We have chosen for discussion in this section some works that we think that are closer to what has been presented in this paper, namely [1, 2, 3], [9], and [4].

A special class of wrappers called L-wrappers (i.e. *logic wrappers*) for tuples extraction, that were inspired by the logic programming paradigm, is studied in papers [1, 2, 3]. L-wrappers i) have a declarative semantics, and therefore their specification is decoupled from their implementation and ii) can be generated using inductive logic programming. The extraction paths introduced in the current paper are just a special class of L-wrappers of arity 1 for which we devised a new and more efficient learning algorithm that runs in polynomial time. For example, the L-wrapper corresponding to the extraction path from figure 4a is:

$$extract(C) \leftarrow next(B,C) \wedge next(A,B) \wedge next(C,D) \wedge last(D) \wedge child(F,C) \wedge a(F) \wedge next(E,F) \wedge first(E) \wedge next(F,G) \wedge last(G) \wedge child(H,F) \wedge child(I,H) \wedge a(I) \wedge last(I) \wedge first(I).$$

In paper [9] tree wrappers for tuples extraction are introduced. A tree wrapper is a sequence of tree extraction paths. There is an extraction path for each extracted attribute. A tree extraction path is a sequence of triples that contain a tag, a position and a set of tag attributes. A triple matches a node based on the node tag, its position among its siblings with a similar tag and its attributes. Extracted items are assembled into tuples by analyzing their relative document order. The algorithm for learning a tree extraction path is quite similar to ours – the composition of two tree extraction paths corresponds to the generalization operator of two extraction paths (our GEN-PATH algorithm). Note also that our extraction paths use a different and richer representation of node proximity and therefore, we have reasons to believe that our wrappers could be more accurate (this claim needs, of course, further support with experimental evidence). Finally, note that our approach can also be extended to tuples extraction as in [9]. However, as future work we are interested to extend the approach presented in our paper to devise an efficient learner for L-wrappers for tuples extraction ([2, 3]), rather than following the approach outlined in [9].

Finally, a generalization of the notion of string delimiters developed for information extraction from string documents ([7]) to subtree delimiters from tree documents is described in paper [4]. This paper introduces a special purpose learner that constructs a structure called candidate index based on trie data structures. Note that the tree leaf delimiters described in this paper are quite similar to our extraction paths and the representation of reverse paths using the symbols $Up(\uparrow)$, $Left(\leftarrow)$ and $Right(\rightarrow)$ can be easily simulated in our approach by 'c' and 'n' arcs.

7 Concluding Remarks

In this note we presented an efficient algorithm for learning to extract single information items from HTML documents. The algorithm is based on the idea of generalizing the extraction paths of example nodes. The generalized extraction path is then mapped to XPath and embedded into an XSLT stylesheet for performing the actual extraction task. As future work we intend to investigate the extension of this algorithm to efficiently learn L-wrappers for tuples extraction.

References

1. Bădică, C., Bădică, A.: Rule Learning for Feature Values Extraction from HTML Product Information Sheets. In: Boley, H., Antoniou, G. (eds): *Proc. RuleML'04*, Hiroshima, Japan. LNCS 3323 Springer-Verlag (2004) 37–48.
2. Bădică, C., Bădică, A., Popescu, E.: Tuples Extraction from HTML Using Logic Wrappers and Inductive Logic Programming. In: Szczepaniak, P.S., Kacprzyk, J., Niewiadomski, A. (eds.): *Proc.AWIC'05*, Lodz, Poland. LNAI 3528 Springer-Verlag (2005) 44–50.
3. Bădică, C., Bădică, A.: Logic Wrappers and XSLT Transformations for Tuples Extraction from HTML. In: Bressan, S.; Ceri, S.; Hunt, E.; Ives, Z.G.; Bellahsene, Z.; Rys, M.; Unland, R. (eds): *Proc. 3rd International XML Database Symposium XSym'05*, Trondheim, Norway. LNCS 3671, Springer-Verlag (2005) 177–191
4. Chidlovskii, B.: Information Extraction from Tree Documents by Learning Subtree Delimiters. In: *Proc. IJCAI-03 Workshop on Information Integration on the Web* (IIWeb-03), Acapulco, Mexico (2003) 3–8.
5. Clark, J.: XSLT Transformation (XSLT) Version 1.0, W3C Recommendation, 16 November 1999, `http://www.w3.org/TR/xslt` (1999).
6. Gottlob, G., Koch, C., Schulz, K.U.: Conjunctive Queries over Trees. In: *Proc.PODS'2004*, Paris, France. ACM Press, (2004) 189–200.
7. Kushmerick, N., Thomas, B.: Adaptive Information Extraction: Core Technologies for Information Agents, In: *Intelligent Information Agents R&D in Europe: An AgentLink perspective* (Klusch, et al. eds.). LNCS 2586, Springer-Verlag (2003).
8. Li, Z., Ng, W.K.: WDEE: Web Data Extraction by Example. In: L. Zhou et al. (Eds.): *Proc.DASFAA'2005*, Beijing, China. LNCS 3453, Springer-Verlag (2005), 347-358.
9. Sakamoto, H., Arimura, H., Arikawa, S.: Knowledge Discovery from Semistructured Texts. In: Arikawa, S., Shinohara, A. (eds.): *Progress in Discovery Science*. LNCS 2281, Springer-Verlag (2002) 586–599.
10. World Wide Web Consortium. XML Path Language (XPath) Recommendation. `http://www.w3c.org/TR/xpath/`, November 1999.

A XSLT Code of the Sample Wrapper

```
<?xml version="1.0" encoding="UTF-8"?>
  <xsl:stylesheet xmlns:xsl="http://www.w3.org/1999/XSL/Transform" version="1.0">
    <xsl:template match="html">
      <result>
        <xsl:apply-templates mode="selhotel" select=
          "//*[not (preceding-sibling::*)][not (following-sibling::*)]/
          tr[not (preceding-sibling::*)][not (following-sibling::*)]/
          td[preceding-sibling::*[1][local-name() = 'td'][not (preceding-sibling::*)]]/
          a[not (preceding-sibling::*)][not (following-sibling::*)]/
          font[not (preceding-sibling::*)][not (following-sibling::*)]/
          text()[not (preceding-sibling::*)][not (following-sibling::*)]"/>
      </result>
    </xsl:template>
    <xsl:template match="node()" mode="selhotel">
      <xsl:variable name="var_hotel"> <xsl:value-of select="normalize-space(.)"/>
      </xsl:variable>
      <hotel><xsl:attribute name="hotel_name"> <xsl:value-of select="$var_hotel"/>
      </xsl:attribute></hotel>
    </xsl:template>
</xsl:stylesheet>
```

Trustworthiness Measurement Methodology for e-Business

Farookh Khadeer Hussain[1], Elizabeth Chang[1] and Tharam S. Dillon[2]

[1] School of Information Systems
Curtin University of Technology
Perth, WA, 6845
Australia
{Farookh.Hussain, Elizabeth.Chang}@cbs.curtin.edu.au

[2] Faculty of Information Technology
University of Technology, Sydney
Sydney, NSW
Australia
tharam@it.uts.edu.au

Abstract The purpose of the Trustworthiness Measure is to (a) to determine the quality of the Trusted Agents and (b) once the trusting agent has determined and recorded the trustworthiness of the trusted agent or the quality of the trusted agent, the trusting agent can use this determined and recorded quality of the trusted agent when some other agent queries it about the quality of the trusted agent. As can be clearly seen, if the trusting agent has not determined the trustworthiness of the quality of the trusted agent and subsequently recorded it, then it will not be in a position to communicate recommendations about the trusted agent. Unfortunately, in the existing literature there is no methodology for quantifying and expressing the trustworthiness of the trusted agent. In this paper we propose a methodology by that the trusting agent needs to following in order to determine the trustworthiness of the trusted agent. This methodology helps trusted business transactions, virtual collaboration and keeps the service-oriented environment trustworthy as well as helping to provide a transparent and harmonious nature to the distributed, heterogeneous, anonymous, pseudo-anonymous, and non-anonymous e-service networks.

1. Introduction

Trustworthiness of an agent / product / service quantifies and expresses the *quality* of an agent or service or product. Trustworthiness of an Agent, service or a product in the service-oriented environment implies the "*quality*" of an Agent, service or product. The quality of a given service in service-oriented environments is determined by determining the correlation between

(a) The *delivered quality* of service (and)
(b) The *mutually agreed quality* of service.

Mark Last et al. (Eds.): Advances in Web Intelligence and Data Mining (SCI) **23**, 21-30 (2006)
www.springerlink.com

We define the delivered service as the set of all functionalities that the trusted agent has delivered to the trusting agent. The delivered quality of service is a numeric value that quantifies and expresses, in commonly used terminology the value of all the delivered functionalities.

We define the mutually agreed service as the set of all functionalities that the trusted agent has promised to deliver to the trusting agent. The mutually agreed quality of service is a numeric value that quantifies and expresses, in commonly used terminology the value of all the mutually agreed functionalities.

The quality of a given product in service-oriented environments is determined by determining the correlation between

(i) The *delivered value of the product* (and)
(ii) The *mutually agreed value* of the product

We define the delivered value of a product as a numeric value that quantifies and expresses, in commonly used terminology the value of all the delivered functionalities of the product.

We define the mutually agreed value of a product as a numeric value that quantifies and expresses, in commonly used terminology the value of all the mutually agreed functionalities of the product.

The trustworthiness of an agent in service-oriented environments is determined by determining the correlation between

(1) The actual behaviour of the trusted agent in the interaction (and)
(2) The mutually agreed behaviour of the trusted agent in the interaction

We define the actual behaviour of the trusted agent as the set of all the functionalities that the trusted agent has delivered to the trusted agent in the interaction.

We define the mutually agreed behaviour of the trusted agent as the set of all the functionalities that the trusted agent has agreed to deliver to the trusted agent in the interaction.

We can see from the above discussion that the mutually agreed behaviour and the actual behaviour of the trusted agent / product and the service are the pivotal factors based on which the the trusting agent can determine the trustworthiness of the trusted agent/ product / service. In this paper we propose a methodology by which the trusting agent and the trusted agent can determine the mutually agreed behaviour and the actual behaviour in the interaction.

This paper in organized as follows,

In Section 2 we propose a conceptual framework comprising of four steps for measuring and quantification of trustworthiness. Section 3-6 explains each of these four steps in detail. Section 7 concludes the paper.

Through this chapter I will explain the proposed methodology using an example. The example is as follows:

Assume that there are two logistic companies namely; East Field and West Field are located in Sydney and Perth respectively. Let us further assume that they have their areas of operation specific to the area that they are located in.

Let us furthermore assume that East Field wants to store some of its consignment of goods in the warehouse belonging to West Field. It sends a request to West Field asking for warehouse space of say 6000 sq feet for duration of 6 days. In this paper we address the sequence of steps that East Field as the trusting agent needs to go through in order to come to the stage of assigning a trustworthiness value to the West Field.

2. Conceptual Framework for Measurement of Trustworthiness Measurement Methodology

The process of *Trustworthiness Measure Methodology Framework* comprises of 4 steps, which are detailed below:

The framework of the Trustworthiness Measure Methodology contains four major steps, namely:

1) Obtain (Determine) Context from associated domain knowledge
2) Identify the Criteria from the knowledge domain
3) Develop Quality Assessment Criteria for each quality aspect
4) Measure the Quality and Trust against Quality Assessment Criteria through CCCI metrics.

In the following section we explain each of the above steps in detail with examples.

3. Determine Context from associated domain knowledge

As discussed in our earlier publication [1], when we refer to the trust that a trusting agent has in a trusted agent, the trust is specific to a specific context/s and the time slot/s. The *'context'*, in which the trust relationship exists, is one major basis based on which the trustworthiness measure by the trusting agent would be carried out. As was pointed in our earlier publication [1], the context can be regarded as a scenario or environment in which the trust exists between the trusting agent and trusted agent.

We define the *context of an interaction*, as a means of representing the set of all the coherently related functionalities that the trusting agent is looking for in an interaction with the trusted agent. As discussed above the context of the interaction can be represented by different terms, as long as they all mean the same. The context of the interaction can be derived only from the knowledge domain.
As an example, based on the service level agreement between East Field and West Field, the context of the interaction could be inferred as *'Storing Goods'* or *'Leasing Warehouse Space' or 'Renting Warehouse Space' or 'Storing Goods'*. As can be seen, from the above discussion there is more than one way in which a given context of interaction can be represented by the trusting agent. It does not matter using which words and terminology how the trusting agent uses to symbolizes or represents the context in an interaction as far as they all mean the same thing semantically. From the above example, East Field could have used either of the above three terms (*'Storing Goods in Warehouse'* or *'Leasing Warehouse Space' or 'Renting Warehouse Space' or 'Storing Goods'*), to represent the context of interaction as semantically they all refer to East Field using the warehouse space of West Field.

Based on the context of interaction, East Field as the trusting agent would determine the trustworthiness of West Field after the interaction. However the context of interaction provides little knowledge that the trusting agent can use to determine the trustworthiness value of the trusted agent. For example, East Field as the trusting agent has little knowledge based on which it could determine the trustworthiness value of West Field.

The context of an interaction has to be determined from the domain knowledge. The domain knowledge would differ for different trusted agents as shown in Table 1, below.

	Trusted Agent	**Knowledge Domain**
1	Human Agent or Software Agent	Mutually Agreed Behaviour(or) Contract(or) Agreement
2	Service Provider or Service	Mutually Agreed Service (or) Service Level Agreement (or) Advertisements(or) Contract(or) Agreement
3	Product	Advertisements (or) Product Catalogue(or) Product Manual(or) Contract(or) Agreement

Table 1: Table showing the knowledge domain associated with trusted agent/ product and service.

We define the **knowledge domain** as a clear and precise natural language description of the mutually agreed behaviour of the trusted agent. The mutually agreed behaviour comprises of a set of functionalities that have been mutually agreed to by both the interacting parties and which the trusting agent expects the trusted agent to perform. Depending on who actually is the trusted agent in the interaction, the knowledge domain is derived from various ways. However an important point to be noted here is that, the knowledge domain specifies the behaviour of the trusted agent in the interaction and this behaviour has to be agreed to by both the trusting agent and the trusted agent. If the trusted agent is a

(a) Human agent or software agent, then the trusting agent and the human / software agent need to enter into a Negotiation Phase. During the Negotiation Phase, the trusting agent and the trusted agent draw up a description of the mutually agreed behaviour. The Negotiation Phase comprises of the following steps

Step 1: Behaviour Proposal: The reason why the trusting agent in engaging the trusted agent in an interaction is to fulfill a certain set of objective/s. The objective/s of the trusted agent in an interaction could be achieved by the trusted agent by performing a set of coherently related activities. The initial behaviour proposal should in practice contain a finite set of clearly and precisely specified activities that the trusting agent expects the trusted agent to perform in the interaction. By clearly specification of the activities, I mean that the activity the trusting agent expects the trusted agent to perform in the interaction should be spelled out in easily comprehensible terminology and in unambiguous terminology during the negotiation phase. By precise specification of the activities, I mean that the if possible, the out put of each activity that the trusting agent expects the trusted agent to perform in the interaction should quantified in terminology that is mutually understandable to both the interacting agents.

Step 2: Revision of Behaviour Proposal: Based on the initial behaviour proposal the trusted agent should revise the *first behaviour proposal* into a behaviour proposal that comprises of a set of clearly and precisely specified set of activities, which the trusted agent feels that it is willing and capable of delivering upon. The trusting agent determines those set of activities specified in the initial behaviour proposal that it feels that it is not capable or is unwilling of carrying out. It would then revise the initial behaviour proposal into a new behaviour proposal that comprises of a set of clearly and precisely specified set of activities, which the trusted agent is capable and willing to carry out in its interaction with the trusted agent. The trusted agent in practice should try it best that the new set of activities would help the trusting agent achieves its set of objectives in the interaction. The trusted agent would then communicate the revised behaviour proposal to the trusting agent.

Step 3: Considering the Revised Behaviour Proposal: Once the trusting agent receives the revised behaviour proposal, it would consider whether the revised behaviour proposal would help it to achieve the objectives in the interaction. If it feels that its objectives that it aims for in the interaction would be achieved by the revised behaviour proposal then it has the option of going ahead and changing the revised behaviour proposal into the *Mutually Agreed Behaviour.*

On the contrary then if it feels that the revised behaviour proposal form the trusted agent would not help it achieve its aims in the interaction, then it has two option as explained below

Step 3(a): Propose an alternative behaviour proposal: The alternative behaviour proposal can be derived by the trusting agent from the revised behaviour proposed communicated to it in Step 2. Alternatively the alternative behaviour proposal may comprise of a new set of clearly and precisely specified activities, that the trusting agent feels would help us in achieving its objectives in the interaction. Irrespective of whether the new proposal is derived by modifying the revised behaviour proposal or comprises of a new set of coherently related set of activities the trusting agent and the trusted agent would need to go through the step 1, step 2 and step 3 in the same order to arrive on the mutually agreed behaviour. We term each iteration through the step 1, step 2 and step 3 as a Negotiation Cycle.

Step 3(b): Consider engaging an alternative trusted agent: Based on the revised proposal from the trusted agent, if the trusting agent feels that the revised proposal would not help it to achieve its objectives in the interaction, then it may choose to consider the engaging an alternative trusted agent.

(b) **Service Provider (or) Service :** If the trusted agent is a service provider, then the knowledge domain , may be one of the following

i. **Contract (and)**
ii. **Agreement(and)**
iii. **Mutually Agreed Service (and)**
iv. **Service Level Agreement:** The trusting agent and the service provider determine the mutually agreed service by going through the Negotiation Phase, as explained above. At the completion of the negotiation phase, the trusting agent will have a specification that clearly and precisely details the service that will be carried out by the trusted agent. The service in turn is specified as a set of clearly, precisely and coherently set of related activities.
v. **Advertisement:** Service Providers usually advertise their services. The advertisement of a service, specifies the Quality of Service. The advertisement of service specifies that the service provider is committed to providing a particular level of service. If the trusting agent is feels that the Quality of Service advertised by the Service Provider could help it in achieving its objectives in the interaction then the trusting agent can go ahead and interact with the service provider on the terms, conditions and the Quality of Service as specified in the advertisement.

(c) **Product:** If the trusted agent is a product, then the knowledge domain , may be one of the following

i. **Product Manual (and)**

ii. **Product Catalogue:** Each product manual and product catalogue is made by the product manufacturer. The product manual or product catalogue specifies the functionality of the product. The functionality of the product should be clearly, precisely specified by the product manufacturer. The product catalogue (or) product catalogue specifies that the product manufacturer assure that the product will accomplish certain functionalities. The Quality of Product will specify the level to which the product will accomplish the given functionality.

iii. **Advertisement:** Provider Manufacturers usually advertise their products. The advertisement of a product, specifies the Quality of Product. The advertisement of a product specifies certain functionalities of the product. Additionally the advertisement of the product specifies quantitatively in commonly used terminology the value that can be obtained by the user. If the trusting agent is feels that the Quality of Product advertised by the Product Manufacturer could help it in achieving its objectives in the interaction then the trusting agent can go ahead and interact with the product on the terms, conditions and the Quality of Product as specified in the advertisement.

The context of an interaction can be additionally described as a high level summarized description of all functionalities that the trusting agent is looking for in its interaction with the trusted agent. The context of an interaction provides little knowledge of the factors based on which the trusting agent would assign a trustworthiness value to the trusted agent. Based on the above discussion of the context of the interaction between East Field and West Field is *'Storing Goods in Warehouse'* or *'Leasing Warehouse Space'* or *'Renting Warehouse Space'* or *'Storing Goods'*. As can be seen East Field does not have enough information to assign a trustworthiness value of West Field based on the above description of context. In order to determine the trustworthiness of the trusted agent, the trusting agent needs to determine the criteria based on which it would assign a trustworthiness value to the trusted agent.

4. Identify the Criteria from the knowledge domain

We define the criteria as *a decisive factor, the performance or the output of which has been mutually agreed by both the trusting agent and the trusted agent, and the trusting agent would evaluate the performance of the trusted agent in that decisive factor in its interaction with the trusted agent.* The Criteria of an interaction can be additionally regarded as a quality dimension for the purpose of quality assessment.

The criterion has to be derived from the domain knowledge. As mentioned before, the domain knowledge comprises of the mutually agreed behaviour / mutually agreed service of the trusted agent from the perspective of the trusting agent. From the above service level agreement between East Field and West Field, the decisive factors the performance of which has been mutually agreed by both the trusting agent and the trusted agent and the performance of which the trusting agent would assign trustworthiness value to the trusted agent is the following

- **The Space of Leased Warehouse:** West Field has agreed to lease a ware house space of 6000 sq feet to East Field.
- **The Duration for which the Warehouse space would be leased:** West Field has agreed to lease the warehouse space for duration of 6 days.

Looking at it from another perspective the above two are the criteria or the decisive factor/s on which East Field will assign a trustworthiness value to West Field.

In this stage the trusting agent, should go through the knowledge domain and determine all the criteria which,

1. have been mutually agreed between the trusting agent and the trusted agent (and)
2. based on which it is going to evaluate the trustworthiness of the trusted agent.

As mentioned before, the mutually agreed behaviour of the trusted agent in a given criterion will be expressed and agreed to by both the interacting parties in commonly used terminologies.

5. Develop the Criteria Assessment Factors

We define Criteria Assessment Factors as those factors based on which the trusting agent would assess whether or not a given criteria in its interaction has been delivered up on. The criteria assessment factors metric can be alternatively defined as *a set of rules, regulations or policies which the trusting agent would use to the whether or not a given criterion has been delivered up on by the trusted agent.*

For the above two criteria in the interaction between East Field and West Field, the criteria assessment factors developed by East Field are as follows,

Criteria Assessment Factor for Size of Warehouse Space Allocated:

If West Field allocate a warehouse space of greater than or equal to 6000 sq ft then this criteria has been delivered upon by the West Field else not.

Criteria Assessment Factor for Duration of Warehouse Space Allocated:

If West Field has allocate the warehouse space for a duration of greater than or equal to 6 days then this criteria has been delivered upon by the West Field else not.

6. Measure the Trustworthiness using CCCI Metrics

Finally in order to determine the trustworthiness of the trusted agent, the trusting would make use of the CCCI Metrics. The CCCI Metrics, as I will explain below determines the correlation value of each individual criterion in the interaction. The correlation of each individual criterion is subsequently weighted by the clarity of the criterion and the importance of the criterion. The correlation values of the all the criteria in the interaction are then combined to determine the correlation of the interaction.

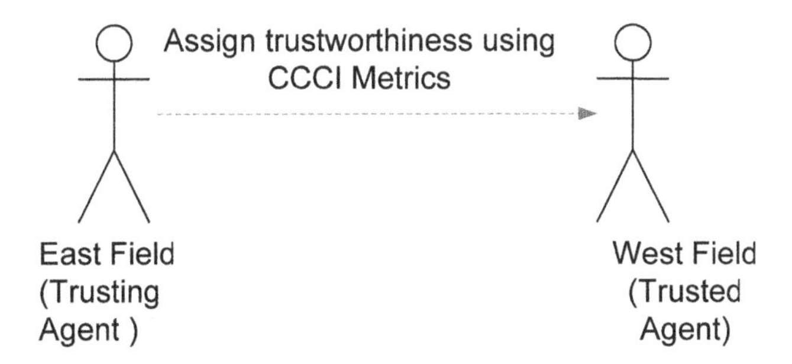

Figure 2. Through the correlation of the *Quality Assessment Criteria* with the *actual delivered service, we* determine the Trustworthiness of Trusted Agent

The Measurement of the Trustworthiness against Quality Assessment Criteria is carried out through CCCI metrics. Due to space constraints we are unable to provide a detailed explanation on the CCCI Metrics. Interested readers are encouraged to refer to [1] for a detailed explanation of the working of the CCCI Metrics.

7. Summary

In this paper we have proposed a methodology, comprising of four steps which the trusting agent can make use of in order to determine the trustworthiness of the trusted agent. The methodology comprises of four steps, namely

(a) Obtain (Determine) Context from associated domain knowledge
(b) Identify the Criteria from the knowledge domain
(c) Develop Quality Assessment Criteria for each quality aspect
(d) Measure the Quality and Trust against Quality Assessment Criteria through CCCI metrics.

We explained each of the first three steps in detail, along with examples. Due to space constraints, we were unable to explain the working other CCCI Metrics in detail. The working however has been explained in great depth, detail and along with examples in [1]. Interested readers are encouraged to refer to [1], for a thorough discussion on the CCCI Metrics.

8. References

[1]Elizabeth Chang, Tharam Dillon , Farookh Khadeer Hussain (2006), *Trust and Reputation for Service Oriented Environment,* John Wiley and Sons, To appear in 2006.

Routing Using Messengers in Sparse and Disconnected Mobile Sensor Networks

Qiong Cheng, Yanqing Zhang, Xiaolin Hu, Nisar Hundewale, Alex Zelikovsky
Dept. of Computer Science, Georgia State University, Atlanta, GA 30303 USA
qcheng1@student.gsu.edu, {yzhang, xhu}@cs.gsu.edu, nisar@computer.org, alexz@cs.gsu.edu

***Abstract*— Sparse mobile sensor networks, such as those in the applications of ecology forest and modern battlefield, can frequently disconnect. Unfortunately, most existing routing protocols in mobile wireless networks mainly address connected networks, either sparse or dense. In this paper, we study the specific problem for dynamic routing in the sparse and disconnected mobile sensor networks utilizing messengers. We propose two routing discovery protocols: Genetic Fuzzy Straight Line Moving of Messengers (GFSLMM) and Genetic Fuzzy Flexible Sharing Policy of Messengers (GFFSPM). A preliminary simulation shows the efficacy of our protocols.**

***Index Terms*— Mobile Sensor Networks(MSN), Minimum Spanning Tree(MST), Dynamical Source Routing Protocol (DSR), Ad Hoc On Demand Distance Vector Routing (AODV), Genetic Algorithm (GA), Fuzzy Inference System, Disjoint Mobile Sensor Networks (DMSN), Straight Line Moving of Messengers (SLMM), Flexible Sharing Policy of Messengers (FSPM), Genetic Fuzzy Straight Line Moving of Messengers (GFSLMM) and Genetic Fuzzy Flexible Sharing Policy of Messengers (GFFSPM).**

1 Introduction

Rapid progress of wireless communication and distributed embedded sensor and actuator technologies has lead to the thriving of the applications of mobile sensor networks (MSN) which range from natural ecosystem to security monitoring, especially in inaccessible terrains or disaster relief operations [1].

Mobile sensor network is a dynamic sensor network with a large number of static sensor nodes and mobile nodes and wireless communication between them. These mobile nodes can be mobile vehicles (cars or buses) which are loaded with sensors. In a habitat monitoring scenario, animals can perform the role of mobile vehicles [5]. The MSN possess the self-organizing and cooperative ability to detect record, collect, process, predict and estimate some events of interest [1].

The first application we consider here is to monitor basic forest ecology. UCLA used infrared imagers to track forest temperatures and heat patterns [8]. But it is not cost-effective to study the entire ecology environment change by deploying lots of sensor nodes in the entire forest to form a dense connected network and maintaining the entire network. So a feasible way is to attach sensors to the trained or observed animals.

The second application we present here is the modern BattleField which demands critical surveillance information system of the enemy site and the most rapid and precise decision support system. Mobile wireless sensors, scattered in targeted zones, form the MSN. They are able to quickly and secretly gather infor-

Mark Last et al. (Eds.): Advances in Web Intelligence and Data Mining (SCI) **23**, 31-40 (2006)
www.springerlink.com

mation about the location and environment and periodically relay this information back to the command, control, and communication center [6].

Since sensor nodes inherently have limited power, short-distance communication and prone to failure, in the above two applications, it is difficult to form a globally connected topology. For sparse/disconnected networks,[2][5][7] proposed mobility-based approaches. [2] assumes network topology is relatively stable and the information about network partitions are conveyed by out-of-band means. [5] assumes that the sensor networks in the bottom tier are fixed. All the assumptions allow ignoring the inherently and highly dynamic nature of the distributed dynamic environment. Our previous work [7] was mostly devoted to implementing the agents based simulator for the problem. We proposed two solutions, one based on straight line moving of messengers and the other based on flexible sharing policy of messengers. However, we did not give the problem formulation for it and show the feasibility topologically. These previous work will be described in details in section 2.

This paper deals with the distributed dynamic environment. We propose solution and show the feasibility. Because the uncertainty of dynamic environment and wireless communication, we propose to employ the genetic fuzzy system in the previous two solutions.

The problem and solution method are as follows. Because the use of a long range radio consumes excessive energy, we employ the mobile vehicle (cars, buses, people, or animals with sensors) as the messenger. Once network partitions are generated, the autonomous routing discovery and maintenance based on messenger will be invoked in delay-tolerant sparse/disconnected mobile sensor networks. Through SLMM and FSPM based on the messengers [7], the information of the available network partitions can be shared. And then, in our problem model, we set every one of network partitions to be a vertex and find the cut-edges in the minimum spanning tree (MST) which connects these partitions and meets the condition of the minimum distance weight. The minimum distance weight represents the minimum total energy consumption in the process of the moving. Furthermore, distributed and dynamic characters of the specific applications make the problem more challenging and interesting. With the advent or moving of targets, the network partitions, as vertices in the MST, are changing. On the consequence, new MST should be built and new cut-edges should be discovered. So the problem is focused on the discovery and rebuilding of the dynamic MST with the tradeoff between the delay and real-time availability of network partitions information. Additionally, after MST is constructed, the problem is simplified and the currently existed techniques can be applied [3]. Our goal is to design an energy efficient, decentralized, scalable and flexible approach to share the information of network partitions and rebuild the MST for the distributed dynamic application environment in the delay-tolerant range when network partitions change. Figure 1 shows Process of building topology in disconnected network partitions. It can be iteratively operated. And every network partition can be composed of the hierarchy of subnetwork partitions no matter whether there has been a MST in it. The iterative operation of the loop results in the decentralized rebuilding of MST for dynamic network partitions in energy efficient, scalable and flexible way for the distributed dynamic application environment.

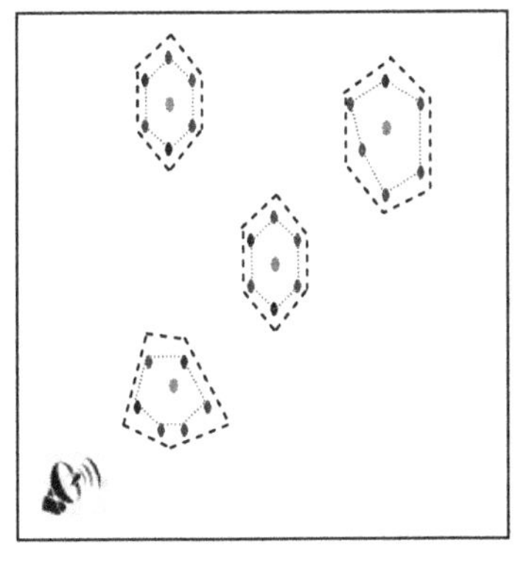

Figure 1.a. Form disconnected MSN with partitions (at T timestep)

Figure 1.c. Establish network topology via MST for T timestep

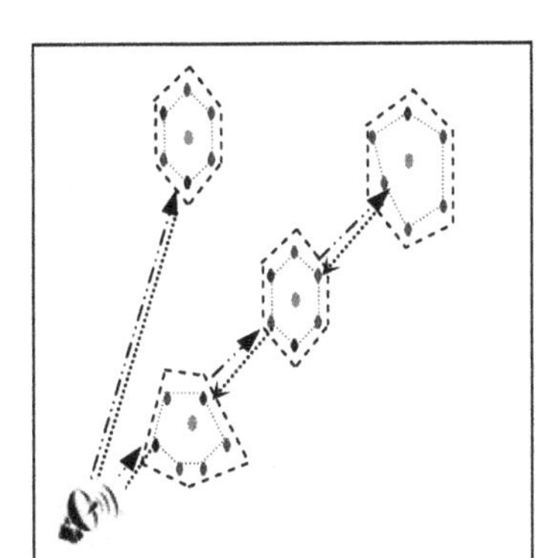

Figure 1.b. Share the information of network partitions by messengers (one of solutions)

Cluster regular member

Messenger

Target

Cluster Leader

Network Partition

Wireless Communication

Messaging

Returning

Cut_edges between networks partitions (MST edges)

Figure 1. Process of building topology in disconnected network partitions.

The remainder of the paper is organized as follows: Section 2 presents previous work. Section 3 introduces genetic fuzzy system routing discovery and maintenance. The autonomous agents' simulation and evaluations on energy consumption are presented in Section 4. Finally, we conclude the paper in Section 5.

2 Previous Work

A message ferrying approach is developed in [2] for sparse mobile ad hoc network. This method employs a set of special mobile nodes called ferries to provide communication service for regular nodes in the deployment area according to a specific route which is generated in out-of-band means so that they are responsible for data delivery in sparse networks. Because authors assume that network partitions are relatively stable and their information is conveyed in the out-of-band means, in [2], they ignore the route discovery and maintenance in the sparse or disjoint dynamic system. Additionally, as a result of the asymmetric roles between ferries and regular nodes, the system will more depend on the robustness of ferries and the consistent coordination between ferries.

R. Shah [5] proposed 3-tier architecture Mules for sparse sensor networks. In the middle tier, the mobile transport agents as messengers collect sensor data from the bottom tier and deliver the data to the top tier. They assumed that the sensor networks in the bottom tier are fixed. The assumption is not suitable for the distributed dynamic application environment.

In our previous work [7], we presented a novel autonomous messenger-based route discovery and routing protocol for disjoint clusters-based MSN. The proposed protocol does not depend upon centralized control or prior global knowledge. It does not depend upon ferries or sensors with special hardware capabilities. The protocol can establish communication in disjoint network of clusters that has not been addressed previously. We provided the framework for the route discovery and routing protocol. And we designed and implemented two route discovery protocols, one based on straight line moving of messengers (SLMM) and the other based on flexible sharing policy of messengers (FSPM). Furthermore, we designed the agent-based modeling and simulator in the application and implemented the prototype. However, our previous work [7] also ignores to describe the highly dynamic problem model and to show the feasibility of the solutions. Additionally, we treated all uncertain elements such as signal strength and the preciseness of direction and distance as crisp elements in the previous version.

3 Genetic Fuzzy System Routing Discovery and Maintenance

3.1 Framework

Based on the framework in previous work, we add a higher tier genetic fuzzy system to SLMM and FSPM in Routing Discovery and Maintenance Component. The whole framework is shown as figure 2.

In routing discovery and maintenance component, we provide three sub components and implement the four routing discovery protocols, one based on straight line moving of messengers (SLMM), one based on flexible sharing policy of messengers, one base on genetic fuzzy flexible share policy of messengers (GFFSPM), and the other based on genetic fuzzy straight line movement of messengers (GFSLMM).

3.2 Genetic Fuzzy System

The genetic fuzzy system prototype is shown as figure 3. The prototype is composed of TSK fuzzy system and genetic algorithm. The TSK fuzzy inference system generates the crisp output which is input into the genetic algorithm as fitness function value. The training feature of genetic algorithm optimizes the TSK fuzzy system. The system input parameters is the distance and angle between the messenger and the sensor which belongs to a full cluster.

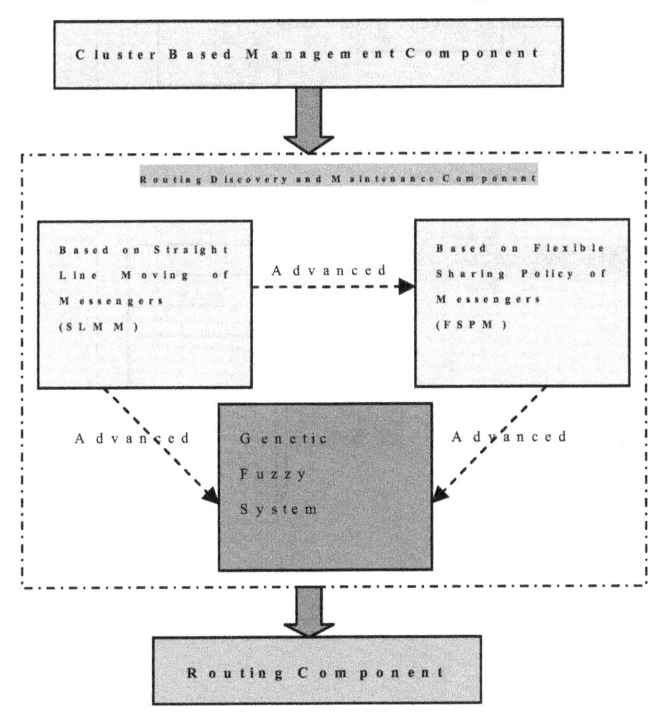

Figure 2. The Framework

3.2.1 The TSK Fuzzy System

Regular fuzzy inference system consists of fuzzy sets, fuzzy if-then rules, and aggregation and defuzzification parts. According to the direction of path planning, we define eight linguistic variables related to direction shown in figure 4: Front Zero (FZ), Front Left (FL), Front Right (FR), Rear Zero (RZ), Rear Left (RL), Rear Right (RR), Left (L), and Right(R). Additionally, we define Signal Strength linguistic variable (SS). The term set for Signal Strength (SS) = {Strong, Medium, Weak}.

Individually the linguistic variables have the membership function as figure 5 and figure 6.

According to a thorough understanding of the system, we design 24 fuzzy rules as the following format:

If direction is FZ and signal is strong, move to FZ;

… …

If direction is FZ and signal is medium, move to the average of FZ and Base Station; … …

If direction is FZ and signal is weak, move to Base Station; … …

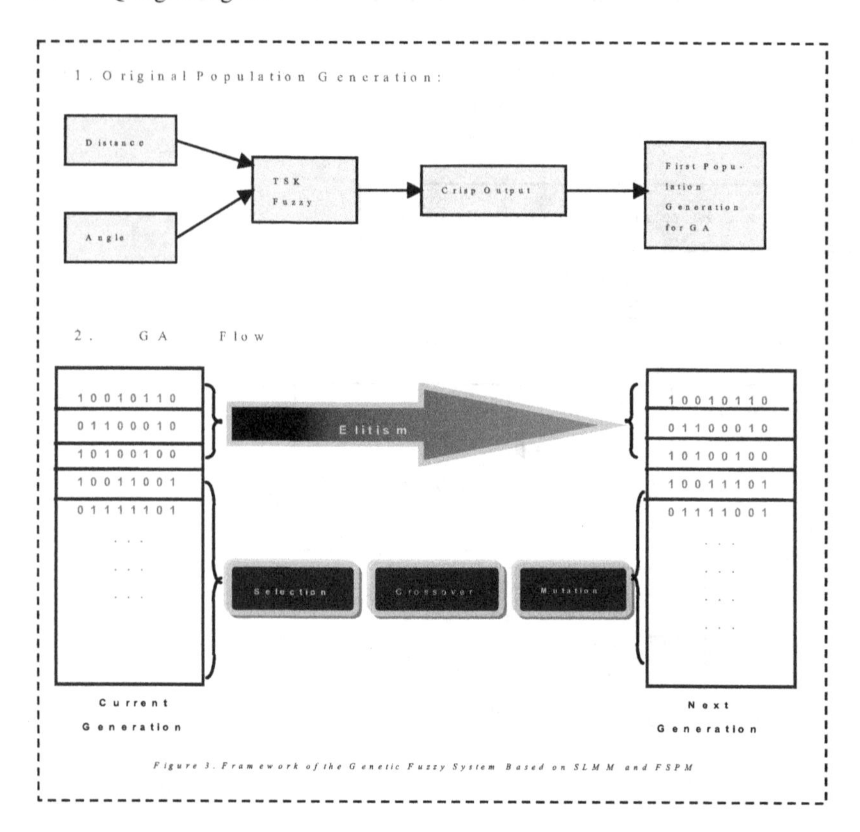

Figure 3. Framework of the Genetic Fuzzy System Based on SLMM and FSPM

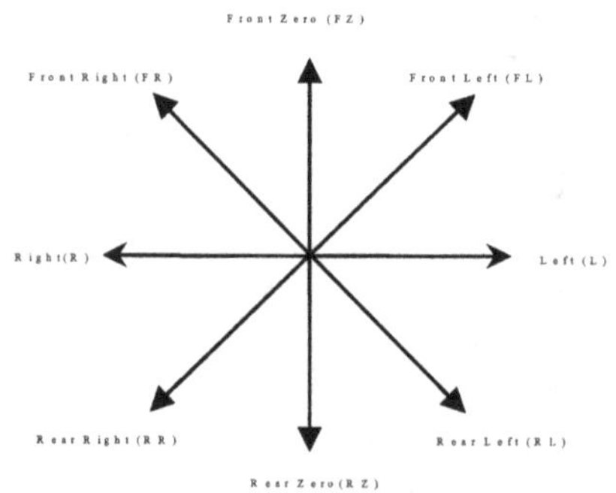

Figure 4. Direction Define

And in the design of the fuzzy rules, we consider the convergence of all messengers' movement towards the nearer available neighbors and the base station.

Based on the distance and angle inputs and the fuzzy rules, we define the angles with the larger certainty as the crisp output.

3.3 Genetic Algorithm (GA)

The second part of figure 3 presents one cycle of the iterative steps of genetic algorithm for generating the offspring generations.

In our GA, chromosomes are stored as floating point numbers. The outputs of the fuzzy inference system are input as the discrete data set and are retrieved as fitness function value.

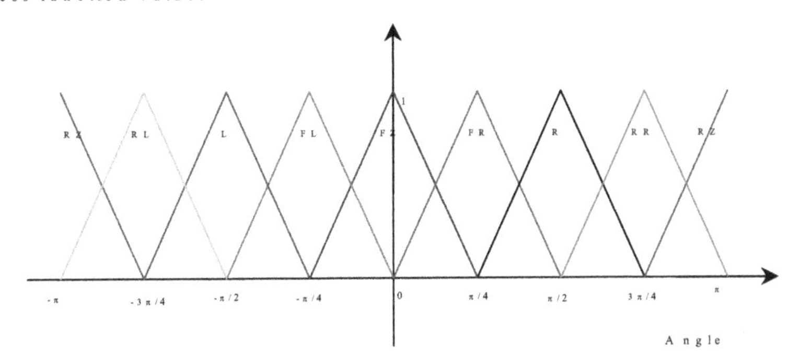

Figure 5. Membership Functions of Path Direction

The same color curve lines represent the membership function of the same linguistic variables

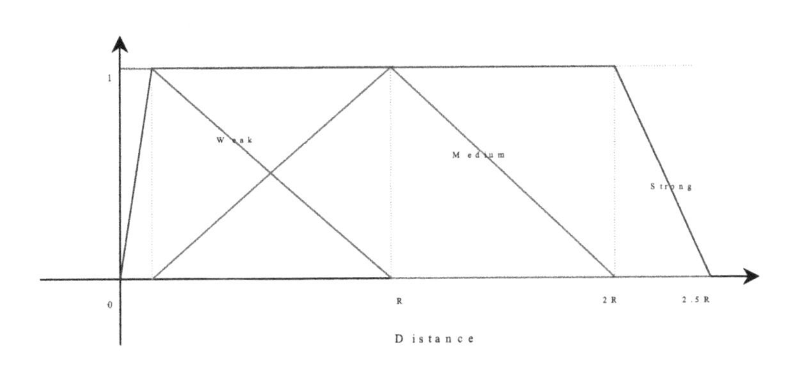

Figure 6. Membership Functions of Signal Strength (R= Sensor Transmission Radius)

The same color curve lines represent the membership function of the same linguistic variables

In the simulation experiment mentioned in section 6, we set the following GA parameters:

Chromosome Dimension (number of genes) = 10;

Population of Chromosomes = 100;

Crossover Probability = 0.7;

Random Selection Chance % (regardless of fitness) = 6

Max Offspring Generations = 30;

Number Prelim Runs (to build good breeding stock for final--full run) = 10;

Max Prelim Generations = 20;

Chromosome Mutation Probing = 0.1;

Crossover Type = Two Points Crossover;

After messengers move for one time step, they broadcast probing signal and receive the response from nearby sensor nodes. First, the distance and angle from the nodes are input to the TSK fuzzy system. Secondly, the crisp outputs are input as fitness function value to GA. After the specified offspring generation, the best fitness value is used as the current optimal direction.

4 Agents Simulation and Evaluation

4.1 Simulation

Due to the dynamic and non-predetermined connection and disconnection among sensor nodes in DMSN, we still use the Autonomous Agents Simulator MSensorSim designed in our previous work [7]. The result will be addressed in the following sub section.

4.2 Evaluation

Based on the agents based modeling and simulation in the application of route discovery and routing protocol in DMSN, we conducted experiments to evaluate and compare the energy consumption of the protocols: genetic fuzzy routing discovery based on SLMM, genetic fuzzy routing discovery based on FSPM, and route discovery based on flexible sharing policy of messengers.

As we know, mobile nodes in MANET have limited battery capacity. So the saving of battery power is a vital issue when determining the network route [4]. We denote the battery power consumption for the entire network as BPC. Because the battery power consumption of a node is caused by the following activities:

1. The roaming of sensor nodes;
2. The moving of messengers;
3. Uploading data or message to cluster head;
4. Messenger assigning of cluster head(because of the routing consumption in the cluster);
5. Data exchanging with messengers;

We define BPC as a linear function shown below:

$$BPC = W_{Roaming} \times M_{Roaming} + W_{Moving} \times M_{Moving} + W_{UploadData} \times M_{UploadData} + W_{AssignMessenger} \times M_{AssignMessenger} + W_{ExchangeData} \times M_{ExchangeData}$$

where the $W_{Roaming}$, W_{Moving}, $W_{UploadData}$, $W_{AssignMessenger}$ and $W_{ExchangeData}$ are the battery power consumed by the network interface when a node does the above mentioned five activities; $M_{Roaming}$, M_{Moving}, $M_{UploadData}$, $M_{AssignMessenger}$ and $M_{ExchangeData}$ are the amount of five types of activities respectively. Therefore, BPC is obtained by averaging the total amount of the power consumption for every T time steps.

In the prototype, we initially define the following constant of energy consumption on different activities: ENERGY_SENSOR__MSGER_EXCHANGE_DATA = 1; SENSOR_FULL_ENERGY = 100000; ENERGY_SENSOR_ROAMING = 10; ENERGY_SENSOR_MEM_UPLOAD_MSG = 1; ENERGY_SENSOR_HEAD_SELECT_MSGER = 1; ENERGY_SENSOR__MSGER_MOVE = 10; SENSOR_NO_ENERGY = 10;

For retrieving the energy consumption trend in the environment of emulating the random event of interest, we employ the average value method. First we do the same experiment ten times using the same parameter set and then calculate the average value. These experiments ran in the Windows XP system of my Pentium ® 4 CPU 2.66 GHz HP laptop. We have 100 sensor nodes and 20 targets. Two hundred mobile nodes were deployed within a 640 m *800 m area. Each node had a radio propagation range of 20m and channel capacity was 0.1Mb/s. Also, 20 targets appear in randomly distributed places that are far enough away from each other to form network partitions. The experiment results are provided in the table and Chart 1. The user interface is shown in figure 7.

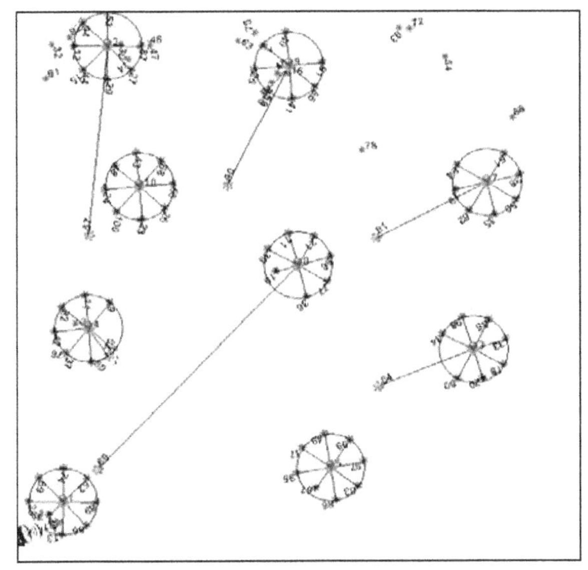

Figure 7. Simulation of route discovery and routing in DMSN

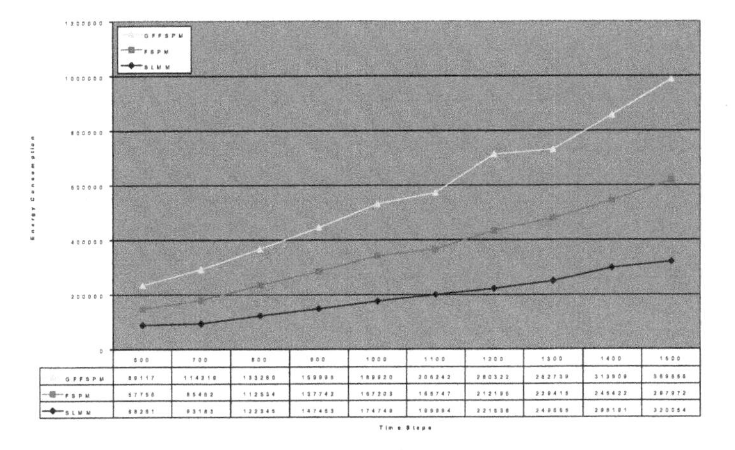

	600	700	800	900	1000	1100	1200	1300	1400	1500
GFFSPM	89117	114219	133260	159995	189920	206242	280322	252739	313909	369555
FSPM	57756	85482	112534	137742	167203	165747	212195	229415	245422	297972
SLMM	88261	93183	122345	147453	174749	199994	221538	249555	298181	320054

From the above stacked line chart and table, we observe that FSPM route discovery protocol is consistently more energy efficient than SLMM and GFFSPM. As the simulation time increases we can see that FSPM becomes more and more energy efficient.

From the theoretic point of view, genetic fuzzy system based on the routing discovery protocols should show better energy consumption. But in my experiment, due to the limited clusters, the population space is not big enough to make efficient use of genetic algorithm. In out next parallel simulation version for the large scale mobile sensor networks, we will see the results of genetic fuzzy system applied to the novel routing discovery protocols.

5 Conclusion

In this paper, according to the highly dynamic property of the distributed dynamic environment, we present the specific problem and the corresponding solution model and show the feasibility from network partitions topology graph's perspective. And due to the uncertainty of dynamic environment and wireless communication, we employ the genetic fuzzy system in the previous two solutions, one based on straight line moving of messengers and the other based on flexible sharing policy of messengers. With varied repetitive experiments based on the simulator MSensorSim, we draw a conclusion that the flexible sharing policy route discovery and routing protocol (FSPM) provides more energy savings than the genetic fuzzy solution based on the straight line moving one (SLMM). Additionally, due to the limited clusters, the population space is not big enough to make efficient use of genetic algorithm. On the other hand, our protocol not only increases network life time but also enables sensor data and cluster knowledge sharing for co-operative efforts of the autonomous agents to perform autonomous monitoring of the terrain and communicating with the base station for prolonged duration in an energy-efficient manner. Our future work will focus on improving the efficiency of routing discovery planning in disjoint network and finishing the routing component.

[1] I. Akyildiz, W. Su, Y. Sankarasubramaniam, and E. Cayirci. *"A Survey On Sensor Networks"* IEEE Commun. Mag. (ASSC02), 40(8):102-114, 2002.

[2] Wenrui Zhao, Mostafa Ammar, Ellen Zegura, *"A Message Ferrying Approach for Data Delivery in Sparse Mobile Ad Hoc Networks"* Proceedings of ACM Mobihoc 2004.

[3] David B. Johnson and David A. Maltz. *"Dynamic Source Routing in Ad Hoc Wireless Networks"*, In Mobile Computing, edited by Tomasz Imielinski and Hank Korth, Chapter 5, pages 153-181, Kluwer Academic Publishers, 1996.

[4] H. Liu, J. Lie, Y. Pan, and Y.-Q. Zhang, *"An adaptive genetic fuzzy multi-path routing protocol for wireless ad-hoc networks"*, Proc. of SAWN 2005, May, 2005.

[5] R. Shah, S. Roy, S. Jain, and W. Brunette. *"Data MULEs: Modeling a three-tier architecture for sparse sensor networks."* In IEEE SNPA Workshop, 2003

[6] C-K Toh. "Ad Hoc Mobile Wireless Networks: Protocols and Systems" 2002 by Prentice Hall PTR

[7] Nisar Hundewale, Qiong Cheng, Xiaolin Hu, Anu Bourgeois, Alex Zelikovsky. *"Autonomous Messenger Based Routing in Disjoint Clusters of Mobile Sensor Networks"* Agent Directed Simulation 2006 ADS'06 (in SCS2006)

[8] http://www.engineer.ucla.edu/stories/2003/nims.htm

Content Consistency Model for Pervasive Internet Access*

Chi-Hung Chi[1], Lin Liu[1], Choon-Keng Chua[2]
[1] School of Software, Tsinghua University, Beijing, China
[2] School of Computing, National University of Singapore, Singapore
Contact email: chichihung@mail.tsinghua.edu.cn

ABSTRACT. In this paper, we propose a new content consistency model for pervasive Internet access. We argue that content retrieved over the Internet consists of not only the data object but also its attributes needed to perform appropriate network or presentation related functions such as caching, content reuse, and content adaptation. With this model, four types of content consistency are defined. To get a deeper insight on the current situation of content consistency over Internet, real content on replica / CDN (Content Delivery Network) was monitored and analyzed. Surprisingly, we found that there are lots of discrepancies in data object and attributes found by comparing the original copy and the retrieved copy of the content. This result is important because they have direct implications to the trustworthiness of information over the Internet.

I. INTRODUCTION

Web caching is a mature technology to improve the performance of web content delivery. To reuse a cached content, the content must be *bit-by-bit equivalent* to the origin (known as data consistency). However, since the internet is getting heterogeneous in terms of user devices and preferences, we argue that traditional data consistency cannot efficiently support pervasive Internet access. There are two problems that have not yet been addressed: i) correctness of function execution in network, and ii) reuse of pervasive content. On Internet, there lies a fundamental difference between "data" and "content". Data usually refers to entity that contains a single value, for example, in computer architecture each memory location contains a word value. On the other hand, content (such as a web page) contains more than just data; it also encapsulates *attributes* to administrate various functions of content delivery. Unfortunately, present content delivery only considers the consistency of data but not attributes. Web caching, for instance, relies on caching information such as expiry time, modification time and other caching directives, which are included in attributes

* This research is supported by the funding 2004CB719400 of China.

Mark Last et al. (Eds.): Advances in Web Intelligence and Data Mining (SCI) **23**, 41-50 (2006)
www.springerlink.com

of web contents (HTTP headers) to function correctly. However, since content may traverse through intermediaries such as caching proxies, replicas and mirrors, the HTTP headers users receive may not be the original. Therefore, instead of using HTTP headers as-is, we question about the *consistency of attributes*. This is a valid concern because the attributes directly determine whether the functions will work properly and they may also affect the performance and efficiency of content delivery. Besides web caching, attributes are also used for controlling the presentation of content and to support extended features such as privacy and preferences.

Under pervasive Internet access, contents are delivered to users in their best-fit presentations (also called variants or versions) for display on heterogeneous devices [5] [6] [9]. As a result, users may get presentations that are *not* bit-by-bit equivalent to each other, yet all these presentations can be viewed as "consistent" in certain situations. Data consistency, which refers to bit-to-bit equivalence, is too strict and cannot yield effective reuse if applied to this pervasive environment. In contrast to data consistency, our proposed content consistency *does not* require objects to be bit-by-bit equivalent. This relaxed notion of consistency increases reuse opportunity, and leads to better performance in pervasive content delivery.

In this paper, we propose a new concept termed *content consistency* and show how it helps to maintain the correctness of functions and improve the performance of pervasive content delivery. With this model, four types of content consistency are defined. To get a deeper insight on the current situation of content consistency over Internet, real content on replica / CDN (Content Delivery Network) was monitored and analyzed. Surprisingly, we found that there are lots of discrepancies in data object and attributes found by comparing the original copy and the retrieved copy of the content. This result is very important because these findings have direct implications to the trustworthiness of information systems over the Internet. Results from this study also provide hints on how the quality of the content provided over the Internet can be improved.

2. CONTENT CONSISTENCY MODEL

In this section, we are going to propose our content consistency model for pervasive Internet access. Its system architecture, content model, and a new classification of content consistency will be given.

2.1. System Architecture

Our vision of the future content consistency is depicted in Figure 1. We begin by describing the pervasive content delivery process in three stages: server, intermediaries, and client. In stage one, server *composes* content by associating an object with a set of attributes. *Content* is the unit of information in the content delivery system where object refers to the main data such as image and HTML while attributes are metadata required to perform *functions* such as caching, transcoding, presentation, validation, etc. The number of functions available is infinite, and functions evolve over the time as new requirements emerge.

In stage two, content travels through zero or more intermediaries. Each intermediary might perform transformation on content to create new *variants* or *versions*. Transformations such as transcoding, translation, watermarking, and insertion of advertisements, are operations that change object and/or attributes. To improve the performance of content delivery, intermediaries might also cache contents (original and/or variants) to achieve full or partial reuse. In stage three, content is received by client. *Selection* is performed to select object and the required attributes from content. The object is then used by user-agents (for display or playback), and functions associated with content are performed. Two extreme cases of selection is to select all available attributes, or to only select attributes for a specific function.

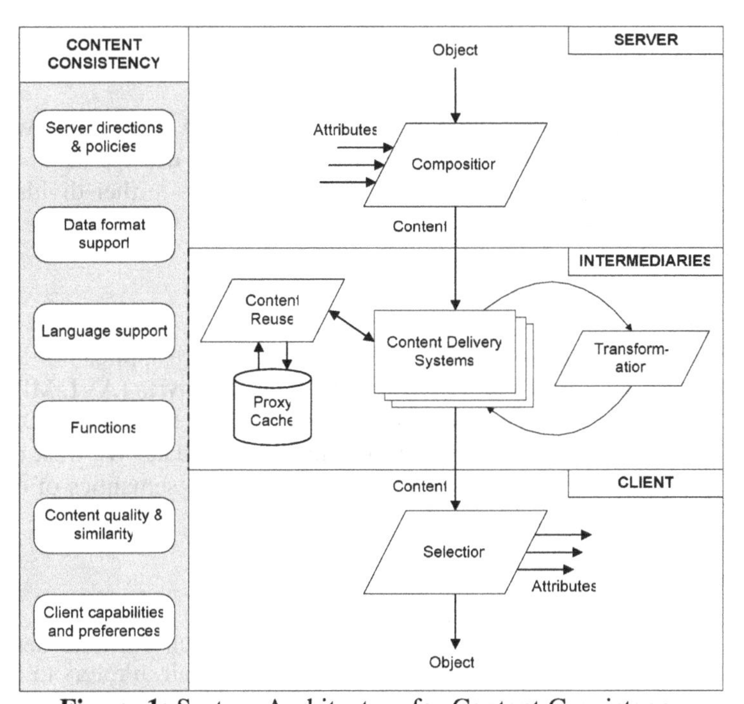

Figure 1: System Architecture for Content Consistency

In addition, if subsequent requests for the same content are found in cache, one particular attribute, called the Time-To-Live (TTL, or data freshness), will be checked. If the TTL has not expired yet, the cached copy will be considered as "fresh" and content will be returned to the client without contacting the original content server. Otherwise, the original content server will be asked to send the content back to the client if necessary (using request like "If-Modified-Since").

2.2. Content Model

Functions require certain knowledge in order to work correctly. Here, we use a simple representation of knowledge by using attributes to illustrate the concept behind. There exists a many-to-many relationship between function and attribute, that is, each function might require a set of attributes while each attribute may associate with many functions.

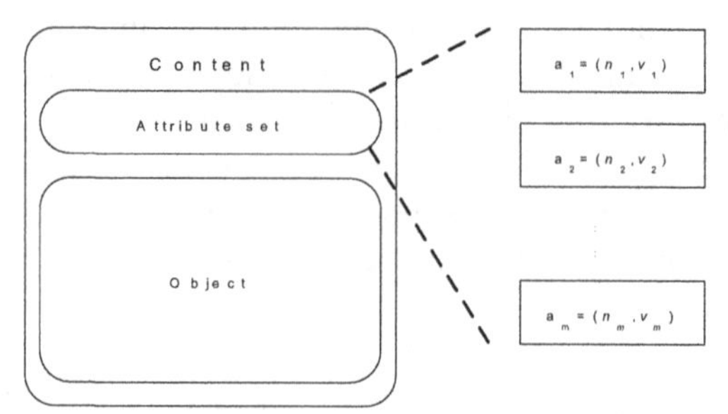

Figure 2: Content Decomposition and Function Mapping

Each object is bundled with relevant attributes to form what we called "content", as shown in Figure 2. Formally, a content C is defined as $C = \{O, A\}$, where O denotes object and A denotes attribute set. We further divide content into two types: *primitive content* and *composite content*, which will be formally defined later in Section 2.4.

2.2.1. Object

O denotes object of any size. Objects we consider here are application-level data such as text (HTML, plain text), images (JPEG, GIF), movies (AVI, MPEG), etc. Objects are also known as resources, data, body and sometimes files. Our model does not assume or require any format or syntax for the data. We treat data only as an opaque sequence of bytes; no understanding of the semantics of objects is required.

2.2.2. Attribute Set

A denotes the attribute set, where $A = \{a_1, a_2, a_3, \ldots\}$ and $a_x = (n_x, v_x)$. The attribute set is a set of zero or more attributes, a_x. Each attribute describes a unique concept and is presented in the form of a (n, v) pair. n refers to the name of the attribute (the concept it refers to) while v is the value of the attribute. Examples are ("Date", "12 June 2004") and ("Content-Type", "text/html"). We assume that there is no collision of attribute names. That is, no two attributes will have the same name but describe different concepts. This is a reasonable assumption and can be achieved in practice by adopting proper naming conventions. Since an attribute set A might associate with a few functions, we denote this set of functions as $F_A = \{f_1, \ldots, f_n\}$. We can also divide the attribute set A into smaller sets according to the functions they serve. We call such an attribute set function-specific attribute set, denoted A^f where the superscript f represents the name of function. Consequently, for any content C, the union of all its function-specific attribute sets is the attribute set A.

Suppose $F_A = \{f_1, \ldots, f_n\}$, then $\{A^{f_1} \cup \ldots \cup A^{f_n}\} = A$

2.2.3. Equivalence

In our discussion, we use the equal sign (=) to describe the relation among contents, objects and attribute sets. The meaning of equivalence in the three cases is defined as follows:

Case 1: $O_1 = O_2$. In this case, O_1 is bit-by-bit equivalent to O_2.

Case 2: $A_1 = A_2$. Here we mean that $\forall a_x \in A_1$, $a_x \in A_2$ and vice versa (set equivalence) or A_1 is semantically equivalent to A_2.

Case 3: $C_1 = C_2$, we mean that $O_1 = O_2$ and $A_1 = A_2$.

2.3. Content Operations

Two operations are defined for content: *selection* and *union*. *Selection* operation is typically performed when a system wishes to do a specific function on content while *union* operation is used to compose content. Both operations might be used in content transformation / adaptation.

2.3.1. Selection

The selection operation, SEL_{FS}, is an operation to filter a content so that it contains only the object and the attributes of a set of selected functions. Let n be the selected functions, represented by the set $FS = \{fs_1, \ldots, fs_n\}$. The selection operation SEL_{FS} on content C is defined as:

$$SEL_{FS}(C) = SEL_{FS}(\{O, A\})$$
$$= \{O, A^{fs_1} \cup \ldots \cup A^{fs_n}\}$$
$$= \left\{ O, \bigcup_{i=1}^{n} A^{fs_i} \right\}$$

2.3.2. Union

The union operation, $\cup$, is an operation to combine m contents $C_1, \ldots, C_m$ into a single content, provided all of them have the same object, that is $O_1 = \ldots = O_m$. The union operation $\cup$ on content $C_1, \ldots, C_m$ is defined as:

$$C_1 \cup \ldots \cup C_m$$
$$= \{O_1, A_1\} \cup \ldots \cup \{O_m, A_m\}$$
$$= \left\{ O_1, \bigcup_{i=1}^{m} A_i \right\}$$

2.4. Primitive and Composite Content

We classify content into two types: primitive content and composite content.

Definition 1: Primitive Content (C)

A *primitive content* C^f is a content that contains only object and attributes of a function, where the superscript f denotes the name of the function. This condition can be expressed as $F_A = \{f\}$. Primitive content is also called function-specific content. A primitive content can be obtained by applying a selection operation on any content C, that is, $C^f = SEL_{\{f\}}(C)$.

Definition 2: Composite Content (CC)

A *composite content*, CC is a content that contains attributes of more than one function. This condition can be expressed as $|F_A| > 1$.

There are two ways to generate a composite content: selection or union. The first method is to apply selection operation on content C with more than one function.

$$CC = SEL_{FS}(C), \text{ where } |FS| > 1$$

The second method is to apply union operation on contents C_1, ..., C_r, if they contain attributes of more than one function.

$$CC = \bigcup_{i=1}^{i=r} C_i \text{ , where } r > 1 \text{ and } \exists i, j; 1 \le i, j \le r; i \ne j; F_{A_i} \ne F_{A_j}$$

3. CLASSIFICATION OF CONTENT CONSISTENCY

Content consistency compares two primitive contents: a *subject* S and a *reference* R. It measures how consistent, coherent, equivalent, compatible or similar is the subject to the reference. Let C_S and C_R represent the set of all subject contents and the set of all reference contents respectively. Content consistency is a function that maps the set of all subject content and reference content to a set of consistency classes:

$$is\_consistent : \{C_S, C_R\} \mapsto \{Sc, Oc, Ac, Wc\}$$

The four classes of content consistency are strong, object-only, attributes-only and weak consistency. For any two content $S \in C_S$ and $R \in C_R$, to evaluate the consistency of S against R, both S and R must be primitive content. That is, they must only contain attributes of a common function; otherwise, content consistency is undefined. This implies that content consistency addresses only one function at a time. To check multiple functions, content consistency is repeated with primitive contents under different functions.

Definition 3: Strong Consistency (*Sc*)

S is strongly consistent with R if and only if $O_S = O_R$ and $A_S = A_R$.

Definition 4: Object-only Consistency (*Oc*)

S is object-only consistent with R if and only if $O_S = O_R$ and $A_S \ne A_R$.

Definition 5: Attributes-only Consistency (*Ac*)

S is attributes-only consistent with R if and only if $O_S \ne O_R$ and $A_S = A_R$.

Definition 6: Weak Consistency (*Wc*)

S is weakly consistent with R if and only if $O_S \ne O_R$ and $A_S \ne A_R$.

The four content consistency classes are non-overlapping classification: given any subject and reference content, they will map to one and only one of the consistency classes. Each consistency class represents an interesting case for research especially since *Oc*, *Ac*, and *Wc* are now common in pervasive content delivery context. Each has its unique characteristics, which poses different implications on functions and content reuse.

4. CASE STUDY: REPLICA / CDN

4.1. Objective

Large web sites usually replicate content to multiple servers (replicas) and dynamically distribute the load among them. A crucial aspect of replication is to ensure all replica hold the exact same content. This is particularly important for end users as replication is usually transparent to them. Any inconsistency can create problems because replicas are collectively viewed as a single server and are expected to behave like one. The objective of this experiment is to study a set of replicated web content to determine whether they are consistent in terms of content attributes.

4.2. Experimental Setup ad Methodology

In our study, the input was the NLANR traces gathered on Sep 21, 2004. For each request, we extracted the site (Fully Qualified Domain Name) from the URL and performed DNS queries using the Linux "dig" command to list all the IP addresses associated with the site. Usually, each site translates to only one IP address. However, if a site has more than one IP address, it indicates the use of DNS-based round-robin load-balancing. This usually means that the content of the website is replicated to multiple servers. An example is shown in Table 1.

Site	www.cnn.com
Replica	64.236.16.116, 64.236.16.20, 64.236.16.52, 64.236.16.84, 64.236.24.12, 64.236.24.20, 64.236.24.28

Table 1: An Example of Site with Replicas

In reality, there might be more than one server behind each IP address. However, it is technically infeasible to find out how many servers are there behind each IP address and there is no way to access them directly. Therefore, in our study, we only consider each IP address as one replica.

The traces originally contained 5,136,325 requests, but not all of them were used in our study. We performed two levels of pre-processing. Firstly, we filtered out URLs not using replica (3,617,571 of 5,136,325). Secondly, we filtered out URLs with query string (those containing the ? character) because NLANR traces were sanitized by replacing each query string with a MD5 hash. URLs with query strings became invalid and we could no longer be able to fetch them for study (227,604 of 1,518,754). After pre-processing, we had 1,291,150 requests to study, as shown in Table 2.

Input traces	NLANR Sep 21, 2004
Request studied	1,291,150
URL studied	255,831
Site studied	5,175

Table 2: Statistics of Input Traces

4.3. Results

We are interested to know the extent of trustworthiness of attributes (i.e. the HTTP header fields) of web content in general. We did this by searching for URL with "critical inconsistency" in caching and revalidation headers, according to the types stated in Table 3. Here, the word "critical inconsistency" means that the discrepancy of the attribute values will result in wrong content presentation to the content requester.

Function in Network	Content Attributes Problems (Based on HTTP Header Fields)
Caching	Expires: missing, Conflict
	Pragma: missing
	Cache-Control: missing, or the directives private, no-cache, no-store, must-revalidate, max-age missing, or max-age inconsistent
	Vary: missing or conflict
Revalidation	ETag: missing, multiple or conflict
	Last-Modified: missing, multiple or conflict

Table 3: Critical Inconsistency / Inaccuracy in Caching and Revalidation Headers

(where Missing – the header appears in some but not all of the replicas; Multiple – at least one the replica have multiple header H ; Conflict – at least 2 of the replica have conflicting and unacceptable values).

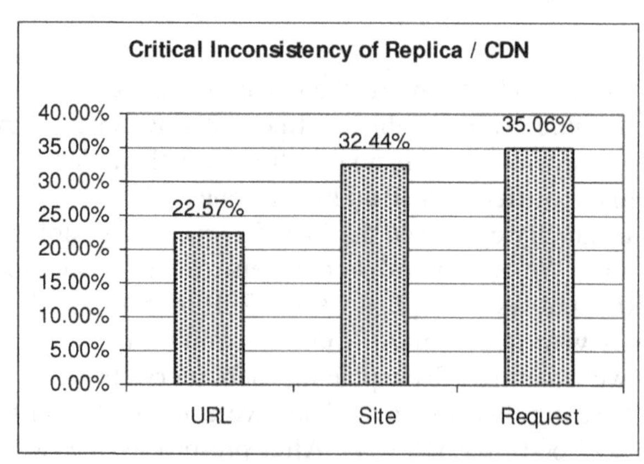

Figure 3: Distribution of Inaccurate/Inconsistency Headers of Replica / CDN

Figure 3 shows that 22.57% of URLs with replica suffer some form of critical inconsistency. In turns, this affects 32.44% of sites and 35.06% of requests. Our study confirms that replica and CDN suffer severe inconsistency problems that result in revalidation failure (performance loss), caching error, content staleness and presentation errors. This definitely puts up a serious trust problem on the content presentation seen by web clients.

5. RELATED WORK

HTTP/1.1 supports basic consistency management using TTL (time-to-live) mechanism. Gwertzman et al. [4] proposed the adaptive TTL which is based on the Alex file system [1]. In this approach, the validity duration of a content is the product of its age and an update threshold. Weak consistency guarantee offered by TTL may not be sufficient for certain applications, such as websites with many dynamic or frequently changing objects. As a result, server-driven approach was proposed to offer strong consistency guarantee [7]. Duvvuri et al. [3] proposed adaptive lease which intelligently compute the optimal duration of leases to balances these tradeoffs.

Many web pages are dynamically generated upon request and are usually marked as non-cachable. This causes clients to retrieve them upon every request, increasing server and network resource usage. Challenger et al. [2] proposed the Data Update Propagation (DUP) technique, which maintains data dependence information between cached objects and the underlying data (eg. database) which affect their values in a graph. MONARCH is proposed to offer strong consistency without having servers to maintain per-client state [8]. The approach achieves strong consistency by examining the objects composing a web page, selecting the most frequently changing object on that page and having the cache request or validate that object on every access.

Caching and replication creates multiple copies of content, therefore consistency must be maintained. This problem is not limited to the web, many other distributed computing systems also cache or replicate content. Saito et al. [10] did an excellent survey of consistency management in various distributed systems. Solutions for consistency management in distributed systems share similar objective, but differ in their design and implementation. They make use of their specific system characteristics to make consistency management more efficient. For example, Ninan et al. [9] extended the lease approach for use in CDN, by introducing the cooperative lease approach. Another solution for consistency management in CDN is [11].

6. CONCLUSION

In this paper, we study the inconsistency problems in web-based information retrieval. Firstly, we redefine content as entity that consists of object and attributes. Then, we propose a novel content consistency model and introduce four content consistency classes. We also show the relationship and implications of content consistency to web-based information retrieval. In contrast to data consistency, "weak" consistency in our model is not necessarily a bad sign. To support our content consistency model, we present a detailed case study of inconsistency in web replica and CDN. Replicas and CDN are usually managed by the same organization, making consistency maintenance easy to perform. In contrast to common beliefs, we found that they suffer severe inconsistency problems, which results in consequences such as unpredictable caching behavior, performance loss, and content presentation errors. This not only raises a big question of trustworthiness of content delivered on Internet but also provides hints to address this problem, content integrity and ownership. This has already been included in the extension of our work here.

REFERENCES

1. V. Cate. Alex – a global filesystem. Proceedings of the 1992 USENIX File System Workshop, May 1992.
2. J. Challenger, A. Iyengar, P. Dantzig. A Scalable System for Consistently Caching Dynamic Web Data. Proceedings of IEEE INFOCOM'99, Mar 1999.
3. V. Duvvuri, P. Shemoy, and R. Tewari. Adaptive leases: A Strong Consistency Mechanism for the World Wide Web. Proceedings of INFOCOM 2000.
4. J. Gwertzman and M. Seltzer. World-wide Web cache consistency. Proceedings of the 1996 Usenix Technical Conference, 1996.
5. M. Hori, G. Kondoh, K. Ono, S. Hirose and S. Singhal. Annotation-based Web Content Transcoding. Proceedings of The Ninth International World Wide Web Conference (WWW9), May 2000.
6. B. Knutsson, H. Lu and J. Mogul. Architecture and Pragmatics of Server-Directed Transcoding. Proceedings of the 7th International Web Content Caching and Distribution Workshop, pp. 229-242, August 2002.
7. C. Liu and P. Cao. Maintaining Strong Cache Consistency in the World-Wide Web. IEEE Transactions on Computers. 1998.
8. M. Mikhailov and C. E. Wills. 2003. Evaluating a New Approach to Strong Web Cache Consistency with Snapshots of Collected Content. Proceedings of WWW 2003, May 2003.
9. R. Mohan, J. R. Smith and C. Li. Adapting Multimedia Internet Content for Universal Access. IEEE Transactions on Multimedia, Vol. 1, No. 1, 1999 pp. 104-114.
10. A. Ninan, P. Kulkarni, P. Shenoy, K. Ramamritham, and R. Tewari. Cooperative Leases: Mechanisms for Scalable Consistency Maintenance in Content Distribution Networks. Proceedings of WWW 2002, May 2002.
11. Y. Saito and M. Shapiro. Optimistic replication. To appear in ACM Computing Surveys. http://www.hpl.hp.com/personal/Yasushi_Saito/survey.pdf

Content Description Model and Framework for Efficient Content Distribution

Chi-Hung Chi[1], Lin Liu[1], Shutao Zhang[2]
[1] School of Software, Tsinghua University, Beijing, China
[2] School of Computing, National University of Singapore, Singapore
Contact email: chichihung@mail.tsinghua.edu.cn

ABSTRACT. In this paper, we propose a content description model and framework for efficient content distribution. The content description model employs ideas from Resource Description Framework and External Annotation, which allow flexible descriptions for Web content. The model also allows a server to efficiently select any subset of the descriptions of any Web page and deliver them to a proxy. The framework consists of algorithms for the proxies to map user preferences and device capabilities to a set of functions to be performed, and for the server to select and deliver necessary content descriptions to the proxy, and for the proxy to efficiently cache and reuse the content descriptions. With our content description model and framework, best-fit content presentation for pervasive Internet access can be made possible.

I. INTRODUCTION

Today, the Web has become a highly heterogeneous environment. To accommodate the needs for heterogeneous users and devices, network nodes between servers and end users start to perform various functions (such as image transcoding, content transformation, and content filtering) on the Web content before it is distributed to the users. These network nodes are often referred to as active web intermediaries or proxies, in the rest of this paper, we call them proxies. In order to perform these functions properly, a proxy usually requires semantic information about the Web content. We will refer this kind of semantic information as content descriptions here.

To support various functions, many content description models and frameworks have been proposed to provide semantic information about different attributes of Web content. For example, the Edge Side Includes (ESI) [4] language was proposed to describe attributes such as expected expiry time or Time-To-Live (TTL) for Web content to support dynamic content caching. Extensible Device Independent Markup Language (XDIME) was proposed by Volantis [5] to describe content layout, image color and others attributes to support content adaptation for mobile devices. Under the heterogeneous environment, efficient content distribution has become a problem. Some of the major challenging issues related to this problem are listed given below.

Mark Last et al. (Eds.): Advances in Web Intelligence and Data Mining (SCI) **23**, 51-60 (2006)
www.springerlink.com

First of all, it is not clear how a proxy should decide which functions to perform given any user preferences and device capabilities. It is because it is not easy, if possible at all, for every proxy to understand every type of devices and users, and the users may not be able to know all the functions provided by proxies, either. If this is not properly handled, we may end up delivering non-acceptable content to users. Secondly, to provide semantic information about different attributes of Web content, the server may need to store a large amount of content descriptions. Delivering all the descriptions about a Web page to a proxy when the Web page is requested may be highly inefficient because the proxy may only need a small fraction of the content descriptions to perform the desirable functions. Thirdly, repeatedly delivering the same content descriptions to the same proxy is unnecessary. But insofar, there lacks a mechanism for a proxy to properly cache and reuse the content descriptions that are already retrieved.

In this paper, we propose a content description model and framework for efficient content distribution. The content description model employs ideas from Resource Description Framework [8] and External Annotation [1] [6], which allow flexible descriptions for Web content. The model also allows a server to efficiently select any subset of the descriptions of any Web page and deliver them to a proxy. The framework consists of several algorithms for the proxies to map user preferences and device capabilities to a set of functions to be performed, and for the server to select and deliver necessary content descriptions to the proxy, and for the proxy to efficiently cache and reuse the content descriptions.

2. RELATED WORKS

Related to the content description models, the two representative ones are InfoPyramid [7] and Resource Description Framework (RDF) [8]. InfoPyramid is a representation scheme for handling Web content (text, image, audio and video) hierarchically along the dimension of fidelity/resolution (in different quality but in the same media type) and modality (in different media type). Content under this model is authored in XML [9], allowing the author to provide more information to the system performing content modification as only limited information about the content can be deducted from an HTML page directly. The content will later be converted to HTML prior to delivery. The authored content is analyzed to extract information that will be useful in adaptation.

Resource Description Framework (RDF) is another general purpose content description framework. This framework is based on XML and uses a collection of triples to provide descriptions. A triple consists of a subject, a predicate and an object. The assertion of an RDF triple says that some relationship, indicated by the predicate, holds between the things denoted by subject and object of the triple. A set of such triples is called an RDF graph. The assertion of an RDF graph amounts to asserting all the triples in it, so the meaning of an RDF graph is the conjunction (logical AND) of the statements corresponding to all the triples it contains. External Annotation [1][6] proposed by W3C has suggested a way to reference to any node of an XML document. For a well formed HTML page, we can parse it into a tree and use External Annotation to create a URI to any node in the HTML parse tree. That means combining RDF and External Annotation can create a very flexible approach to provide descriptions about any node in a well formed HTML Web page.

To deliver the best-fit presentation of content to the users, W3C proposes the Composite Capability and Preference Profile (CC/PP) [2] to describe user preference and device capabilities. Wireless Application Protocol (WAP) Forum also proposes a similar approach named User Agent Profile (UAProf) [10] to handle user descriptions. Both CC/PP and UAProf are based on Resource Description Framework (RDF) [8]; it aims at describing and managing software and hardware profiles. In our framework for efficient content distribution, we can use CC/PP or UAProf to provide descriptions about user preferences and device capabilities.

Besides descriptions about the clients, there are also approaches on the server side to address the issue of customized content delivery. Approaches in this category fall into two main streams: providing web content descriptions or giving instructions on how to process web content from the web server. W3C has proposed a working draft on content selection for web contents for device independence [3]. It specifies a processing model for general purpose content selection. Selection involves conditional processing of various parts of an XML information set according to the results of the evaluation of expressions. Using this mechanism some parts of the information set can be selected for further processing and others can be suppressed. ESI [4] uses a similar mechanism as W3C's content selection. Logical expressions are embedded with ESI markups into HTML object and evaluated at run to determine which fragment will be selected. But the main purpose of the ESI selection is for dynamic content assembly for different users.

3. A GENERAL CONTENT DESCRIPTION MODEL

As mentioned in Section 1, we need a general content description model to provide content semantic information to support different functions by different parties. In this section, we give the general settings for the content description model. Under our setting, we discuss the design considerations of the model. In particular, we look at how the content should be described, and how the descriptions should be organized and associated with the content.

3.1. General Settings

As illustrated in Figure 1, without loss of generality, we assume that there are three entities, namely, a Web server, a proxy, and a user. A user may send requests for Web objects to the Web server through the proxy. Each time, the user may choose to use a difference device for a different application, and sends a list of device capabilities and user preferences to the proxy to indicate the requirements on the Web objects imposed by the device and application for the current request.

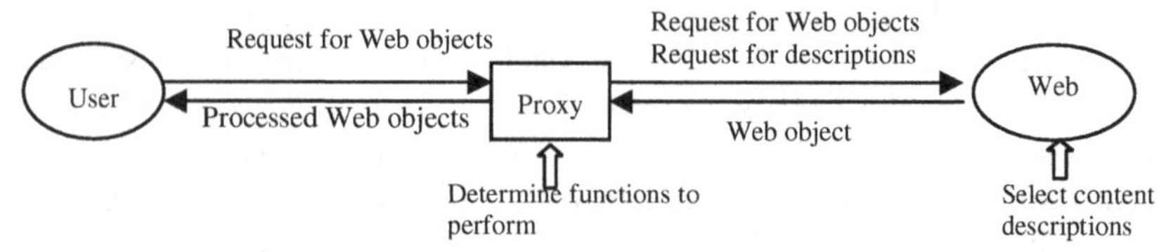

Figure 1: General Settings

The proxy, based on the user's preferences and device capabilities as well its local policies, determines a set of functions to perform on the content before delivering it to the user. Each function here refers to a set of logical operations to be performed on a Web object, e.g., caching and image transcoding. Next, the proxy requests from the server for the Web objects and the content descriptions that are necessary for the corresponding functions. After that, the proxy performs the desired functions on the Web objects and delivers the results to the user. Note that whether some functions are performed on a certain Web object may not depend on the capabilities and preferences provided by the user at all, e.g., in the case of caching, but for some other functions, such decisions may in deed depend on the capabilities and preferences, e.g., image transcoding.

3.2. Proposed Content Description Model

3.2.1. Web Objects

Without loss of generality, we assume that the Web content that the user is interested in is always stored in the form of a mark-up language that is well-formed, in the sense that (1) the tags always appear in pairs, and (2) after removing any pair of tags and the content between them, the remaining content is still well-formed. Note that the second requirement implies that the tags are properly nested, so that we can always compute a parse-tree from a document encoded using such a mark-up language. XML and XHTML are examples of such mark-up languages. In the rest of this paper, we will use XHTML as an example, but our content description model applies to any well-formed mark-up language. We also assume that each Web object is all the content enclosed by a pair of tags, and is always represented by a parse-tree.

Recall that we require that every Web object is uniquely identified by some identifier (ID). For any XHTML page *P* which includes a unique pair of tag "<html>" and "</html>", let its URL be *Up*, then the ID for the Web object that represents *P* would be *Up#root()*, where *root()* represents the root node of the parse tree of the page. Similarly, we can give the identifier for each element in *P*. For example, *Up#root().child(1)* is the ID for first child node of the parse tree, while *Up#root().child(1).child(5)* would be the ID for the fifth child node of the first child node of the parse tree. Note that the above notation is similar to the use of annotation scheme proposed by W3C for content transcoding.

3.2.2. Object Description Scheme

In the proposed model, we give an **object description scheme** (**ODS**) to describe Web objects. Under our framework, every Web object, which is uniquely identified, is associated with a number of descriptions. Each description of a Web object is a tuple <ID, attribute, value>, where ID is the unique identifier associated with the Web object, whose attribute, which is a string, is specified by value, which is another string. The "attribute" is the property of the Web object we want to describe. In particular, the description takes the form of the following.

```
<ods object=ID>
<attribute>value</attribute>
</ods>
```

For example, for a Web object that is an XHTML page, its identifier could be its URL, which is clearly unique among all other Web pages, and the attribute could be a string "Author", and the value is the name of the author. We can write this description as below.

```
<ods
object='http://mywebsite/mywebpage.html#root()'>
<Author>Anonymous Author</Author>
</ods>
```

For the ease of selection of different descriptions, descriptions for a Web object are organized according to the attributes of the Web object, such as expected time of expiry of a Web object. However, to accommodate different types of descriptions for an attribute, we define a *type* to differentiate various types of descriptions for the same attribute. In this way, all the descriptions for a Web object are organized as a set of XML documents, each document stores descriptions for a particular attribute of a particular type. Such an XML document then consists of three parts:

- **XML and XML Name Space Declarations**
  Since all the descriptions are in the form of XML, we need a XML declaration "<?xml version='1.0'?>" to indicate the beginning of a XML document. Further more, all the name spaces used in this XML document needs to be declared too. These name spaces include the scheme's default name space as well as name spaces for property descriptions.
- **Attribute Meta-data Definition**
  This is to specify meta-data about the attribute we are describing in this XML document. These meta-data is specified as different attributes of the tag "<ods:attribute>". The attribute "attr" indicates what attribute we are describing in this document and attribute "type" is for the type of the

description for the attribute. The attribute "isDefault" is an indication on whether this is the default description for this attribute about the Web object and attribute "mode" is mode of the description.

- **Description Tuples**
  As shown above, a description tuple consists of an object ID, a property and its value. Note that the property can be from another namespace other than the default; this allows reusing descriptions by existing content description frameworks. The value can be both literals and markups from a XML namespace. All the description tuples must be enclosed in between the root element "ods:Desp". In Figure 2, there is an example of the descriptions of the Web objects in a simple XHTML page, where the property of the objects is their time-to-live (TTL).

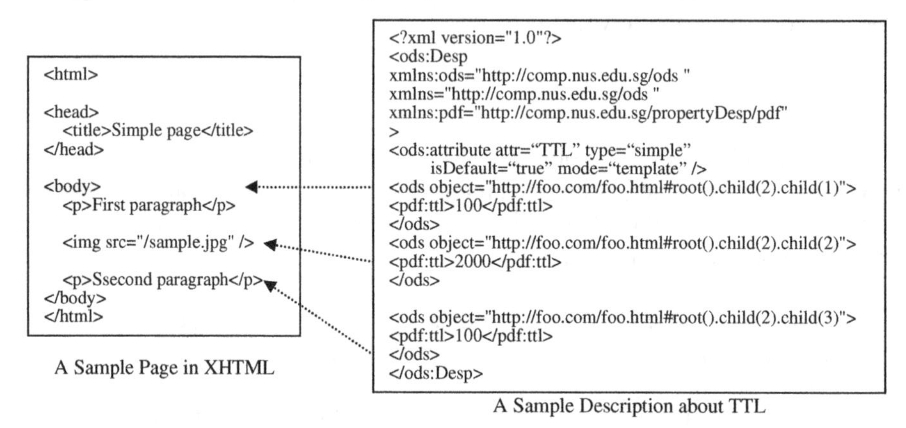

Figure 2: Description for a Simple XHTML Page

3.2.3. Attribute Rules

As one Web object can contain other Web objects, we define some rules to follow when determining the descriptions for a particular Web object:

- *Rule one: By default, attribute descriptions associated with the parent node apply to all Web objects in the child nodes.* This makes description simpler when we specify attribute descriptions for multiple Web objects.
- *Rule two: Ignore irrelevant attribute descriptions.* Sometimes attribute descriptions to a parent node does not apply to child nodes. We can show this example. A web page contains a news article in English with a few images among the text. The author would like to specify an attribute to show that the text information is in English. This description is put at the "root" node level and obvious it doesn't apply on the image Web objects. So we ignore them during processing.
- *Rule three: Local attributes have a higher priority than global attributes.* If attribute descriptions are attached to the child and parent nodes, then those

attached to the child nodes will overwrite the ones in the parent node. This is a complement to rule one.

With the three rules defined above, it is clear on which attributes apply on a particular piece of content.

4. A FRAMEWORK FOR EFFICIENT CONTENT DISTRIBUTION

In previous sections, we have introduced a content description model for Web objects. We focus on how to use the model to provide content descriptions and facilitate efficient selection of content descriptions. In this section, we propose a framework to improve efficiency of content distribution in a heterogeneous environment.

4.1 Design Objectives

The following are the objectives we would like to achieve by the framework:

- **System architecture**
  The functions discussed in previous sections can be performed by the server, the proxy, and even the user. To improve efficiency, we need to have a system architecture where responsibilities of the server, the proxy, and the user are clearly defined.
- **Select the right functions to perform**
  A proxy may have a set of functions for different purposes. For a user with certain preferences and device capabilities, we need to select the right functions to perform on the content before delivering the result to the user. Otherwise, we may end up delivering unacceptable content to the user.
- **Transfer necessary content descriptions only**
  Due to variety of user preferences and device capabilities, web servers need to maintain a large set of content descriptions to support various functions on Web objects. Different proxies may require different portions of content descriptions on the server to support different functions. Hence we need a mechanism for a server to transfer only the necessary descriptions desired by a proxy.
- **Reuse existing content descriptions**
  Repeatedly transferring the same set of descriptions to a proxy is a waste of network bandwidth, especially when the volume of descriptions is large. Under the condition that descriptions for a Web object may expire, we need a mechanism to properly cache, validate and reuse existing descriptions retrieved from a server.

4.2. System Architecture

When we design the architecture to handle content delivery from the Web server to the user in a heterogeneous environment, we need to address the following issues: who provides the content descriptions, who performs the functions, and who provides descriptions about user preferences and device capabilities. In the following, we would like to discuss the considerations for the system architecture first, followed by the overall architecture design.

There are some design considerations we would like to take into account for the system architecture, namely scalable service and transparency to end users. For scalable service, due to rapid increase of Internet hosts and end users, the way to provide content delivery needs to be *scalable.* It means that relatively maintaining the same response time for content delivery with the increasing number of users requiring customized delivery from the same website. For the transparency to users, due to the large amount of users access the Web, it is very inconvenient or even impractical if there is a dramatic change to the software (e.g., browsers) used to access the Web by the users. In our system architecture, users only need to express their preferences and device capabilities and no other change on the software for end users is required.

Having these design considerations, we will address the issues about how to allocate the responsibilities on the Web server, the proxy and the end user. For content descriptions, the Web server is the most appropriate to host them. Since content authors provide Web content, they know their content better than others. It is very natural for the content authors to provide both the Web content and the content descriptions and let all of them hosted on the Web server. As for the question on who to perform the functions on Web content, we have a few choices, namely the Web server, the proxy and even the end user. There are different benefits and constraints for different choices.

For the Web server, it hosts Web content and content descriptions. If it is going to perform functions on Web content, the benefits is that it can refers to content descriptions locally and performs the necessary functions. At the same time, it also has to get the user descriptions (preferences, capabilities) to make the decision on what functions to perform. Since the user descriptions vary, the server may need to perform different sets of functions on the content for different users. When there is an increase in the number of users accessing the server, the server's needs to increase its capability (processing power, network bandwidth) to maintain a relatively constant response time. So this solution has a problem of scalability.

For the proxy to perform those functions, the disadvantage is that additional bandwidth is needed to transfer content descriptions from the server. The advantage is the proxy only handles a small portion of users accessing the server because number of proxies is usually much more the number of servers. So scalability is not an issue. We do not consider performing those functions by the end user here, as this will require dramatic change of the user software and violates our design considerations. Since the proxy is in a better position to perform the functions, there is another argument on who should make the decision on what functions to perform. If the Web server is to make the decision, the benefit is that the server can refer to the content descriptions locally and make the decision. Then it needs to convey the server's decisions to the proxy to perform the functions. However, there is no way for the server to ensure that the proxy will follow the decision. As the proxy has its own configuration and policies, it may or may not follow the decisions by the server. In this case, it is more appropriate for the proxy to make the decision on what functions to perform.

From the above, we have separated the responsibilities into the Web server, the proxy and the end user in the system architecture. The Web server host Web content and content descriptions. The proxy performs functions on Web content and retrieves necessary content descriptions from the server. The end user sends its preferences and device capabilities along with its request for the Web content. In this framework we assume *data integrity* from the Web server to the user. In other words, the content is modified according to their intended use from the server to the user and no malicious entities can jam the content.

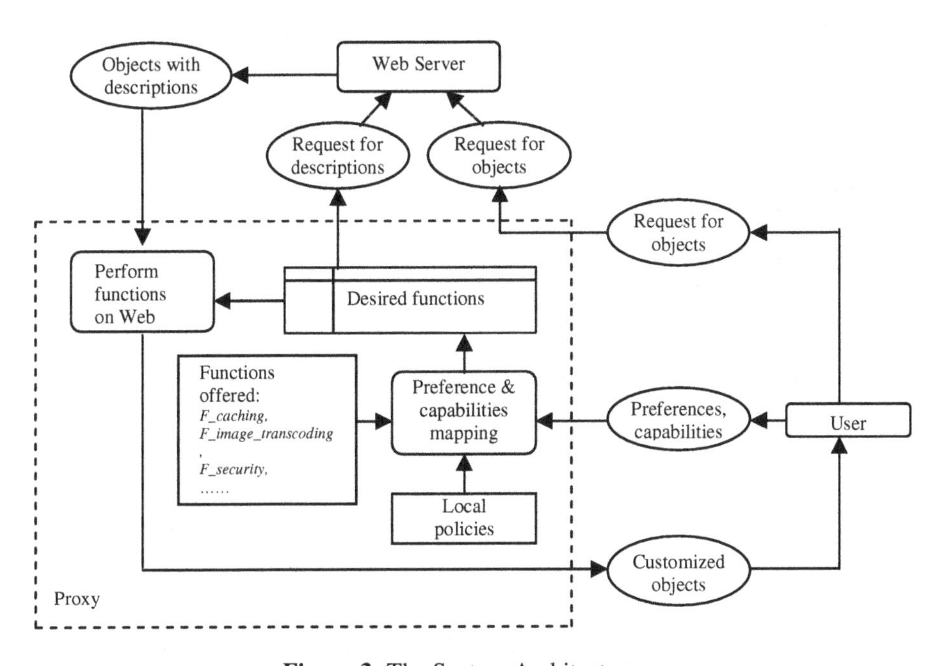

Figure 3: The System Architecture

This architecture is shown in Figure 3. In this figure, the user issues a request to the web server through a proxy for a set of Web objects. At the same time, his preferences and device capabilities are also sent along with the request. The proxy that offers a set of functions examines the user's preferences and device capabilities and some other information to determine what functions need to be performed on the Web objects. It then determines the descriptions needed to support these functions and send a request to the web server hosting these Web objects and the descriptions. The Web server, upon receiving the request, delivers the Web objects and the required descriptions to the proxy. The proxy then performs the functions on Web objects based on the descriptions and deliver the customized Web objects back to the user. In the following sections, we will cover detailed operations for each of the key entities for Web content delivery: server, proxy, and user.

5. CONCLUSION

In this paper, we study the problem of efficient content distribution in a heterogeneous environment. We propose a content description model and framework. The content description model defines how the Web objects are described and how these descriptions are organized. Our model supports a wide range of content descriptions and allows a server to easily select any subset of content descriptions of a given Web page. The framework consists of several algorithms and guidelines for a proxy to easily and correctly map user device capabilities and preferences to a set of functions to be performed, and to determine which content descriptions are necessary to perform these functions. We also define HTTP protocol extensions and local rules for a proxy to select, cache and validate content descriptions from a server. With the proposed model and framework, we can improve the efficiency of content distribution by sending only the desired content to a user, requesting only the necessary content descriptions from a server, and by caching and reusing existing content descriptions at the proxy.

REFERENCES

1. Annotation of Web Content for Transcoding, http://www.w3.org/1999/07/NOTE-annot-19990710/
2. Composite Capability and Preference Profile, http://www.w3.org/Mobile/CCPP/
3. Content Selection for Device Independence, http://www.w3c.org/TR/cselection/
4. ESI Language Specification 1.0, 2000, http://www.esi.org.
5. The Extensible Device Independent Markup Language, http://www.volantis.com
6. H. Masahiro, K. Goh, O. Kouichi, S. Hirose, S. Sandeep, "Annotation-Based Web Content Transcoding," Proceedings of the 9th *WWW Conference*, 2000.
7. R. Mohan, J. R. Smith, C. S. Li, "Adapting Multimedia Internet Content for Universal Access," *IEEE Transactions on Multimedia*, *Vol. 1*, *No. 1*, March 1999, pp. 104–114.
8. Ora L., Ralph S, "Resource Description Framework (RDF) Model and Syntax Specification", *World Wide Web Consortium Recommendation*, 1999, http://www.w3.org/TR/1999/REC-rdf-syntax-19990222/.
9. Resources for the Extensible Markup Language (XML), http://www.w3.org/XML/.
10. User Agent Profile, WAP forum.

Exploiting Wikipedia in Integrating Semantic Annotation with Information Retrieval

Norberto Fernández-García, José M. Blázquez-del-Toro, Luis Sánchez-Fernández and Vicente Luque

Telematic Engineering Department. Carlos III University of Madrid.
{berto,jmb,luiss,vlc}@it.uc3m.es

Summary. The Semantic Web can be seen as an extension of the current one in which information is given a formal meaning, making it understandable by computers. The process of giving formal meaning to Web resources is commonly known in the state of the art as semantic annotation. In this paper we describe an approach to integrate the semantic annotation task with the information retrieval task. This approach makes use of relevance feedback techniques and exploits the information generated and maintained by Wikipedia users. The validity of our approach is currently being tested by means of a Web portal, which also uses the annotations defined by users in providing basic semantic search facilities.

1 Introduction

In order to give to the content of Web resources a formal, computer understandable, meaning and make possible the Semantic Web vision [1], we need to add semantic metadata to such Web resources. The process of adding semantic metadata to Web resources is commonly referred in the state of the art as semantic annotation. In general, the semantic annotation of a Web resource requires to relate, to link, its whole content or a part of it with a certain concept identifier taken from an ontology or other knowledge source.

Semantic annotation has a critical importance in order to make the Semantic Web become a reality. In consequence, this topic is an important field of research in the Semantic Web area. In the literature different approaches to this problem can be found [7, 8, 9, 10, 11, 12, 13, 14, 15, 16, 17, 18, 19], but no one of them takes advantage of the effort of the millions of users who every day look for information on the Web.

So, in [2] we introduced the SQAPS, *Semantic Query-based Annotation, P2P Sharing*, system. The main idea behind this system was to exploit keyword-based user queries in semantic annotation of Web resources. The annotation was performed by the user adding semantic metadata to his query and giving relevance feedback over the results of the query in a Web search

Mark Last et al. (Eds.): Advances in Web Intelligence and Data Mining (SCI) **23**, 61-70 (2006)
www.springerlink.com

engine. The concepts in the annotated query were used to annotate relevant Web resources for such query. Annotations were stored and shared by means of a peer to peer infrastructure.

But the system we presented in [2] had a set of limitations: due to the nature of the peer to peer infrastructure, it only allowed users to define new annotations and to get the annotations related with a certain Web resource. Semantic search facilities, that is, obtaining a set of Web resources talking about a certain concept or set of concepts, were not provided. Additionally, the knowledge used to annotate the queries was taken from a static source: WordNet[1]. The evolution of knowledge with time was left outside of the scope of that work.

In this paper we present an evolution of the system in [2], which adopts the architecture of a Web portal and tries to address these limitations. A prototype of this portal is implemented and is currently being tested. It is publicy available at [3]. The rest of this paper describes the current state of this portal. It is organized as follows: section 2 introduces some basic definitions of terms that we will use in the paper. Section 3 briefly describes the architecture of our system, introducing its main components, their functionality and the way these components have evolved from our previous system. Section 4 describes the current system working model. Section 5 briefly describes some related work in the state of the art. Concluding remarks and future lines, in section 6, finalize this paper.

2 Basic definitions

Term Word or set of words between quotes typed in by the user as part of a query. For instance, the query: *Java tutorial "class definition"*, has three terms: *Java*, *tutorial*, and *"class definition"*.

Query Concept Term or set of terms in a query which, as a result of a query annotation process, have a certain concept associated. This concept is represented by an identifier in a certain knowledge source. For instance, in the previous example, if the user annotates the term *Java* with the identifier *http://monet.nag.co.uk/owl#Java* taken from an ontology about programming available on the Web, we get the query concept *[Java, http://monet.nag.co.uk/owl#Java]*.

Query Literal A term which has not any concepts associated.

Query A sequence of one or more query concepts and/or query literals.

$$query := (query\_concept | query\_literal)+ \qquad (1)$$

Simple Query A query which consist only of a query concept.

$$simple\_query := (query\_concept) \qquad (2)$$

[1] http://wordnet.princeton.edu/

Complex Query A query which has at least a query concept, but also other query concepts and/or query literals.

$$complex\_query := (query\_concept, (query\_concept|query\_literal)+) \quad (3)$$

Annotated Query A query which has at least a query concept. Simple queries and complex queries are both annotated queries.

$$annotated\_query := (simple\_query|complex\_query) \quad (4)$$

Literal Query A query which is only composed of query literals.

$$literal\_query := (query\_literal)+ \quad (5)$$

3 System Architecture

Figure 1 shows the intended architecture of our current system. Its main components are:

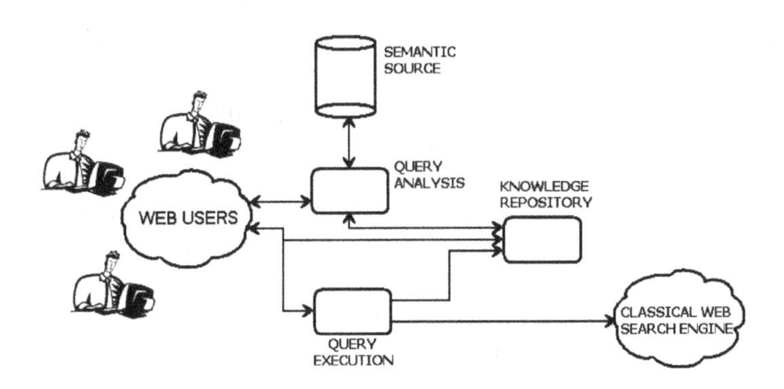

Fig. 1. System Architecture

Query Analysis Its main purpose is to allow the annotation of a keyword-based query by the user. In order to do so, the Query Analysis component divides the query into its terms, and looks into a Semantic Source for candidate concepts to be associated to concrete terms or combinations of terms. In our original system, the process of query annotation was performed by generating all possible term combinations and looking into the Semantic Source for concepts whose labels matched the terms in such combinations. In our current prototype the process of query annotation has been divided into two separate steps. The first one tries to match the query with a single concept (simple query). The second one tries to match the query with several concepts (complex query) and is similar to the original annotation process. We have taken this decision after observing that most popular queries in Google Web Search engine, though sometimes contain more than one term, usually refer to a single concept [4]. So, instead of going directly to the complex query definition process, which requires more effort from users because different terms and combinations need to be considered, we allow the users to directly specify a single concept of interest. If this does not fit the interests of the user, the complex query definition process is still available. It is expected that this decision will make the system more usable, which is a crucial issue in a manual annotation system as the one described here.

Semantic Source It is intended to provide concrete concepts to be used in the query annotation process. For each concept the Semantic Source should provide at least a label, a human readable definition of the concept and a unique identifier. In the first version of SQAPS, we proposed the usage of WordNet as Semantic Source. The main problem with that approach comes from the fact that WordNet is a static Semantic Source and knowledge changes with time: new concepts appear, others can change their meanings, etc. Of course, new versions of WordNet can appear with time, but we should rely on WordNet providers to get new Semantic Source versions and knowledge evolves faster than WordNet. Additionally, the WordNet approach has also other problems: specific vocabularies of concrete domains are not well covered and concrete named entities (persons, locations, etc) are not always inside WordNet. In order to address these problems, we can think of using a more dynamic Semantic Source, allowing users to define their own entries. This leads us to the approach we are currently exploring: using Wikipedia[2] as our Semantic Source. As Wikipedia is based on Wiki[3] technology, users can define their own entries and maintain the information in a collaborative manner. Additionally, the election of Wikipedia as our Semantic Source is supported by other reasons: it provides us with the information we require in our model: label, definition and identifier; it covers a broad range of topics and it has thousands of users collaborating in maintaining the information. Furthermore, there exist external analysis that have shown that its contents seem to be of reasonable quality [5]. On the negative side, the semantics offered by Wikipedia is not very formal. It can be seen as a semantic network of linked topics which, for instance, makes difficult the usage of reasoning mechanisms. In any case, in the state of the art we can find proposals [6] for providing semantics to Wikipedia, making it a more formal source. Our work may benefit from such approaches.

Knowledge Repository It stores all the knowledge generated by the system operation. In our current prototype this includes not only the seman-

[2] http://www.wikipedia.org/
[3] http://en.wikipedia.org/wiki/Wiki

tic annotations defined by system users, as in [2], but also information about the users themselves, about the documents which are annotated and about the concepts which are used in the annotation process. All this information is represented in RDF[4] format according to the RDFS[5] schema available at [3]. This model provides basic annotation characterization based on Annotea schema[6], as in [2], but also takes advantage of well-known metadata vocabularies as Dublin Core[7], PRISM[8], SKOS[9] or FOAF[10] to describe the other entities involved in the system process: persons, concepts and documents.

Query Execution It receives a user query (literal query or annotated query) as input and looks for relevant Web resources for such query. In the case the query is a literal query, this module just sends the query to a Web search engine and shows results to user. In the case the query is an annotated query, it searches both into the Knowledge Repository using the concepts in the query (semantic search) and into Web search engine using the keywords (syntactic search). The results of both processes are combined and shown to the user, in order to allow him to annotate them by giving relevance feedback.

4 System Working Model

The current system working model is described in figure 2. Basically it consists of the following steps:

Authentication In order to gain access to the system the user needs to be authenticated. We require this, because we are interested in associating a certain annotation with the user who has created it. By doing so, the system can learn about the interests of the users. This information could be later exploited in implementing personalized search services or automatic suggestion services. At this moment, the authentication process is performed by a classical login/password scheme. If the user has not an account in the system, he can create a new one simply by providing and e-mail address. When creating a new account, the user can optionally provide an URL linking to his FOAF profile information. At the moment this FOAF information is not used in the system process, but it could be used in the future to implement services exploiting, for instance, the social network links provided by the FOAF profile. When a new account

[4] http://www.w3.org/RDF/
[5] http://www.w3.org/TR/rdf-schema/
[6] http://www.w3.org/2000/10/annotation-ns
[7] http://dublincore.org/
[8] http://www.prismstandard.org/
[9] http://www.w3.org/2004/02/skos/
[10] http://www.foaf-project.org/

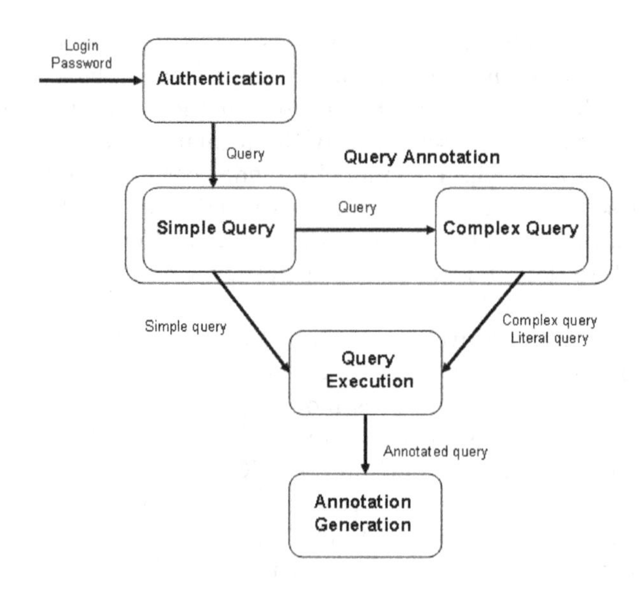

Fig. 2. System Working Model

is created, the information about the user: login, e-mail and FOAF profile URL, is stored, in RDF format, inside the Knowledge Repository.

Query Annotation Once the user has been authenticated, he can access the system and start querying and annotating. In order to do so, the system provides the user with an HTML form where he can type in his query. This query is sent to the Query Analysis component, where the process of query annotation begins. As we have said in previous sections, the query annotation process is divided into two steps:

1. The first step allows the user to look for a single concept associated to his query (simple query definition). In order to do this, a list of possible candidate concepts inside the Semantic Source (Wikipedia) matching the user query needs to be shown. This list is obtained by the system by searching in Google Web Service[11] with a query which consist of the original user query plus the restriction *site:wikipedia.org*. Every Wikipedia entry shown to the user in this list has a button. By clicking on one of these buttons the user can say which one of the entries matches his interests. Given the URL entry, the system knows which concept is the relevant one. Then a simple query is defined and send to the Query Execution component. If no one of the shown entries fits with the user intention, the user can go to the second query annotation step, the complex query definition process.
2. In this step the system splits the original user query into its terms. For each single term the system looks into Wikipedia using Google Web

[11] http://www.google.com/apis/

Service. With the results, a list of Wikipedia entries for all terms is generated and shown to the user. Every entry in the list has a checkbox that the user can click on to add the entry to his query. By clicking on a "More about this" button, the user defines a complex query which is sent to the Query Execution component. Of course, it can also happen that the user does not find any entry of interest. In such a case, the user defines a literal query, which is also sent to the Query Execution component. It has to be noted that term combinations are not generated in our current prototype, in order to reduce the number of possibilities to be shown to the user and make the query annotation process simpler. In any case, the user can always define a term with several words simply by reformulating the query using quotes.

Another process which occurs during Query Annotation is the concept information capture. Every time the user expresses his interest in a certain Wikipedia page representing a concept, the system looks for the URL of the Wikipedia page into the Knowledge Repository to know if the concept which is represented in a certain language for such page is already there. If the concept is not inside the repository, the Wikipedia page describing it is processed to obtain the concept label and definition in a certain language and the links to Wikipedia pages in other languages talking about the same topic. All this information is represented in RDF and stored in the Knowledge Repository. As Wikipedia pages can change with time, the processing of the Wikipedia pages of concepts already inside the Knowledge Repository is repeated periodically to refresh the information.

Query Execution This process takes a query as input, both literal query or annotated query, and shows to the user a set of relevant Web resources for such query. In the case of a literal query, the query execution process just consists in redirecting the query literals to Google Web Service and showing results to the user. No semantic annotations can be generated, as the query has no semantics. Let us assume this is not the case, and the system receives an annotated query. In such a case the system performs two operations. First, it uses the terms in the query to search in Google. Second, it takes the URLs of Wikipedia entries relevant to the query, uses them to obtain the concepts in the query and looks into the Knowledge Repository for Web resources annotated with such concepts. The results obtained from Knowledge Repository could be ranked attending to many different criteria such as annotation freshness (the more recent the annotations which link the resource to the concepts are, the better), number of different users which have annotated the resource with a certain concept (the more, the better), matching between concepts annotating the resource and user interests, etc. In our current implementation a simple ranking approach is used: the more annotations from different users a resource has, the better. Additionally, if a resource is in both Google results and Knowledge Repository results it is given more relevance. More complex ranking mechanisms may be explored in the future. Once obtained the most relevant results from Google and the Knowledge Repository, these are shown to the user. By clicking on a button, he could say if a certain resource is relevant for his query. By doing so, the annotation generation process starts. It has to be noted that the user explicitly needs to give relevance feedback in order to generate an annotation. So he decides at any moment which information is shared with the system and its users, minimizing privacy problems.

Annotation Generation This process just associates an annotated query with a Web resource, generating an RDF representation of the annotation and storing it inside the Knowledge Repository. As can be seen in the schema available at [3], every annotation includes information about its author, a creation timestamp, a link to the resource being annotated and the query literals and query concepts of the query which has been used to generate the annotation. During this process it is also checked if information about the Web resource is available at the Knowledge Repository. If not, RDF triples describing the document are generated and inserted into the Knowledge Repository. Basically these triples contain information about the URL of the Web resource, its title and a snippet.

5 Related Work

In the state of the art in semantic annotation, we can find tens of proposals. Attending to the degree of automation of the annotation process, we could classify such proposals as:

Manual Annotation Systems It is the case of systems as, for instance, Annotea [7], the SHOE Knowledge Annotator [8], SMORE [9] or the CREAM framework [10]. Typically these systems provide a user interface which allows human annotators to view and browse both ontologies and Web resources, using the knowledge in the ontologies to add annotations to the Web resources. Also of interest to the topic of manual annotation are the proposals of extending content creation tools with semantic annotation capabilities, like Semantic Word [11]. These systems allow the addition of annotations to the contents of the resources while these contents are composed.

(Semi)automatic Annotation Systems It is the case of systems as, for instance, AeroDAML [12], SemTag [13], S-CREAM [14], PANKOW [15], C-PANKOW [16], KIM [17] or MnM [18]. Basically these systems exploit natural language processing techniques, like pattern discovery and matching or machine learning techniques, in order to extract the references in text to certain concepts in ontologies. Another work which deserves special attention in this section is [19] where the authors propose a system where a Web site manager can annotate manually SQL queries to a database used to generate dynamic Web pages. Then, the pages generated by the system are automatically annotated using the annotations on SQL.

But no one of the systems described here explores the possibility of integrating the semantic annotation task with the keyword-based information retrieval task. Additionally, exploiting the information maintained collaboratively by Wikipedia users as knowledge source for the annotation process is also not suggested by none of these works.

6 Conclusions and Future Lines

In this paper we have described an approach for the manual semantic annotation of Web resources that could exploit the effort of the millions of users who every day look for information on the Web by integrating the semantic annotation task with the keyword-based information retrieval task.

The system introduced here exploits the information generated and maintained by Wikipedia users in the annotation process. This has the advantage that the source of knowledge evolves with time instead of being static.

On the negative side, one of the main limitations of our system comes from the dependence on user collaboration. But, from our point of view, integrating the annotation activities with habitual user actions, as Web search, should be a point in favor. We expect that the work finally done by the users will imply a low overhead compared to classical keyword-based search.

A Web portal showing the functionalities of the system described in the paper is currently implemented and being tested. It is publicly accessible from [3]. Future lines of development of this prototype could include:

- Integrate some trustness approach which could determine the degree of trust of a certain annotation. This would allow us to deal with users inserting into the system wrong annotations both by error or malice.
- Perform extensive usability analysis in order to test the validity of our approach.

Acknowledgements

This work has been partially funded by the *Ministerio de Educación y Ciencia de España*, as part of the Infoflex Project, TIC2003-07208.

References

1. Berners-Lee T, Hendler J, Lassila O (2001) The Semantic Web: A new form of Web content that is meaningful to computers will unleash a revolution of new possibilities. Scientific American, May 2001.
2. Fernández-García N, Sánchez-Fernández L, Blázquez-del-Toro J, Larrabeiti D (2005) An Ontology-based P2P System for Query-based Semantic Annotation Sharing. In Ontologies in P2P Communities Workshop colocated with ESWC 2005, Heraklion, Greece, May 2005.

3. SQAPS Homepage, http://www.it.uc3m.es/berto/SQAPS/
4. Google ZeitGeist, http://www.google.com/press/zeitgeist.html
5. Wikipedia Evaluations, http://en.wikipedia.org/wiki/Wikipedia#Evaluations
6. Krtzsch M, Vrandecic D, Vlkel M (2005) Wikipedia and the Semantic Web - The Missing Links. Proceedings of Wikimania 2005: The First International Wikimedia Conference, Germany, August 2005.
7. Kahan J, Koivunen MR, Prud'Hommeaux E, Swick RR (2001) Annotea: An Open RDF Infraestructure for Shared Web Annotations. In 10th International World Wide Web Conference, Hong Kong, May 2001.
8. SHOE Knowledge Annotator http://www.cs.umd.edu/projects/plus/ SHOE/KnowledgeAnnotator.html.
9. Kalyanpur A, Hendler J, Parsia B, Golbeck J; SMORE - Semantic Markup, Ontology, and RDF Editor, http://www.mindswap.org/papers/SMORE.pdf.
10. Handschuh S, Staab S (2002) Authoring and Annotation of Web Pages in CREAM. In 11th International World Wide Web Conference, Honolulu, Hawaii, May 2002.
11. Tallis M (2003) Semantic Word Processing for Content Authors. In 2nd International Conference on Knowledge Capture, Sanibel, Florida, USA, 2003.
12. Kogut P, Holmes W (2001) AeroDAML: Applying Information Extraction to Generate DAML Annotations from Web Pages. In 1st International Conference on Knowledge Capture. Workshop on Knowledge Markup and Semantic Annotation, Victoria, B.C. October 2001.
13. Dill S, Eiron N, Gibson D, Gruhl D, Guha R, Jhingran A, Kanungo T, Rajagopalan S, Tomkins A, Tomlin JA, Zien JY (2003) SemTag and Seeker: Bootstrapping the semantic web via automated semantic annotation. In 12th International World Wide Web Conference, Budapest, Hungary, May 2003.
14. Handschuh S, Staab S, Ciravegna F (2002) S-CREAM: Semi-automatic CREAtion of Metadata. In European Conference on Knowledge Acquisition and Management, Madrid, Spain, October 2002.
15. Cimiano P, Handschuh S, Staab S (2004) Towards the Self-annotating Web. In 13th International World Wide Web Conference, New York, USA, May 2004.
16. Cimiano P, Ladwig G, Staab S (2005) Gimme' The Context: Context-driven Automatic Semantic Annotation with C-PANKOW. In 14th International World Wide Web Conference, Chiba, Japan, May 2005.
17. Popov B, Kiryakov A, Kirilov A, Manov D, Ognyanoff D, Goranov M (2003) KIM, Semantic Annotation Platform. In 2nd International Semantic Web Conference, ISWC 2003, LNCS 2870, pp. 835-849.
18. Vargas-Vera M, Motta E, Domingue J, Lanzoni M, Stutt F, Ciravegna F (2002) MnM: Ontology Driven Semi-Automatic and Automatic Support for Semantic Markup. In 13th International Conference on Knowledge Engineering and Management, Springer Verlag, 2002.
19. Handschuh S, Staab S, Volz R (2003) On Deep Annotation. In 12th International World Wide Web Conference, Budapest, Hungary, May 2003.

Ontology based Query Rewriting on Integrated XML based Information Systems

Jinguang Gu and Yi Zhou

College of Computer Science and Technology, Wuhan University of Science and Technology, Wuhan 430081, China simon@wust.edu.cn

As an extension to database information system, XML based information system (XIS) plays a key role in web information community. This paper develops a semantic query rewriting mechanism on Integrated XISs with complex ontology mapping technology to enable accessing web information at semantic level. It discusses the patterns of complex ontology mappings at first, and discusses the ontology-based query mechanism in mediated integrated environment, which includes the extension of XML query algebra and XML query rewriting mechanism secondly.

1 Introduction

Due to its ability to express semi-structured information, XML based information System (XIS), which is an extension to Database Information System, plays a key role in web information community, and XML is rapidly becoming a language of choice to express, store and query information on the web, other kinds of web information such as HTML-based web information can be transferred to XML based information with annotation technologies. Users can query information with XML languages, XPath based languages such as XQuery, XUpdate are suitable for retrieving information in distributed integration systems. Problems that might arise due to heterogeneity of the data are already well known within the distributed database systems community: *structural heterogeneity* and *semantic heterogeneity*. Structural heterogeneity means that different information systems store their data in different structures. Semantic heterogeneity considers the content of an information item and its intended meaning[1]. How to accessing distributed information with a consistent semantic environment and how to make the XML query mechanism with semantic enabled are the main problems that should be discussed in distributed XISs.

The use of ontologies for the explication of implicit and hidden knowledge is a possible approach to overcome the problem of semantic heterogeneity.

Mark Last et al. (Eds.): Advances in Web Intelligence and Data Mining (SCI) **23**, 71-80 (2006)
www.springerlink.com

Ontologies can be used to describe the semantics of the XIS sources and to make the content explicit. With respect to the data sources, they can be used for the identification and association of semantically corresponding information concepts.

This paper focuses on how to use ontology technology to enable semantic level querying on integrated XISs. It uses ontology mapping technology to get a consistent semantic environment, and extends the XML based querying technologies to enable semantic querying on XISs. The remainder of this paper is structured as follows. Section 2 gives the general discussion about ontology enabled integrated XML based information systems (XISs). Section 3 defines the patterns of ontology mapping with semantic similarity enhanced, we use description logic style formalism to introduce the notions used in this paper. Section 4 discusses ontology enabled querying mechanism on integrated XML based semi-structured information systems, such as ontology enhanced XML algebra and XML query rewriting. Section compares with the related works. Section 6 summarizes the whole paper.

2 The Integrated XML based Information Systems

Because XML has been the standard language to represent web information and semantic web resources, using XML to represent web based or semi-structured information systems is a good choice. In the distributed environment, every local site contains the local ontology based structured or semi-structured information source, this information source may be a relational database, native XML database, web site, XML based application or other autonomous system. From the point of view of web based or semi-structured information processing, all local information sites can be expressed as collections of XML instances. An XML instance can be described as a structure $X_d := (V_d, E_d, \delta_d, \mathcal{T}_d, \mathcal{O}_d, t_d, oid_d, root_d)$[2]. By an XML based information system (XIS) we mean $S = (\{X_d\}, W)$, where $\{X_d\}$ is a finite set of XML instances, W is the ontology based wrapper or mediator[3].

By a integrated XML based information system we mean $PS = (\{S_i\}_{i \in I}, M)$, where I is a set of sites, S_i is an XIS for any $i \in I$, M is mapping relation on the set I which can be expressed as $M : (S_1, S_2, \ldots, S_n) \rightarrow S_0$ while S_i $(1 \leq i \leq n)$ denotes the local XIS sites, S_0 denotes the global XIS site acted as the mediator site, M denotes an integrated procedure.

Ontologies are used for the explicit description of the information source semantics. We employ web service based components named **Semantic Adapter** to perform the task of semantic information processing on every local site. Semantic adapter acts as the wrapper of local site information, different local site has different semantic adapter. The function of the semantic adapter can be described as follows[3]:

- **Ontology Establishing**. With the help of domain expert, semantic adapter creates the local ontology to supply a local semantic view to express the semantic of local information source;
- **Semantic Mapping**. The semantic adapter maintains a mapping table, which maps the local semantic to the semantic of other sites using method introduced in paper [3];
- **Query Processing**. The semantic adapter accepts the query request from the other sites, transfers it to the form which the local information source, the local information source executes the query and the semantic adapter transforms the result to the form which the other XIS sites need using XSLT technology;
- Some other functions will be added in the future.

The main components of semantic adapter can be described as following:

1. **SKC** (semantic knowledge construction). SKC constructs semantic mapping knowledge between schemata, it uses the results of schema extraction and concept matching to establish mapping between local and global semantic schemata. The mapping knowledge is saved in VMT;
2. **MDD** (Meta data dictionary). It could include some description of information source, such as schema, storage path, type and provider etc.;
3. **SKB** (semantic knowledge base). It includes the knowledge needed to understand Ontology concept and their attributes, they are synonymous words, comparison of Chinese and English etc. these knowledge is crucial to concept matching. SKB can expand automatically in the process of matching;
4. **VMT** (vocabulary mapping table).The VMT contains the mapping list of local ontology and its instances, one item of the list can be described as follows:

```
<TItem>
  <STerm>Ontology.Term</STerm>
  <Description>Description of the Concept or Ontology</Description>
  <MappingList>
    <MapItem Type="M"> %Direct Mapping
      %IP Address, Port or Semantic Adapter Service Descrption
      <Source>Source Description</Source>
      <MTerm>Source1.Term</MTerm>
      <Relation>Map1.Relation</Relation>
      <CValue>Confidence Value></CValue>
    </MapItem>
    <MapItem Type="S"> %Subsumption Mapping
      <Source>Source Description></Source>
      <MTerm>Source2.Term1</MTerm>
      ....
      <MTerm>Source2.Termn</MTerm>
```

```
      <Relation>Map2.Relation</Relation>
      <CValue>Confidence Value></CValue>
    </MapItem>
    <MapItem Type="C"> %Composition Mapping
      <Source>Source Description></Source>
      <MTerm>Source3.Term1</MTerm>
      <MConcatenate>Term1.Concatenate</MConcatenate>
      ....
      <MTerm>Source3.Termn</MTerm>
      <MConcatenate>Termn.Concatenate</MConcatenate>
      <Relation>Map3.Relation</Relation>
      <CValue>Confidence Value></CValue>
    </MapItem>
  </MappingList>
</TItem>
```

5. **MQW** (Mapping and Querying Wrap), act as the wrap for ontology mapping and information retrieval.

3 The Patterns of Ontology Mapping

The major bottleneck of semantic query answering on XISs is ontology mapping discovery, that means how to find the similarities between two given ontologies, determine which concepts and properties represent similar notions, and so on. Many technologies, such as heuristics-based, machine learning based or Bayesian network based methods, were discussed in recent years, a survey of ontology mapping is discussed in paper [4]. The patterns of ontology mapping can be categorized into four expressions: direct mapping, subsumption mapping, composition mapping and decomposition mapping[5], a mapping can be defined as:

Definition 1 *A* ***Ontology mapping*** *is a structure* $\mathcal{M} = (\mathcal{S}, \mathcal{D}, \mathcal{R}, v)$, *where* $\mathcal{S}$ *denotes the concepts of source ontology,* $\mathcal{D}$ *denotes the concepts of target ontology,* $\mathcal{R}$ *denotes the relation of the mapping and* v *denotes the confidence value of the mapping,* $0 \leq v \leq 1$.

A direct mapping relates ontology concepts in distributed environment directly, and the cardinality of direct mapping could be one-to-one. A subsumption mapping is used to denote concept inclusion relation especially in the multiple IS-A inclusion hierarchy. The composition mapping is used to map one concept to combined concepts. For example, the mapping *address=contact(country, state, city, street, postcode)* is a composition mapping, in which the concept *address* is mapped to combined concept "*contact, country, state, street, and postcode*" of local schema elements. The decomposition mapping is used to map a combined concept to one local concept. The example for the decomposition is the reverse of the composition. These four mapping patterns can be described in the figure 1.

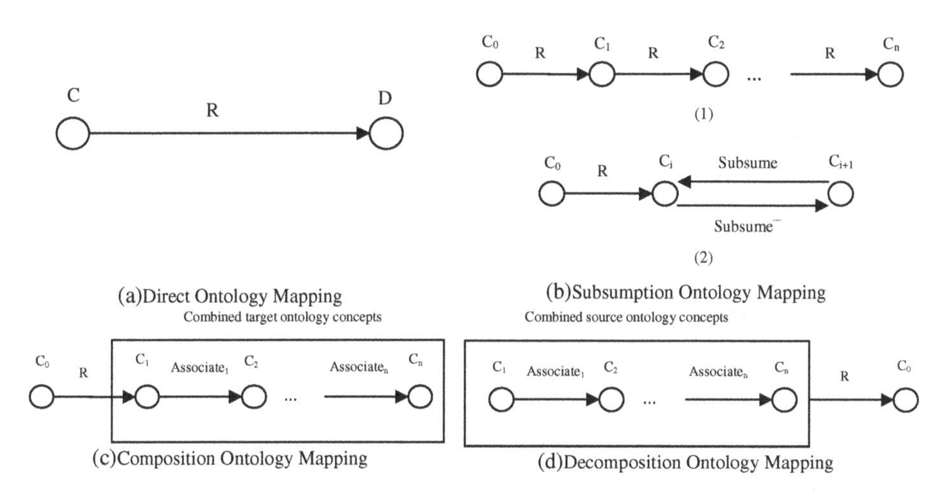

Fig. 1. The Patterns of Semantic Mapping

We define the some properties of semantic mapping which are useful in the task of ontology fusion. The first property is transitivity, for the mapping $\mathcal{M}_{i-1,i} = (C_{i-1}, C_i, \mathcal{R}, v_{i-1,i})$ and $\mathcal{M}_{i,i+1} = (C_i, C_{i+1}, \mathcal{R}, v_{i,i+1})$, a new mapping $\mathcal{M}_{i-1,i+1} = (C_{i-1}, C_{i+1}, \mathcal{R}, v_{i-1,i+1})$ can be created to satisfy the mapping relation R. The second property is symmetric, which means that the mapping $\mathcal{M} = (\mathcal{S}, \mathcal{D}, \mathcal{R}, v)$ is equal to the mapping $\mathcal{M}' = (\mathcal{D}, \mathcal{S}, \mathcal{R}, v)$. The third property is strong mapping property, it can be described as follows.

Definition 2 *A set of mappings $\mathcal{M}_i$ $(0 \leq i \leq n)$are* ***strong*** *if they can satisfy the following conditions:*

i). They share the same mapping relation R, and the mapping relation is transitivity;

ii). For $\forall(i,j,k)$,v_i, v_j, v_k are the confidence value of mapping $\mathcal{M}_i, \mathcal{M}_j, \mathcal{M}_k$, then $v_i \leq v_j + v_k$.

4 XML Query Rewriting with the Extension of Ontology Mapping

4.1 The extension of XML algebra with semantic query enhanced

We extended XML algebra TAX[6] to enable semantic querying on mediated XISs, TAX uses Pattern Tree to describe query language and Witness Tree to describe the result instances which satisfy the Pattern Tree. The definition of pattern tree with ontology extension can be described as follows:

Definition 3 *An Ontology Enhanced Pattern Tree is a 2-tuple $SPT := (T, F)$, where $T := (V, E)$ is a tree with node identifier and edge identifier. F is a combination of prediction expressions.*

The prediction expression F supports the following **atomic condition** or **selection condition**[7]. Atomic condition have the form of X *op* Y, where:

- $op \in \{=, \neq, <, \leq, >, \geq, \sim,$ **instance of**, **isa**, **is_part_of**, **before**, **below**, **above**$\}$
- X and Y are conditional *terms*, which are attributes ,types,type values $v : \tau$ and $v \in dom(\tau)$, ontology concepts and so on;
- $\sim$ stands for the estimation of semantic similarity.

The selection condition is:

- Atom conditions are selection conditions;
- If c_1 and c_2 are selection conditions, then $c_1 \wedge c2$, $c_1 \vee c_2$ and $\neg c_1$ are both selection conditions;
- No others selection conditions forms.

4.2 XML Query Rewriting

In order to simplify the discussion, this paper just pays attention to the rewriting mechanism of the selection operation. Briefly, a selection operation can be expressed as $\sigma(X, Y)$ $\{X \subseteq P_i \cup P_o, Y \subseteq PE\}$, where P_i is the input pattern tree, P_o is output pattern tree, PE is predication list. We define two operators $\cup$ and $\bowtie$ to represent *Union* and *Join* operation separately, and define the operator $\Rightarrow$ to represent the query rewriting operation.

Firstly, we propose how to rewrite pattern tree (which is the X element of expression $\sigma(X, Y)$), there maybe several cases as follows:

1. X is one of the elements of input pattern tree or output pattern tree, and it is also a concept in the global ontology hierarchy. $X_i (1 \leq i \leq n)$ are the concepts for different local ontologies. X and X_i were combined into one concept in the integrated global ontology with strong direct mappings, which means that X and X_i can match each other, then rewrite X as $X \cup \bigcup_{1 \leq i \leq n} X_i$. The responding selection rewriting can be expressed as:
$$\sigma(X, Y) \Rightarrow \sigma(X, Y) \cup \sigma(X_1, Y) \cup \sigma(X_2, Y) \ldots \cup \sigma(X_n, Y) \quad (1)$$
2. The concept of X is generated by the subsumption mapping or composition mapping of $X_i (1 \leq i \leq n)$, then rewrite X as $\bigcup_{1 \leq i \leq n} X_i$. The responding selection rewriting can be expressed as:
$$\sigma(X, Y) \Rightarrow \sigma(X_1, Y) \cup \sigma(X_2, Y) \ldots \cup \sigma(X_n, Y) \quad (2)$$

And then, we propose how to rewrite the predication expressions (which is the Y element of the expression $\sigma(X, Y)$, there are also several cases, which can be described as follows:

1. If there are lots of concept $Y_i (1 \leq i \leq n)$ combined in the concept Y of global Ontology, rewrite Y as $Y \cup \bigcup_{1 \leq i \leq n} Y_i$. The corresponding selection rewriting can be described as:

$$\sigma(X,Y) \Rightarrow \sigma(X,Y) \cup \sigma(X,Y_1) \cup \sigma(X,Y_2) \ldots \cup \sigma(X,Y_n) \tag{3}$$

2. If the concept Y is generated by the subsumption mapping of $Y_i (1 \leq i \leq n)$, rewrite Y as $\bigcup_{1 \leq i \leq n} Y_i$. The corresponding selection rewriting can be described as:

$$\sigma(X,Y) \Rightarrow \sigma(X,Y_1) \cup \sigma(X,Y_2) \ldots \cup \sigma(X,Y_n) \tag{4}$$

3. If the concept Y is generated by the composition mapping of $Y_i (1 \leq i \leq n)$, suppose the composition condition is F, rewrite Y as $(Y_1 + Y_2 + \ldots Y_n) \cap F$. The corresponding selection rewriting can be described as:

$$\sigma(X,Y) \Rightarrow \sigma(X,Y_1 \wedge F) \bowtie \sigma(X,Y_2 \wedge F) \ldots \bowtie \sigma(X,Y_n \wedge F) \tag{5}$$

It is worth to point out that rewriting process may require a recursion in the transitivity property of semantic mapping. The process of rewriting pattern tree and predication expressions can be described as algorithm 1 and 2.

Algorithm 1: SEL_Rewrite_X(X)

```
   Input: X is the pattern tree of selection query σ(X,Y).
1  foreach x ∈ X do
2      switch Mappings of X node do
3          case funsion_node
4              x ← x ∪ ⋃_{1≤i≤n} x_i;
5              σ(X,Y) ⇒ σ(X,Y) ∪ σ(X_1,Y) ∪ σ(X_2,Y) ... ∪ σ(X_n,Y);
6              foreach x_i do
7                  SEL_Rewrite_X(x_i);
8              end
9          case subsumption or composition
10             x ← ⋃_{1≤i≤n} x_i;
11             σ(X,Y) ⇒ σ(X_1,Y) ∪ σ(X_2,Y) ... ∪ σ(X_n,Y);
12             foreach x_i do
13                 SEL_Rewrite_X(x_i);
14             end
15         end
16     end
17 end
```

Algorithm 2: SEL_Rewrite_Y(Y)

```
   Input: Y is the predication list of selection query σ(X,Y).
1  foreach y ∈ Y do
2      switch Mappings of Y concept do
3          case funsion_node
4              y ← y ∪ ⋃_{1≤i≤n} y_i;
5              σ(X,Y) ⇒ σ(X,Y) ∪ σ(X,Y_1) ∪ σ(X,Y_2) ... ∪ σ(X,Y_n);
6              foreach y_i do
7                  SEL_Rewrite_Y(y_i);
8              end
9          case subsumption
10             y ← ⋃_{1≤i≤n} y_i;
11             σ(X,Y) ⇒ σ(X,Y_1) ∪ σ(X,Y_2) ... ∪ σ(X,Y_n) ;
12             foreach y_i do
13                 SEL_Rewrite_Y(y_i);
14             end
15         case decomposition
16             y ← (y_1 + y_2 + ... y_n) ∩ F;
17             σ(X,Y) ⇒ σ(X,Y_1 ∧ F) ⋈ σ(X,Y_2 ∧ F) ... ⋈ σ(X,Y_n ∧ F) ;
18             foreach y_i do
19                 SEL_Rewrite_Y(y_i);
20             end
21         end
22     end
23 end
```

Now we discuss the problem of reducing redundant in the process of ontology query, A selection is redundancy if it satisfy

$$\exists(i,j)\{X_i \in P_o \wedge X_j \in P_o \wedge X_i \cap X_j \neq \emptyset\} \tag{6}$$

and corresponding rewriting of selection can be described as:

$$\sigma(X,Y) \Rightarrow \sigma(X_i,Y) \cup \sigma(X_j - (X_i \cap X_j), Y) \tag{7}$$

The advantage of complex ontology mapping with semantically enhanced similarity can be expressed as follows:

- It can match the semantic similar concepts more exactly, especially for the concepts which are part of concept hierarchy;
- It can reduce the semantic inconsistent by solving problem semantic absent. For example, both the concept $O_1 : C_1(a,b,c,d,f)$ and $O_2 : C_2(a,c,d,e,f)$ represent part of the real concept C with a,b,c,d,e and f attributions, the complex mapping mechanism can supply complete view of concept C at the user view;

- It can reduce the redundancy of the querying by finding more semantic matching in subsumption and composition (decomposition) mappings;
- The complex mapping mechanism refines the process of querying, and it makes the result more precisely.

5 Comparison with related works

In this section, we compare this mechanism with other four approaches. The comparison is based on five different categories, named *mapping method, complex mapping support, pattern support, semantic similarity support*, and *query rewriting support.* As shown in table 1, different approach uses different method to map and integrate ontology, IF-Map[8] and FCA-Merge[9] are the most mature methods which have been accepted in knowledge management community widely, OBSA focuses on a mediator-wrapper based distributed environment, just like FCA-Merge approach, it uses the bottom-up mapping method to integrate different local ontologies. Most of the approaches do not define the ontology mapping patterns, and they can not define the relationship between the local ontologies and the integrated global ontologies with formal method, and this is the reason why they can not apply ontology integration to support semantic level query rewriting, MBL[5] approach defines the mapping patterns, but it uses logic mapping method, and does not discuss how to combine the ontology reasoning with structured or semi-structured information query rewriting. TOSS[7] is the most similar approach with our proposed approach, but it only support one-one mapping and the query rewriting algorithm is simple. Another problem is the definition of ontology mapping, it is not flexible, and can not express the relationship between mapping or integrated ontologies formally, which makes users can not use the global ontology to enable semantic level query information over integrated systems.

Table 1. Comparison of proposed approach with related works.

	mapping method	complex mapping	pattern	semantic similarity	query rewriting
IF-MAP	Information Flow	NA	NA	Support	Not Support
FCA-Merge	Bottom-Up	NA	NA	Support	Not Support
MBL	Logic	Support	Support	Not Support	Not Support
TOSS	Fusion List	Not Support	Not Support	Support	Support
OBSA	Bottom-Up	Support	Support	Support	Support

6 Discussion and Conclusion

The paper mainly discusses the extension of querying on integrated XISs with wrapped ontologies. It discusses the complex ontology mapping patterns with semantically enhanced similarity, such as subsumption mapping, composition mapping and so forth. It also discusses the semantic query mechanism, which primarily extends XML query algebra based on TAX, on the XISs wrapped with local ontologies. Because common XML query languages such as XQuery and XUpdate can be transferred into XML query algebra based on TAX, so the extension is manageable. Complex ontology mapping ensures distributed querying can solve the problem of the inconsistency of semantic and increases the efficiency by refining on the querying and reducing redundancy.

References

1. Wache, H., Vögele, T., Visser, U., Stuckenschmidt, H., Schüster, G., Neumann, H., Hubner, S.: Ontology-based integration of information - a survey of existing approaches. In: Proceedings of IJCAI-01 Workshop: Ontologies and Information Sharing, Seattle, WA, Springer (2001) 108–117
2. Lü, J., Wang, G., Yu, G.: Optimizing Path Expression Queries of XML Data(in chinese). Journal of Software **14** (2003) 1615–1620
3. Gu, J., Chen, H., Chen, X.: An Ontology-based Representation Architecture of Unstructured Information. Wuhan University Journal of Natural Sciences **9** (2004) 595–600
4. Kalfoglou, Y., Schorlemmer, M.: Ontology Mapping: The State of the Art. The Knowledge Engineering Review **18** (2003) 1–31
5. KWON, J., JEONG, D., LEE, L.S., BAIK, D.K.: Intelligent semantic concept mapping for semantic query rewriting/optimization in ontology-based information integration system. International Journal of Software Engineering and Knowledge Engineering **14** (2004) 519–542
6. H.V.Jagadish, L.V.S.Lakshmanan, D.Srivastava, et al: TAX: A Tree Algebra for XML. Lecture Notes In Computer Science **2379** (2001) 149–164
7. Hung, E., Deng, Y., V.S.Subrahmanian: TOSS: An Extension of TAX with Ontologies and Simarity Queries. In G.Weikum, ed.: Proceedings of the 2004 ACM SIGMOD international conference on Management of data, Paris, France, ACM Press (2004) 719–730
8. Kalfoglou, Y., Schorlemmer, M.: IF-Map: An Ontology-Mapping Method Based on Information-Flow Theory. Journal on Data Semantics, LNCS **2800** (2003) 98–127
9. Stumme, G., Maedche, A.: FCA-MERGE: Bottom-up merging of ontologies. In: Seventeenth International Joint Conference on Artificial Intelligence, Seattle, WA (2001) 225–234

Visually Exploring Concept-Based Fuzzy Clusters in Web Search Results

Orland Hoeber and Xue-Dong Yang

University of Regina, Regina, SK S4S 0A2, Canada
{hoeber, yang}@uregina.ca

Abstract. Users of web search systems often have difficulty determining the relevance of search results to their information needs. Clustering has been suggested as a method for making this task easier. However, this introduces new challenges such as naming the clusters, selecting multiple clusters, and re-sorting the search results based on the cluster information. To address these challenges, we have developed Concept Highlighter, a tool for visually exploring concept-based fuzzy clusters in web search results. This tool automatically generates a set of concepts related to the users' queries, and performs single-pass fuzzy c-means clustering on the search results using these concepts as the cluster centroids. A visual interface is provided for interactively exploring the search results. In this paper, we describe the features of Concept Highlighter and its use in finding relevant documents within the search results through concept selection and document surrogate highlighting.

1 Introduction

Users of web search systems commonly have difficulties determining the relevance of the document surrogates that comprise web search results. While some of these difficulties can be attributed to poorly crafted queries, even when the users provide a query that adequately describes their information needs, the search results are often a mixture of documents with varying degrees of relevance. This inability of web search engines to provide highly relevant search results for users' queries can be attributed to the generality of the collection of documents being searched, the ambiguity of language, and the word mismatch problem [5].

Because most searches result in a combination of relevant and irrelevant documents to the users' information needs, the users are required to make relevance decisions on a document-by-document basis. This can be time consuming, and can result in users giving up when a large portion of the search results are irrelevant. The end result is that users of web search systems often view only one to three pages worth of search results [14, 15].

One possible method for addressing this problem is to cluster the search results such that documents that are similar to one another are grouped together [11]. In such a system, the users navigate the clusters in order to narrow down the search results and avoid clusters of irrelevant documents. In the best case

Mark Last et al. (Eds.): Advances in Web Intelligence and Data Mining (SCI) **23**, 81-90 (2006)
www.springerlink.com

scenario, the users will select the relevant clusters and view lists of document surrogates in which a large portion are relevant to the users' information needs.

One of the primary challenges in clustering is determining an adequate name or description of the clusters. If this information does not correctly describe the document surrogates contained in the cluster, the users will either choose clusters that are not relevant to their information need, or will entirely miss the clusters that contain the relevant documents. Further problems with clustering web search systems include an inability to select multiple clusters simultaneously, and a lack of sorting or re-ordering the search results.

To address these drawbacks of the clustering in web information retrieval, we have developed a tool for visually exploring concept-based clusters in web search results called Concept Highlighter. This tool makes use of a concept knowledge base [9] in order to automatically generate a set of concepts related to the users' queries. The concepts generated using the concept knowledge base are used as the centroids for a single-pass fuzzy c-means clustering algorithm [2, 11] that is applied to the search results as they are retrieved via the Google API [6]. A visual representation of the fuzzy membership scores allows the users to interactively select concepts of interest and identify how this affects the clustering of the search results.

Preliminary studies have shown that this method for providing a visual representation of the fuzzy clustering results can be very effective in allowing users to narrow down the search results. Further, since the users can interactively select and un-select concepts, as well as sort and re-sort the search results, the outcome is an exploration of the search results. This ability to explore the search results allows the user to take an active role in the evaluation of the results of their web search, and is a step towards Yao's vision for web information retrieval support systems [19].

The remainder of this paper is organized as follows: An overview of clustering in web information retrieval is provided in Section 2. In Section 3, we describe our methods for obtaining the concepts and generating fuzzy c-means clusters of the search results using these concepts. Section 4 describes the methods by which the cluster membership scores are visually represented. The process for interactively exploring the results of a web search is presented in Section 5. Conclusions and future work are provided in Section 6.

2 Background

Clustering can be defined as the unsupervised classification of data objects into groups of similar objects (called clusters) [11]. Clustering has been explored for a number of years both for browsing text collections [4, 8] and for organizing web search results [20, 21]. Hearst and Pedersen validated the cluster hypothesis, showing that relevant documents tended to be more similar to each other than non-relevant documents [8]. This research provides evidence that clustering can be used to support the users' tasks of finding relevant groups of documents from a collection of search results.

Recently, a number of publicly available web search systems have been developed that provide clusters of web search results, and allow the users to browse the clusters to narrow down the set of search results. Many of these systems use hierarchical clustering algorithms, and primarily differ in the representation and interaction with the clusters. Two such systems are Vivisimo [17] and Grokker [7].

The hierarchical clustering algorithms used by these systems partition the search results at various levels of similarity [11]. The result is a tree-like structure representing the clusters, where parent clusters contain all the objects of their children clusters.

In Vivisimo, these hierarchical clusters are represented as a tree. The nodes in the tree can be expanded and collapsed in a manner similar to file directory navigation. When a tree node is selected, the document surrogates contained within that cluster are displayed in a separate frame.

In addition to providing a tree-like navigation scheme, Grokker uses a visual representation of the hierarchical cluster structure. This visual representation uses nested circles to represent the clusters and their children, and provides the ability for the users to see the sizes of the clusters and whether they contain additional children or document surrogates. Like Vivisimo, when a cluster is selected, the document surrogates that are contained within that cluster are displayed in a separate frame.

One of the challenges in any clustering system is to provide meaningful names for the clusters. Commonly, the names are generated by choosing the most frequent terms or phrases within the cluster (ignoring very common terms such as "is", "and", "the", etc.) [21]. The ability to choose meaningful descriptions of clusters in a web search system has a direct impact on the ability for the users to correctly navigate the clusters to find relevant document surrogates. If vague or misleading names are chosen for the clusters, this can lead to users choosing clusters that are not relevant to their information needs (resulting in the evaluation of documents that are likely not relevant), or not choosing clusters that are relevant to their information needs (resulting in missing documents that are relevant).

Commonly, the documents that are relevant to a users information needs will be distributed among multiple clusters. However, these systems do not easily support the exploration of multiple clusters. While it is possible to view an intermediate cluster that contains all the document surrogates of its children clusters, viewing the union of an arbitrary set of clusters is not possible. This means that if users wish to explore multiple clusters, they must do so separately.

A final difficulty with these web search clustering techniques is that they do not provide any additional information regarding the organization of the document surrogates within the clusters. When a cluster is selected, the documents are listed in the same order as provided by the underlying search engine. There is no indication of which documents are most similar to the cluster centroid, or the degree of membership to the cluster.

To address these shortcomings of the web search clustering systems, we have developed Concept Highlighter, a tool that provides a visual and interactive interface to concept-based fuzzy clusters of web search results. In this tool, the clusters are named using the concept names; multiple clusters can be selected generating a union of the clusters; and the search results are re-sorted based on their membership score. Information visualization techniques are used to visually represent the fuzzy membership scores of the document surrogates in an abstract and compact form, allowing the users to visually process and interpret this information.

3 Fuzzy Clustering Using Concepts

The first step in generating the concept-based fuzzy clusters is to obtain a set of concepts associated with the users' queries. The source of the conceptual information is a concept knowledge base that was originally devised for query expansion [9, 10]. This concept knowledge base contains relationships between concepts and the terms have been used to describe them. The ACM Computing Classification System [1] was used as the source of the conceptual knowledge for the prototype tool, resulting in a concept knowledge base specifically for the computer science domain.

The process for obtaining the concepts that are related to the users' queries is similar to the process for generating the query space as described in [10]. The query terms are first processed using Porter's stemming algorithm [12], which removes the prefixes and suffixes from terms to generate the root words, called stems. These stems are matched to the stems in the concept knowledge base, and the nearest concepts are selected. For each of these concepts, the set of stems that are nearest to the concept are selected from the knowledge base. Each of these sets will contain one or more of the original query term stems, plus additional stems that are not present in the query.

In our previous work, the resulting query space was used to allow the users to interactively refine their queries. In this work, we instead use this query space to identify potential cluster centroids that may be relevant to the users' information needs. For each concept, a vector is created using the set of stems that were selected from the concept knowledge base. The weight of the link between the concept-stem pair is used to set the magnitude of the concept vector in the dimension associated with the stem.

Therefore, as a result of this query space generation, a set of concept vectors $C = \{c_1, c_2, \ldots, c_m\}$ are generated. If the total number of unique stems that were selected from the concept knowledge base is p, then the dimension of all vectors c_i $(i = 1 \ldots m)$ is p. Further, the magnitude of the vector c_i $(i = 1 \ldots m)$ on dimension j $(j = 1 \ldots p)$ is given by the concept knowledge base weight between concept i and term j.

After the concepts have been obtained from the concept knowledge base, and the concept vectors have been created, the users' queries are sent to the Google API [6]. As each of the document surrogates are retrieved, a single-pass fuzzy c-means clustering algorithm [2, 11] is performed. The title and snippet from the document surrogate are processed using Porter's stemming algorithm [12], and the frequency of each unique stem is calculated. These frequencies are used to generate vectors for each of the document surrogates. Although some argue against using term frequencies (TF) as the sole source of information in a text retrieval system [13], using other global information such as the inverse document frequency (IDF) is not feasible when the document surrogate vectors need to be generated as each document surrogate is retrieved (to achieve a near real-time web information retrieval system).

Given a set of concept vectors $C = \{c_1, c_2, \ldots, c_m\}$ and a document surrogate vector d_i, the fuzzy membership of document surrogate d_i with respect to concept c_j is given by:

$$u_{i,j} = \frac{1}{\sum_{k=1}^{m} (\frac{sim(d_i, c_j)}{sim(d_i, c_k)})^2}$$

In this calculation, the similarity between a document surrogate vector and a concept vector is given by the Euclidean distance metric [11]:

$$sim(x_i, x_j) = (\sum_{k=1}^{p} (x_{i,k} - x_{j,k})^2)^{1/2}$$

Normally, when evaluating the document surrogates, all unique stems would contribute to the construction of the document surrogate vector. However, since the distance calculations in this single-pass fuzzy clustering algorithm are always between concept vectors and document surrogate vectors, we only need to consider the stems that are already present in the concept vectors. This reduction in the dimension of the document surrogate vectors results in an increase in the speed at which the fuzzy clusters are generated. In our prototype system, the fuzzy cluster membership scores are calculated as quickly as the underlying search engine can provide the document surrogates to the system.

While it is common to run the fuzzy c-means clustering algorithm in multiple passes, each time re-calculating the centroids of the clusters, we only run the algorithm in a single pass resulting in a fuzzy membership score for each concept-document surrogate pair. This ensures that the fuzzy clusters remain centred around the concepts.

Since the concepts represent the centroids of the clusters, the clusters can be named using the concept names. This is a valuable benefit since the names of the concepts are derived from the source knowledge upon which the concept knowledge base was constructed (in this case, the ACM Computing Classification System). Further, since the clusters always remain centred on the concepts, they are independent of the search results. Therefore, while two similar queries will result in two different sets of search results, they will commonly result in a very similar set of concepts. This can be beneficial as the users learn which concepts are of interest to their general information seeking needs (and can be extended in the future to support personalized concept selection).

4 Visual Representation of Membership Scores

Information visualization takes advantage of the human visual information processing systems by generating graphical representations of data or concepts [18]. The cognitive activity involved in viewing and processing a visual representation allows users to gain understanding or insight into the underlying data. With respect to the visualization of fuzzy clusters, the ultimate goal is to allow users to *see* the clusters without limiting their ability to view the entire set of document surrogates.

Concept Highlighter provides a compact list-based representation at two levels of detail: the overview map shows the membership scores for the first 100 documents returned by the Google API in a single compact list; the detail view shows approximately 25 document surrogates at a time. A screenshot of these two levels of detail are shown in Figure 1.

Our preliminary studies have shown that most users have a preference for a compact representation of web search results, which can more easily be visually scanned. As such, the only persistent information from the document surrogate provided in the detail view is the title. The snippet and URL associated with each document surrogate can be accessed as needed via a tool tip. Additionally, the detail view provides the document surrogate number, allowing the user to easily identify the degree of importance placed on this document surrogate by the underlying search engine algorithms.

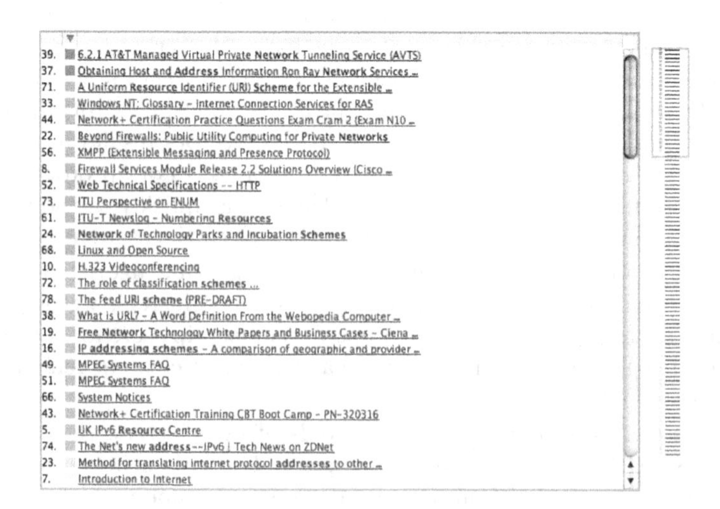

Fig. 1. The visual representation of the web search results consists of an overview map (right) and a detail window (left). These search results were returned from the query "addressing schemes resources networks", and show the fuzzy membership score when the concepts "computer-communication networks: network architecture and design" and "operating systems: communications management" are enabled. The document surrogates with the purple links are those that have been viewed by the user in this search session.

Since the spatial position of an object and its colour can be perceptually separated, colour coding of the fuzzy membership scores can be used without interfering with the spatial layout of the data [18]. In many cases, colour can be pre-attentively processed, allowing the information to be absorbed by the users faster than if they were required to read the corresponding numerical values [18]. While identifying specific values in the colour scale used in Concept Highlighter may not be pre-attentively processed, identifying a relative ordering as well as a few high values from many low values will be processed faster than reading the numerical values.

The choice of a colour scale is not as simple as it might seem. Since we need to represent an ordered sequence of values, a colour sequence that varies monotonically on at least one colour channel is required [16, 18]. A set of nine perceptually distinct colours on a yellow-green-blue colour scale were chosen to represent the fuzzy membership scores. This colour scale varies on all three colour channels: luminance, yellow-blue, and red-green. The ColorBrewer application [3] was used to select this colour scale.

In order to allow the users to remain aware of the location of the detail view with respect to the larger set of documents represented in the overview map, a grey box is used to indicate the correspondence between these two coordinated views. Together, these views allow the user to both investigate the document surrogates in detail, as well as gain insight into the features of the entire set of search results displayed.

5 Interactive Search Results Exploration

Users of Concept Highlighter can interactively explore the search results in a number of different ways. A list of the concepts matched to the users' queries is provided at the top of the display. Beside each concept is a checkbox which can be used to enable or disable the corresponding fuzzy cluster.

When the user checks a cluster, the fuzzy membership scores for all the document surrogates is visually represented in both the overview map and detail view. The user may check multiple concepts, the result of which generates a summation of the fuzzy membership scores corresponding to the selected concepts. Therefore, as multiple concepts are selected, the document surrogates that are nearer to both clusters are represented with a darker colour on the colour scale, indicating their higher fuzzy membership score.

As the documents that belong to the selected clusters are highlighted, the user may visually inspect both the overview map and the detail view to find relevant documents. Clicking on any location in the overview map will automatically scroll the detail view to that location. Therefore, the users can easily scan the entire 100 documents shown in the overview map, and jump to locations of interest based on the fuzzy membership score visualization.

To make it easier for the users to systematically view the set of documents that are contained within the selected fuzzy clusters, a sorting mechanism is supported in the detail view, and is enabled by default. Clicking the column header above the colour codes for the fuzzy membership scores will disable the sort. Any changes to the sorting will be instantly reflected in the overview map as well detail view.

The interactive nature of concept cluster selection, and the sorting of the documents based on the total fuzzy membership score allows the users to interactively explore the search results. Using the fuzzy clusters as a means for organizing the search results in this exploration process can help in bringing documents that are relevant to the users' information needs into focus, even if these documents are deep in the search results.

An example of a scenario in which a user performs a search for "addressing schemes resources networks" is provided in Figure 2. A video showing this scenario is provided on the author's web site [1].

6 Conclusions & Future Work

Even for well crafted queries, the results of web searches often contain document surrogates of varying degrees of relevance to the users' information seeking goals. Clustering of the search results allows the users to navigate the clusters in order to narrow down the set of search results to a smaller collection containing a larger ratio of relevant documents. However, most web search clustering systems use simple keyword-based cluster naming techniques; do not allow the users to select multiple clusters simultaneously; and do not organize the search results once a cluster is selected.

In this paper, we described Concept Highlighter, a tool for generating concept-based fuzzy clusters of web search results, and an interface for visually representing the fuzzy membership scores and interactively exploring web search results. The visual exploration of the concept-based fuzzy clusters allows the users to interactively select the concepts they think may be relevant to their information seeking goal, and see the results of these concept selections in the highlighting of the document surrogates that belong to the corresponding fuzzy clusters.

The ability of Concept Highlighter to allow the users to find relevant document surrogates depends on the ability of the tool to match the users' queries to the concept knowledge base. Further, if there are few concepts returned, or if all the concepts returned are relevant to the users' information needs, the ability to assist the users in narrowing down the search results is diminished. This can occur when the users' queries are very specific, and will often lead to very specific search results.

More often, the users' queries are less specific. This results in multiple concepts being selected from the concept knowledge base, and a more general collection of documents being returned from the search engine. It is in these situations Concept Highlighter can assist the users in finding relevant documents. The interactive exploration of the web search results using the concept-based fuzzy clusters can lead the users to groups of document surrogates that are relevant, and away from groups of document surrogates that are less relevant.

---

[1] `http://www.cs.uregina.ca/~hoeber/ConceptHighlighter/`

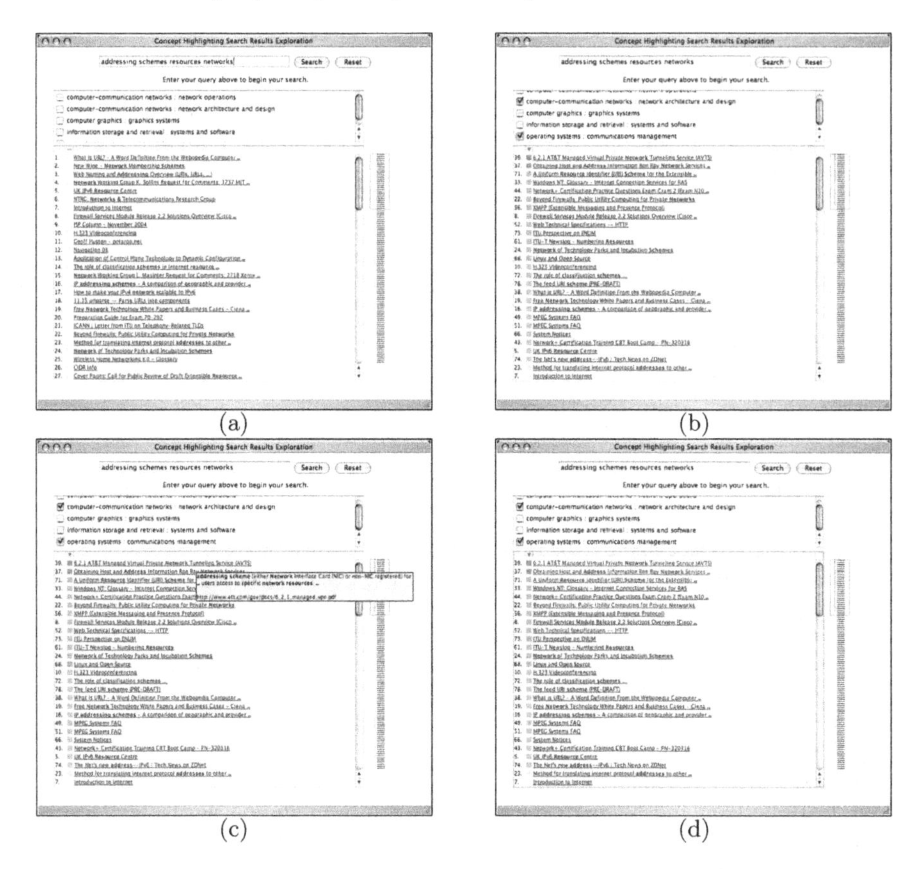

Fig. 2. A common usage scenario would begin with the user entering their query and viewing the search results (a). The user may then check the concepts that are relevant to their query which sorts the document surrogates based on their fuzzy membership score (b). The users may view the snippet and URL contained in the tool tip (c). The link colour changes as documents are viewed, allowing the users to easily identify what they have previously seen (d).

Preliminary investigations have shown this tool to be quite effective in bringing relevant documents to the users' attention; a more systematic study is currently underway to determine the benefits of this work over other clustering methods and simple list-based representations. Since the concept knowledge base used in this work is specific to the computer science domain, the usefulness for general web searching is somewhat limited. The development of a more general concept knowledge base would broaden the applicability of this tool to more general web searching. Other future work includes the integration of this tool with our larger research project of developing a complete framework for a visual and interactive web information retrieval support system.

References

1. ACM. ACM computing classification system. http://www.acm.org/class/.
2. James C. Bezdek. *Pattern Recognition with Fuzzy Objective Function Algorithms.* Plenum Press, New York, 1981.
3. Cynthia A. Brewer. www.colorbrewer.org, 2005.
4. Douglass Cutting, David Karger, Jan Pedersend, and John Tukey. Scatter/gather: A cluster-based approach to browsing large document collections. In *Proceedings of the ACM SIGIR Conference on Research and Development in Information Retrieval*, 1992.
5. G. W. Furnas, T. K. Landauer, L. M. Gomez, and S. T. Dumais. The vocabulary problem in human-system communication. *Communications of the ACM*, 30(11), 1987.
6. Google. Google web API. www.google.com/apis/, 2005.
7. Grokker. http://www.grokker.com/.
8. Marti Hearst and Jan Pedersen. Reexamining the cluster hypothesis: Scatter/gather on retrieval results. In *Proceedings of the ACM SIGIR Conference on Research and Development in Information Retrieval*, 1996.
9. Orland Hoeber, Xue-Dong Yang, and Yiyu Yao. Conceptual query expansion. In *Proceedings of the Atlantic Web Intelligence Conference*, 2005.
10. Orland Hoeber, Xue-Dong Yang, and Yiyu Yao. Visualization support for interactive query refinement. In *Proceedings of the IEEE/WIC/ACM International Conference on Web Intelligence*, 2005.
11. A.K. Jain, M.N.Murty, and P.J. Flynn. Data clustering: A review. *ACM Computing Surveys*, 31(3), September 1999.
12. Martin Porter. An algorithm for suffix stripping. *Program*, 14(3), 1980.
13. S. E. Robertson and K. Sparck Jones. Simple proven approaches to text retrieval. Technical Report TR356, Cambridge University Computer Laboratory, 1997.
14. Craig Silverstein, Monika Henzinger, Hannes Marais, and Michael Moricz. Analysis of a very large web search engine query log. *SIGIR Forum*, 33(1), 1999.
15. Amanda Spink, Dietmar Wolfram, B. J. Jansen, and Tefko Saracevic. Searching the web: the public and their queries. *Journal of the American Society for Information Science and Technology*, 52(3), 2001.
16. Edward Tufte. *Envisioning Information.* Graphics Press, 1990.
17. Vivisimo. http://www.vivisimo.com/.
18. Colin Ware. *Information Visualization: Perception for Design.* Morgan Kaufmann, 2004.
19. Yiyu Yao. Information retrieval support systems. In *Proceedings of the 2002 IEEE World Congress on Computational Intelligence*, 2002.
20. Oren Zamir and Oren Etzioni. Web document clustering: A feasibility demonstration. In *Proceedings of the ACM SIGIR Conference on Research and Development in Information Retrieval*, 1998.
21. Oren Zamir and Oren Etzioni. Grouper: A dynamic clustering interface to web search results. In *Proceedings of the Eighth International World Wide Web Conference*, 1999.

A Grid Scheduling Optimization Strategy Based on Fuzzy Multi-Attribute Group Decision-Making*

Jin Huang, Hai Jin, Xia Xie, and Jun Zhao

Cluster and Grid Computing Lab, Huazhong University of Science and Technology, Wuhan, 430074, China hjin@hust.edu.cn

In grid environment, the grid scheduling technique is more complex than the conventional ones in high performance computing system. Grid scheduling is one of the major factors that would affect the grid performance. In order to optimize grid scheduling, we have to consider various factors. By combining the analysis and prediction methods that are of different principles and approaches, we would be able to make comprehensive decisions on different scenarios and provide reference for scheduling optimization. In this process, a method of fuzzy multi-attribute group decision-making is proposed, which introduces fuzzy set and its operations into decision-making process, and reflects a group or collective ranking of alternatives based on the individual preferences of those alternatives.

1 Introduction

Grid is not only an aggregation of resources, but also a system with some dynamic characteristics. Grid computing evolves from the concept of integrating a collection of distributed computing resources to offer performance which is unattainable by any single machine. So, it is one of the goals for a grid to provide performance as high as possible. The optimization is an approach to get high performance.

In the grid, applications share various resources with others. Therefore, how to make these applications get high performance is a problem that the grid scheduler has to deal with. Since grid has some unique features, i.e., the resources in the grid are always dynamic, heterogeneous and diverse, the schedulers have to deal with local issues, the grid scheduling technique is more complex than those conventional ones that have been applied in most high

* This paper is supported by ChinaGrid project funded by Ministry of Education of China, National Science Foundation under grant 90412010, and China CNGI project under grant CNGI-04-15-7A.

Mark Last et al. (Eds.): Advances in Web Intelligence and Data Mining (SCI) **23**, 91-100 (2006)
www.springerlink.com

performance computing system. Grid scheduling is one of the major factors that may affect the grid performance. If a scheduler can work well in resource selection and task scheduling, each task may get the most suitable resource. Thus the mean response time of the tasks will decrease, which will guarantee each task is completed within the limited period.

In the grid scheduling, scheduler always needs to select a group of the relevant grid resources or services from some available ones to perform the tasks that are submitted by the users. To perform resource selection and task scheduling efficiently, analyzing and predicting the possible status of certain objects in the grid is very necessary. The grid monitoring and accounting tools record the historical data of objects in the grid environments. So we can utilize these data to construct the prediction models that accord with system regulations, and then predict the possible status of certain objects in the future. With the increase of monitoring and accounting information and the factors that have effect on system status, selection and decision-making have become a more and more complex problem. Any single decision method can neither reflect the overall system status nor obtain all necessary information needed in the prediction process. So it is difficult to get the satisfied decision results. Therefore, multiple prediction methods will be combined, which have applied the different prediction principles. It takes advantage of multiple techniques to make up for the shortage of single method. The above process is, by nature, a multi-attribute group decision-making problem. This approach is not only to increase the prediction methods but also to construct a synthetic information entity that applies the different prediction methods.

2 Related Work

At present, many grid scheduling optimization techniques have been suggested [1, 2]. In current grid scheduling research, most schedulers fetch predicted resource parameters from NWS (*Network Weather Service*) [3] predictions. NWS is an agent system deployed on the grid to periodically monitor resource and performance. The overall goal of the PRAGMA project is to realize a next-generation adaptive runtime infrastructure capable of support self-managing, self-adapting and self-optimizing applications on the grid [4, 5]. These methods can predict the performance of the system according to current grid monitoring data. However, the prediction methods that have been used right now merely focus on a single element, not considering the role the historical monitoring information has played in this regard.

Autopilot integrates application and system instrumentation with resource policies and distributed decision procedures. The resulting closed loop adaptive control system can automatically configure resources based on application request patterns and system performance. The goal of GrADS (*Grid Application Development Software*) [6] framework is to provide good resource allocation for grid applications and to support adaptive reallocation if performance degrades because of changes in the availability of grid resources. The methods of this kind can reconfigure accurately the execution of the applications in time. But application programming needs to be changed for grid scheduling. Consequently, the developers will have to be involved in the process of performance optimization.

In multi-attribute decision-making problems under group decision environment, several methods have been proposed for drawing consensus from opinions of decision-makers. Methods in [7] are based fuzzy preference relation. Ishikawa et al. [8] proposed that judgements of decision-makers are represented by an interval-value and a group consensus judgement is derived from cumulative frequency distribution. Bardossy et al. [9] advocated that opinions of decision-makers should be represented by fuzzy numbers. Hsu [10] proposed a method called SAM (*Similarity Aggregation Method*) to aggregate individual opinions of decision-makers.

3 Related Concept

3.1 Fuzzy Multi-Attribute Group Decision-Making

The basic model of multi-attribute decision-making is that given a set of alternatives, corresponding attributes in each alternative and the attribute weights, the aim of decision is to find out the best alternative. Due to the conflicts and constraints between multiple attribute criteria, in common sense, the best solution does not exist, and a set of effective solution is often used in place of it.

In the real world, the number of the decision-makers in the decision-making process is not only one. So, the model of multi-attribute decision-making always combines with the model of group decision-making, and constructs a new kind of selection model. From one decision-maker to more, the complexity of model is increasing, and a series of new problems need to be solved. As decision-makers may have different or even conflict views on the problem, how to form a group preference that reflects all the group members' preferences, that is, to form a group priority relationship or group efficiency function in accordance with the individual relationship or function becomes the key to solve the multi-attribute group decision-making problem.

Frequently, real world decision-making problems are ill defined, i.e., their objectives and parameters are not precisely known. These obstacles of lack of precision have been dealt with using the probabilistic approach. But, due to the fact that the requirements on the data and on the environment are very high and that many real world problems are fuzzy by nature and not random, the probability applications have not been very satisfactory in a lot of cases. On the other hand, the application of fuzzy set theory in real world decision-making problems has given very good results [11]. Its main feature is that it provides a more flexible framework, where it is possible to solve satisfactorily many of the obstacles of lack of precision [12].

3.2 Group Choice Function

Group choice typically involves an aggregation of everyone's preferences. For the special case of two alternatives, the majority rule group choice function uniquely satisfies the criteria of anonymity, monotonicity, and neutrality. But, for larger than two alternatives, the choice outcome may violate the transitivity property of group choice functions, or pairwise independence and so on. It is instance of a more general problem, which is impossible to find a group choice function to satisfy all the criteria we would like to. The foundations of group choice theory lie in the "Paradox of Voting". To avoid the wrong conclusion, many criteria are proposed, which reflect a group or collective ranking of alternatives based on the individual preferences of those alternatives [13].

4 Description of Decision-Making Process

4.1 Data Preprocessing

Generally speaking, the original decision data cannot be utilized directly and needs to be preprocessed. Data preprocessing mainly includes transforming the various attribute values, removing the dimension and doing normalization. In the decision-making process, it is better to be larger for some metrics and smaller for others. Still, there are other metrics that should neither be too large nor too small. All these different metrics should not be put together to judge the advantage of the alternatives. Therefore, it is necessary to transform the decision data so that the better the option, the larger the value of metric. In addition, if a metric has different dimension, comparing each method would be impossible. When the values of different attributes have a big difference, these values need to be normalized for the evaluation and decision-making, that is, the decision data are transformed to interval [0,1]. In the most situations, the essence of data preprocessing is providing real effects of the values of attributes when applying different decision method to rank the alternatives.

4.2 Decision-Making Process

Before describing the algorithm in detail, three related theorems are presented. First, considering the operations of fuzzy set, the following two theorems are true.

Theorem 1. *Let $\tilde{P}$ and $\tilde{W}$ be any two triangular fuzzy numbers, denoted by (p_1, p_2, p_3) and (w_1, w_2, w_3) respectively, then the product of $\tilde{P}$ and $\tilde{W}$, denoted by $\tilde{P} \otimes \tilde{W}$, is a parabola fuzzy number and its member function can be expressed as:*

$$\mu_{\tilde{P}\otimes\tilde{W}} = \begin{cases} \frac{-\delta_2 + \sqrt{\delta_2^2 - 4\delta_1(\delta_3 - x)}}{2\delta_1} \quad , & \delta_3 \le x \le \bar{d} \\ \frac{\lambda_2 - \sqrt{\lambda_2^2 - 4\lambda_1(\lambda_3 - x)}}{2\lambda_1} \quad , & \bar{d} \le x \le \lambda_3 \end{cases}$$

where
$\delta_1 = (p_2 - p_1)(w_2 - w_1), \delta_2 = p_1(w_2 - w_1) + w_1(p_2 - p_1), \delta_3 = p_1 w_1, \bar{d} = p_2 w_2,$
$\lambda_1 = (p_3 - p_2)(w_3 - w_2), \lambda_2 = p_3(w_3 - w_2) + w_3(p_3 - p_2), \lambda_3 = p_3 w_3.$
For convenience, we utilize the related parameters to express parabola fuzzy number in brief as:

$$(\delta_1, \delta_2, \delta_3 / \bar{d} / \lambda_1, \lambda_2, \lambda_3)$$

Theorem 2. *Let $\tilde{M}$ and $\tilde{N}$ be any two parabola fuzzy numbers, are denoted by $(\delta_1, \delta_2, \delta_3/\bar{d}_M/\lambda_1, \lambda_2, \lambda_3)$ and $(\alpha_1, \alpha_2, \alpha_3/\bar{d}_N/\beta_1, \beta_2, \beta_3)$ respectively, then sum of $\tilde{M}$ and $\tilde{N}$, are denoted by $\tilde{M} \oplus \tilde{N}$, is also a parabola fuzzy number and its parameters can be expressed as:*

$$(\delta_1 + \alpha_1, \delta_2 + \alpha_2, \delta_3 + \alpha_3 / \bar{d}_M + \bar{d}_N / \lambda_1 + \beta_1, \lambda_2 + \beta_2, \lambda_3 + \beta_3)$$

Next, considering the definition of mean-value area method for ranking of fuzzy numbers, the following theorem is presented.

Theorem 3. *The defuzzification values of the triangular and parabola fuzzy numbers are calculated as follow:*

(i) Let $\tilde{M}$ be a triangular fuzzy number and be denoted by (l, m, r), then the defuzzification value of $\tilde{M}$ is

$$s(\tilde{M}) = \frac{1}{4}(l + 2m + r)$$

(ii) Let $\tilde{M}$ be a parabola fuzzy number and be denoted by $(\delta_1, \delta_2, \delta_3/\bar{d}/\lambda_1, \lambda_2, \lambda_3)$, then the defuzzification value of $\tilde{M}$ is

$$s(\tilde{M}) = \frac{1}{6}(\delta_1 + \lambda_1) + \frac{1}{4}(\delta_2 - \lambda_2) + \frac{1}{2}(\delta_3 + \lambda_3)$$

In the following algorithm, we assume all the fuzzy rating can be expressed by triangular fuzzy number. The algorithm consists of the following steps:

STEP 1: Transform the language rating and the other imprecise rating into the form of triangular fuzzy number according to the semantic functions.

STEP 2: Combine the fuzzy rating $\tilde{p}_{ij}^{(k)} = (p_{1ij}^{(k)}, p_{2ij}^{(k)}, p_{3ij}^{(k)})$ of decision method k with the fuzzy attribute weight $\tilde{w}_j^{(k)} = (w_{1j}^{(k)}, w_{2j}^{(k)}, w_{3j}^{(k)})$ to construct a weighted evaluation matrix $\tilde{D}^{(k)}$, $i = 1, 2, \ldots, M$ and $j = 1, 2, \ldots, L^{(k)}$, using the multiplication operation of fuzzy numbers. $\tilde{D}^{(k)}$ is calculated as follows:

$$\tilde{d}_{ij}^{(k)} = \tilde{w}_j^{(k)} \otimes \tilde{p}_{ij}^{(k)}$$

where M is the number of alternatives and $\tilde{L}^{(k)}$ is the number of attributes of decision method k. According to Theorem 1, $\tilde{d}_{ij}^{(k)}$ is a parabola fuzzy number $(\delta_{1ij}^{(k)}, \delta_{2ij}^{(k)}, \delta_{3ij}^{(k)}/\bar{d}_{ij}^{(k)}/\lambda_{1ij}^{(k)}, \lambda_{2ij}^{(k)}, \lambda_{3ij}^{(k)})$, where
$\delta_{1ij}^{(k)} = (p_{2ij}^{(k)} - p_{1ij}^{(k)})(w_{2j}^{(k)} - w_{1j}^{(k)}), \delta_{2ij}^{(k)} = p_{1ij}^{(k)}(w_{2j}^{(k)} - w_{1j}^{(k)}) + w_{1j}^{(k)}(p_{2ij}^{(k)} - p_{1ij}^{(k)}),$

$\delta_{3ij}^{(k)} = p_{1ij}^{(k)} w_{1j}^{(k)}, \bar{d}_{ij}^{(k)} = p_{2ij}^{(k)} w_{2j}^{(k)}, \lambda_{1ij}^{(k)} = (p_{3ij}^{(k)} - p_{2ij}^{(k)})(w_{3j}^{(k)} - w_{2j}^{(k)}),$
$\lambda_{2ij}^{(k)} = p_{3ij}^{(k)}(w_{3j}^{(k)} - w_{2j}^{(k)}) + w_{3j}^{(k)}(p_{3ij}^{(k)} - p_{2ij}^{(k)}), \lambda_{3ij}^{(k)} = p_{3ij}^{(k)} w_{3j}^{(k)}.$

STEP 3: Calculate a fuzzy decision number $\tilde{A}_i^{(k)}$ corresponding alternative A_i for decision method k, $i = 1, 2, \ldots, M$, using the fuzzy add operation. $\tilde{A}_i^{(k)}$ is calculated as follows:

$$\tilde{A}_i^{(k)} = \tilde{d}_{i1}^{(k)} \oplus \tilde{d}_{i2}^{(k)} \oplus \cdots \oplus \tilde{d}_{iL^{(k)}}^{(k)}$$

According to Theorem 2, $\tilde{A}_i^{(k)}$ is a parabola fuzzy number

$$(\alpha_{1i}^{(k)}, \alpha_{2i}^{(k)}, \alpha_{3i}^{(k)} / \bar{A}_i^{(k)} / \beta_{1i}^{(k)}, \beta_{2i}^{(k)}, \beta_{3i}^{(k)})$$

where

$$\alpha_{ti}^{(k)} = \sum_{j=1}^{L^{(k)}} \delta_{tij}^{(k)} \quad \beta_{ti}^{(k)} = \sum_{j=1}^{L^{(k)}} \lambda_{tij}^{(k)}, \quad t = 1, 2, 3, \quad \bar{A}_i^{(k)} = \sum_{j=1}^{L^{(k)}} \bar{d}_{ij}^{(k)}$$

STEP 4: Calculate a decision measurement $s(\tilde{A}_i^{(k)})$ of each alternative for decision method k by the mean-value area method, $i = 1, 2, \ldots, M$. According to Theorem 3(ii), $s(\tilde{A}_i^{(k)})$ is calculated as follows:

$$s(\tilde{A}_i^{(k)}) = \frac{1}{6}(\alpha_{1i}^{(k)} + \beta_{1i}^{(k)}) + \frac{1}{4}(\alpha_{2i}^{(k)} - \beta_{2i}^{(k)}) + \frac{1}{2}(\alpha_{3i}^{(k)} + \beta_{3i}^{(k)})$$

STEP 5: For decision method k, its decision weight is $\tilde{v}^{(k)}$. Rank the alternatives in terms of the decision measurements. Assume to set index over again, ranking result is expressed as follows:

$$A_{k_1} \succ_k A_{k_2} \succ_k \cdots \succ_k A_{k_M} : \tilde{v}^{(k)}$$

STEP 6: Aggregate the ranking results of all the decision methods to form the final outcome using Condorcet function value $\tilde{f}_C(x)$. $\tilde{f}_C(x)$ is calculated as follows:

$$\tilde{f}_C(x) = \min_{y \in A \setminus \{x\}} \sum_{k=1}^{K} \tilde{v}^{(k)} | x \succ_k y$$

where K is the number of decision methods and $\tilde{f}_C(x)$ is the smallest sum of weight satisfying priority relationship while comparing x with any other candidate. Because $\tilde{f}_C(x)$ consists of the triangular fuzzy number corresponding the decision weight, it is calculated according to Theorem 3(i). Then, the ranking of alternatives is confirmed by $\tilde{f}_C(x)$ values.

STEP 7: Select the final decision alternative. Assume the subscripts of the alternatives to be reordered, and aggregation outcome is following:

$$A_{1'} \succ A_{2'} \succ \cdots \succ A_{M'}$$

Thus, $A_{1'}$ is the final decision alternative.

5 Application Example

The grid scheduling decision service based on the above method has been applied by grid scheduling optimization engine base on knowledge discovery (GSOE-KD). Through predicting the object performance in the grid, analyzing the behaviors of grid users and mining the partial period association of grid activities in time domain, the system considers many decision factors and makes the grid scheduling decision finally. In the next, we take the data in Table 1 as an example to illuminate the decision-making process in detail. In this example, two services must be requested to perform the task. In the decision-making process, the state of grid services and local networks is considered. The decision methods to be used include the prediction based on time series analysis of utilization states of grid objects, the prediction based on utilization smoothing values in the recent periods and the analysis of reliability of grid services. There are three alternatives in this example. Each decision method considers different attributes. Due to prior two prediction methods need to divide the states of objects, the imprecise metrics are used for prediction.

Table 1. The original decision data

Decision Method	Alternative	Decision Attribute		
Prediction based on time series analysis of utilization state of grid objects		Utilization state of service 1	Utilization state of service 2	Utilization state of local network
	A_1	Middle	High	Low
	A_2	Low	Very High	Middle
	A_3	Middle	Low	Middle
Prediction based on utilization smoothing value in the recent periods		Utilization value of service 1	Utilization value of service 2	Utilization value of local network
	A_1	78%	55%	46%
	A_2	23%	79%	32%
	A_3	61%	40%	64%
Analysis of reliability of grid services		Evaluation of reliability of service 1	Evaluation of reliability of service 2	
	A_1	High	High	
	A_2	Middle	High	
	A_3	High	Low	

First, transform the language rating in Table 1 into the form of triangular fuzzy number according to the semantic functions. In Table 1, because the decision attributes of prior two prediction methods are cost-like metrics, they must be transformed properly. Meanwhile, the fuzzy weights are given to each decision attribute respectively to describe the degree of their importance. In this way, the fuzzy weights are set for each decision method. After constructing the weighted evaluation matrix of each decision method respectively, calculate the fuzzy decision number corresponding to each alternative for each decision method. The results are calculated in Table 2.

Table 2. The fuzzy decision number and its decision measurement

Decision Method	Alternative	Fuzzy Decision Number	Value
J_1	A_1	(0.04,0.24,0.22/0.5/0.03,0.29,0.76)	0.489
	A_2	(0.03,0.18,0.22/0.43/0.05,0.28,0.66)	0.428
	A_3	(0.04,0.24,0.28/0.56/0.03,0.35,0.88)	0.564
J_2	A_1	(0.01,0.107,0.259/0.376/0.025,0.261,0.612)	0.403
	A_2	(0.015,0.138,0.375/0.528/0.025,0.296,0.799)	0.554
	A_3	(0.015,0.149,0.304/0.468/0.02,0.245,0.693)	0.480
J_3	A_1	(0.03,0.25,0.42/0.7/0.01,0.19,0.88)	0.672
	A_2	(0.03,0.21,0.34/0.58/0.01,0.17,0.74)	0.557
	A_3	(0.03,0.21,0.3/0.54/0.01,0.19,0.72)	0.522

Then, calculate decision measurement of each alternative for each decision method by the mean-value area method. The calculation results are also included in Table 2. For each decision method, rank all the alternatives in terms of their decision measurements. The ranking of each decision method corresponds to its decision weight. Finally, rank the alternatives using Condorcet function and its decision measurement. The result of ranking is expressed in Table 3. The ranking result is $A_2 \succ A_3 \succ A_1$. So, select A_2 as the final decision alternative.

Table 3. The aggregation of ranking results

Alternative	Ranking Result	Value	$s(\tilde{f}_C)$
A_1	$A_1 \succ A_2 : (0.2, 0.3, 0.5) \oplus (0.1, 0.2, 0.2) = (0.3, 0.5, 0.7)$	0.5	0.175
	$A_1 \succ A_3 : (0.1, 0.2, 0.2)$	0.175	
A_2	$A_2 \succ A_1 : (0.3, 0.5, 0.6)$	0.475	0.475
	$A_2 \succ A_3 : (0.3, 0.5, 0.6) \oplus (0.1, 0.2, 0.2) = (0.4, 0.7, 0.8)$	0.65	
A_3	$A_3 \succ A_1 : (0.2, 0.3, 0.5) \oplus (0.3, 0.5, 0.6) = (0.5, 0.8, 1.1)$	0.8	0.325
	$A_3 \succ A_2 : (0.2, 0.3, 0.5)$	0.325	

6 Experiment

We have realized the prototype of GSOE-KD based on grid simulator JFreeSim. This system gets a large amount of information in the simulation process, and provides reference for scheduling optimization through knowledge discovery based on data mining and some different ways of analysis and prediction.

GSOE-KD is integrated into grid system as a grid service. The following experiments illustrate the performance of decision service. First, we analyze the mean response time of decision service. For different numbers of tasks, the mean response time of service is shown in Figure 1. The mean response time of service maintains in the low level, and there is no significant wave while the number of tasks is increasing. It indicates that the effect of overhead of decision service upon the performance of system is limited.

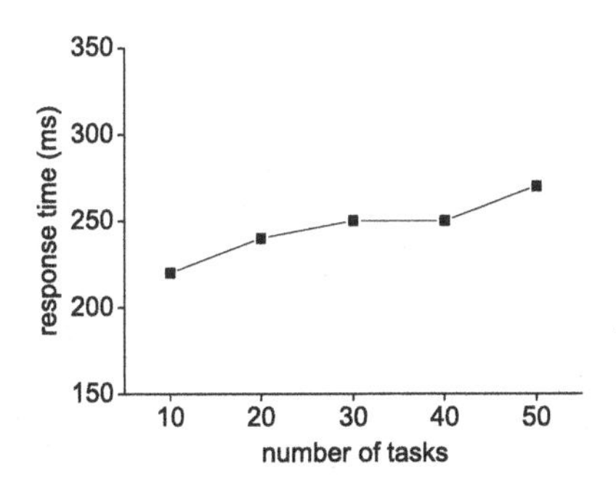

Fig. 1. The mean response time of decision service

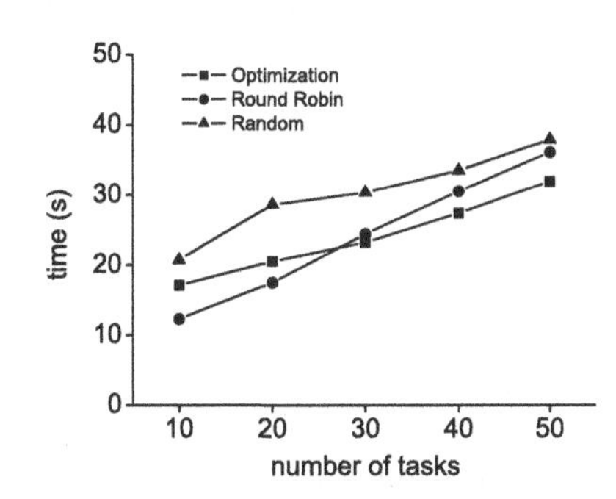

Fig. 2. The execution time of tasks using different allocation policies

Considering the same grid simulation configuration, we compare random and round robin allocation policy with the one based on decision service. Figure 2 shows the execution time in the condition of different numbers of tasks. When the number of tasks is small, the execution time of optimization policy exceeds the one of round robin policy because of the cost of requesting decision service. When the number of tasks is increasing, the optimization scheduling policy will come up with higher execution efficiency, making the total execution time much less than that using other scheduling policies.

7 Conclusion

In grid environments, the grid scheduling technique is more complex than the conventional ones in high performance computing system, and grid scheduling is one of the major factors that would affect the grid performance. In order to optimize grid scheduling, we have to consider the various factors. By combining the analysis and prediction methods that are of different principles and approaches, we would be able to make comprehensive decisions on different scenarios and provide reference for scheduling optimization. In this paper, a method of fuzzy multi-attribute group decision-making is proposed, which introduces fuzzy set and its operations into decision-making process, and reflects a group or collective ranking of alternatives based on the individual preferences of those alternatives. The flexible selection models heighten the expressive force and adaptability greatly. The experiments show that the grid scheduling with this method has high performance.

It should be pointed out that the decision-making approach in this paper is built on the compensability between the decision attributes. But in some cases, the compensability between the decision attributes is conditional, and even non-compensable. Therefore, the other comprehensive decision-making approaches are needed for these features. These approaches will be our further research focus.

References

1. Kapadia N, Fortes J, Brodley C (1999) Predictive application-performance modeling in a computational Grid environment. In: Proceedings of the 8th International Conference on High Performance Distributed Computing, 47–54
2. Bacigalupo D, Jarvis S, He L, Nudd G (2004) An investigation into the application of different performance prediction techniques to e-Commerce applications. In: Proceedings of the 18th International Conference on Parallel and Distributed Processing
3. NWS Project. http://nws.cs.ucsb.edu/
4. Zhu H, Parashar M, Yang J, Zhang Y, Rao S, Hariri S (2003) Self-adapting, self-optimizing runtime management of Grid applications using PRAGMA. In: Proceedings of the International Conference on Parallel and Distributed Processing Symposium
5. Alkindi A, Kerbyson D, Papaefstathiou E, Nudd G (2000) Run-time optimisation using dynamic performance prediction. In: Proceedings of the International Conference on High Performance Computing and Networking, 280–289
6. GrADS Project. http://www.hipersoft.rice.edu/grads/
7. Kacprzyk J, Fedrizzi M, Nurmi H (1992) Group decision making and consensus under fuzzy preferences and fuzzy majority. Fuzzy Sets and Systems, 49(1):21–31
8. Ishikawa A, Amagasa M, Shiga T, Tomizawa G, Tatsuta R, Mieno H (1993) The max-min Delphi method and fuzzy Delphi method via fuzzy integration. Fuzzy Sets and Systems, 55(3):241–253
9. Bardossy A, Duckstein L, Bogardi I (1993) Combination of fuzzy numbers representing expert opinions. Fuzzy Sets and Systems, 57(2):173–181
10. Hsu H, Chen C (1996) Aggregation of fuzzy opinions under group decision making. Fuzzy Sets and Systems, 79(3):279–285
11. Cheng C (1999) A simple fuzzy group decision making method. In: Proceedings of the 10th International Conference on Fuzzy Systems, 910–915
12. Lan J, Xu Y, Liu J (2003) Multiple attributes group decision making under fuzzy environment. In: Proceedings of the International Conference on Systems, Man and Cybernetics, 4986–4991
13. Prodanovic P, Simonovic S (2003) Fuzzy compromise programming for group decision making. In: Proceedings of the International Conference on Systems, Man and Cybernetics, 358–365

An Adaptive PC to Mobile Web Contents Transcoding System Based on MPEG-21 Multimedia Framework

Euisun Kang, Daehyuck Park, JongKeun Kim, Kunjung Sim, Younghwan Lim
[1]Department of Media, Soongsil University, Seoul, Korea
{kanges86}@naver.com
{hotdigi, *jongni*, *inyourhands*, yhlim}@ssu.ac.kr

Abstract. The purpose of this paper is to supply web contents for PC to various multi platform device as PDA or portable device. The conventional studies could not consider these devices, so it created mobile contents beforehand, and then transmitted to the limited mobile device. In this point, the critical problem is to generate mobile contents which are suitable for all kinds of mobile device from PC Web content. This paper propose a service system for transmitting wire web content to various portable device by using MPEG-21 Multimedia Framework. It does not create mobile contents for each device. It just uses DIDL of MEPG-21 as intermediate language to express the structure, resource and description of mobile contents. In DIDL, the multimedia resource is transcoded in off-line previously. The description part is converted in real time as soon as service is requested by end-user. Mobile contents integrate adapted resource with appropriate description and then are transmitted. In addition, this paper proposes the Multi-level caching for reusing mobile contents and describes the result in experiment system.

1 Introduction

During the past few years, due to the tremendous development in the field of wire and wireless communication, users have been demanding the useful information more conveniently; whenever and wherever. To satisfy these needs, ubiquitous has emerged. Supported by the MPEG-21 Multimedia Framework, Ubiquitous provides the necessary service "Any time, any where, any device" overcoming the existent limits. MPEG-21 is a multimedia framework unlike the concept of the existing MPEG series for transmitting and consuming multimedia data efficiently. The gold of the MPEG-21 is to offer a multimedia resource variously, safely, and clearly in extensive network and equipment [1]. Recent research involving MPEG-21 is being undertaken in the field of DMB (Digital Multimedia Broadcasting)[7][8]. Wireless Internet is another area for the application of MPEG-21. It is the technology which provides internet service to portable devices (e.g., mobile phones, PDA, and etc) in a wireless environment at the same time providing services based on the traditional Web pages[9].

* This work was supported by the Korea Research Foundation Grant. (KRF-2004-005-D00198)

Mark Last et al. (Eds.): Advances in Web Intelligence and Data Mining (SCI) **23**, 101-110 (2006)
www.springerlink.com

However, to offer wireless internet, we should consider the following issues. One is how to display contents including various multimedia such as image, animation, movie and picture on a mobile device which has less powerful hardware than PC. The other is how to create adapted contents for various mobile devices. The purpose of this paper is to find a solution to the problem of displaying conventional web contents accessible from PCs to a mobile. In the following chapter, we describe problems that arise in the process of creating mobile contents from PC web contents. We will also propose a system which applies the MPEG-21 Multimedia Framework to provide optimized contents to the mobile device. As a method of expressing optimized contents on the various portable devices, this paper will suggest DIDL(Digital Item Declaration Language) of MPEG-21, written by XML, and proposes Multi-level caching as a method of providing faster response rate.

2 Problems and Solution

2.1 Problem and Related work

Our objective is to enable web contents originally constructed for PC to display on the various multi platform devices such as the PDA or mobile phones. However, there lie several problems in achieving its objective.

First of all, it is difficult to reconstruct original web contents for wireless device. Namely, it is impossible to display the same general web contents to the mobile device. In addition, reconstructing the web contents to specific mobile device will be ineffective. The reasons are that the resolution and the multimedia processing capability of the mobile device are very weak, and there are many types of device. Secondly, it is impossible to transcode multimedia data in real time. Generally, There are diverse sorts of multimedia data such as image, audio, animation, and video in Web contents. The transcoding time for multimedia costs too much. Accordingly, it is inconsistent to transmit in real time. Finally, it is difficult to determine its demands for transformation regarding the various mobile devices. Since there are many varieties of mobile device, it is difficult to know what kind of contents should be transcoded before a mobile device requires contents to server.

There are 3 methods according to the scope of automatization. The methods are about converting the construction of existing web contents into mobile contents. The manual approach considers the characteristic of each device to create new mobile web pages manually. This method is efficient when developing new wireless internet service and contents. An advantage is to generate an optimized mobile page [2][13][14]. Semi automatic approach, page filtering, is a method which selects and transforms specific parts of the web page[2]. Finally, the fully-automatic approach is efficient when there is existing web service and contents. It services to mobile device by converting existing HTML documents into mobile internet format. Since it transforms all web pages automatically, it is the most convenient and the most efficient method [2][12]. However, the above mentioned methods have some difficulties. One is to express optimized multimedia data according to the specifications of various mobile devices, and the other is to create the appropriate page for the new devices that are being produced.

2.2 Solution

In a process of servicing general web contents to a wireless device, the most significant problem is the reuse of PC web contents for mobile device, the possibility of transcoding service for multimedia data in real time, and the method of transforming contents according to various multi platforms. To solve these problems, we first convert and store multimedia data which will be used by mobile device. Server analyses information of device and delivers the most similar data when end-user connects with server. More specifically, after reconstructing various mobile contents referring PC web content, the result, including multimedia data and design, is saved as an intermediate language. When a multi-platform device requests service to the server, multimedia data and description are automatically created and then are transmitted. Here, multimedia data have been pre-transformed according to each device characteristic such as media processing capability or browser information and description has been optimized according to each mobile device. This paper proposes an adaptive PC to Mobile Web contents Transcoding system using MPEG-21 multimedia framework and describe improvement of in the caching method for a faster response rate.

3. A Mobile Gate System based on MPEG-21

3.1 Mobile Gate System Architecture

Mobile Gate System based on MPEG-21Multimedia Framework help service Web content for PC to various mobile devices. The entire structure is as [Fig. 1].

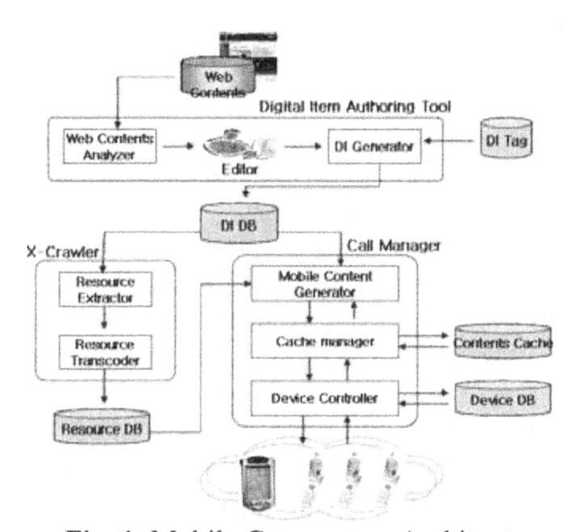

Fig. 1. Mobile Gate system Architecture

To service mobile content to wireless terminal referring wired web contents, this system first edits the existing web contents consisting of HTML or XML by using the Digital Item authoring Tool. The following step is to define the newly constructed mobile contents using DIDL. Among contents defined by DIDL, multimedia data that converting time is too long is transformed beforehand utilizing information about mobile browse by X-Crawler r. When the portable device requests service, the Call Manager analyzes the accessed device information and searches for appropriate contents (i.e., WML, mHTML, xHTML, and etc) in the Contents Cache. If proper content for device exists, Call Manager will service the contents, and if it does not, it requests transformation of contents suitable to the respective device to the Mobile Content Generator, and then services to the device along with the pre-transcoded Resource. In this process, Device Controller saves information about new mobile terminal to the Device Database.

3.2 Components of Mobile Gate System

Digital Item Authoring Tool. As a type of editor, it assists in reconstructing various web contents for mobile use by selecting only resource which will be displayed on the PDA or mobile phone from web content displayed on the PC. This component is done off-line. The produced mobile web contents are automatically created in the type of DI which satisfies MPEG-21 standard. Specific information regarding DI will be referred to in Chapter 4. It consists of three major modules.

Web Contents Analyzer. It parses PC Web pages wrapped by HTML or XML, and then divides the web page into resource and expressive specification and assists users in choosing the resource more conveniently.
Editor. Using the extracted and arbitrary resource, it helps the user construct the mobile web contents more simply through a copy and paste.
DI Generator. it creates the reconstructed mobile page into MPEG-21 DI format using DI Tag Table and saves it in DIDL DB.

X-Crawler. This is a part which carries out transformation in advance regarding actual resource in DI edited by administrator. This component is performed in Off-line. Specific modules are the following.

Mobile Digital Item Parser. It parses DIDL generated by authoring tool, extracts information relative to the resource from DIDL.
Transcoder. Among resources which need transformation, it performs the actual transcoding using the transformation information[3][4]. It is a part of the Resource Adaptation Engine of DIA (Digital Item Adaptation) in MPEG-21[6].
Resource DB. It saves and manages the transcoded resources and these information

Call Manger and Cache Server. In case that a mobile terminal has accessed to server, this module finds out the characteristic of device in Device DB, reconstructs web documents which browser device is able to be recognizable and then sends these. The following are the specific modules.

Mobile Contents Generator. By applying the multimedia data which has been converted according to each device platform, it creates documents suitable for each device in XSL. It is similar to the Description Adaptation Engine of DIA in MPEG-21 DIA[6].

Cache Manager. It provides data appropriate for each device platform. Contents accessed frequently are saved to the Contents Cache and it is serviced immediately in case that the requested document exists in the Contents Cache.

Device Controller. It analyzes information about mobile device which is requesting service and manages the Device DB.

4. Mobile Content Definition Language based on MPEG-21

4.1 Digital Item for expressing Mobile contents

DI is a structured digital object with a standard representation, identification and metadata within the MPEG-21 Framework[5].

Definition : Digital Item within Mobile Gate System

In Mobile Gate, Digital Item is a kind of intermediary file generated by Digital Item Authoring Tool. This help not only transcode Web contents created by inputted HTML, XML, XSL into wireless one such as WML, mHTML, HDML but also make multimedia data playable in every wireless environment.

The following [Fig. 2] indicates how DI is used in the Mobile Gate System.

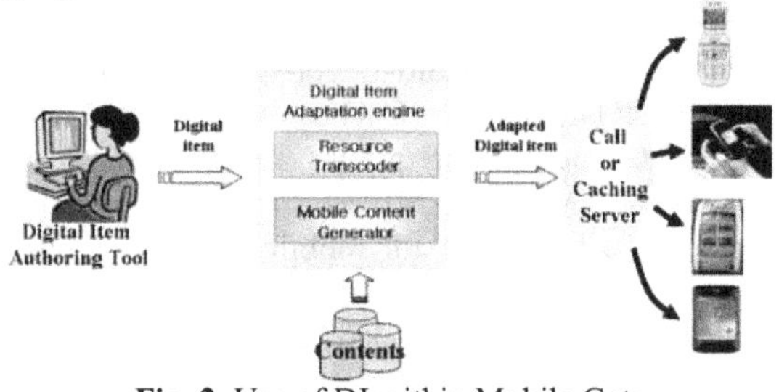

Fig. 2. Use of DI within Mobile Gate

In Mobile Gate, DI consists of resource part and description part. Resources part of DI includes local path or URL of multimedia data filtered from general Web contents. X-Crawler converts resources in Database with a couple of information described in Resource part such as image size or clipping size. Description part of DI is to describe how to present contents in mobile device and short information about resource. Description part is adapted by description transcoder and then transmits contents to mobile device with transcoded data by resource transcoder when service is required by end-user. Hence, although mobile device has been rapidly developed and varied, digital item generated by Authoring Tool help server perform flexible transformation according to properties of device and overload of server may be minimize, once device information is extracted.

4.2 A Structure of Digital Item

In MPEG-21, DI is described in DIDL as represented in XML[5]. The <DIDL> element is the root element of a DIDL instance document and may contain an optional element, followed <Container> or <Item>. <Container> element is a grouping of items and arranges items by means of package shelf. In contrast, <Item> element is intended to be the lowest level of visible to an end-user and systematically construct digital items.

We describe that entire attributes of reconstructed mobile page is in <Container> element and metadata of individual objects in contents and resources is in <Item> elements. Besides, we describe that actually transcoded resource URL is in <Resource> of <Item> element. However, in a stage that express DIDL about mobile pages with Internet contents, we have known that defining a variety of objects within mobile page using DIDL elements within MPEG-21 has limitation. To solve the boundary, we redefine and append necessary elements and their attributes such as text, image, audio, video, flash, form, table or cell in table by namespace in DIDL.

5 Multi-Level Caching Scheme for DIA of PC web-to-Mobile web Service

Cache is to store frequently requested document to a side closer than server and to help transmit faster service to wireless device. The performance of cache is determined by the Response Time when client gains the desired web page and Hit Ratio which is the probability of desired web page to exist within cache. Accordingly, various researches have been doing in order to enhance Response Time and Hit Ratio[10][11]. Our system suggests Multi-Level-Cache method to provide clients of fast response time. Mobile Gate has two Caches inside. First Level Cache is Content Cache and Second Level Cache is Description Cache. Content Cache stores adaptive contents such as WML, mHTML, xHTML, etc. including information of resources. Description Cache converts style sheet for mobile content of DIDL into XSL style, suitable to various mobile terminal, Markup Language by using Mobile Content Generator and stores the result. Transcoded document does not involve resource information.

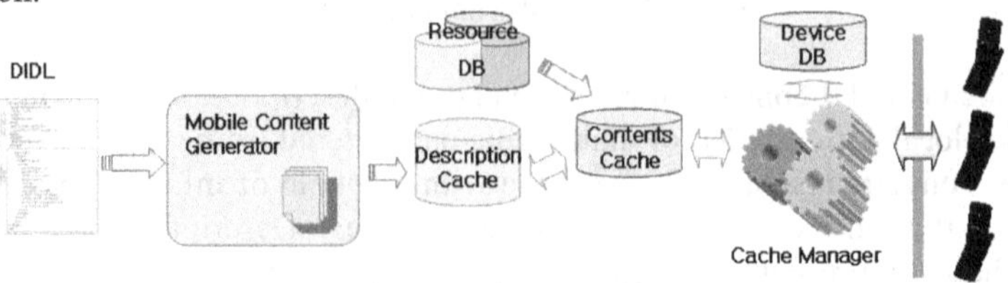

Fig. 3. Multi-level Cache in Mobile Gate system

Algorithm : Multi-level Cache Algorithm

Step 1. A Mobile terminal requests service to Web Server.

Step 2. Cache Manager receives device information from Device Controller.

Step 3. Cache Manager finds out contents that suit the device information in Contents Cache.

Step 4. If it does exist, corresponding content is provided to mobile terminal. If not, content that fits device information is requested to Description Cache.

Step 5. If the requested content exists in the Description Cache,
① Copy the content into Content Cache.
② Provide the content to mobile terminal.
If not,
① Convert it to correspond to the device information in real-time. During the converting process, convert the content of terminal that is similar to present device information and store it to Description Cache and Content Cache.
② Integrate it with converted resource and create desired content.

Multi-level Cache is a method for reducing system overload brought by content conversion in real time. This not only operates conversion for accessed terminal information, but also extracts terminals with high similarity within Device DB. As the standard with high similarity, we select terminals that have identical Markup and browser with lower specification like resolution than accessed terminal has. Hence, Pre-transcoding of Multi-level Cache will be able not only to enhance the Hit Rate of Contents Cache, but also to minimize the number of content conversion.

6 The Result of Implement and Experiment

6.1 The Result of Implement

We have implemented and tested the system in environment that is Windows 2000 Professional, Intel Pentium IV 1.8 GHz, and 512MB. First, we have generated DIDL file that fits rules of [table 2] by using Digital Item Authoring Tool, and have created wireless contents in real-time by Call Manager and X-Crawler. [Fig. 4] shows the result in displaying on portable terminal. We used to two simulators. Openwave is expressed as WML and KTF supports mHTML. Following [Fig. 4] is simulated, but the actual result shows no difference.

Fig. 4. Sample page generated by XSL Generator

6.2 The Result of Experiment

In order to check the response time of a mobile terminal requesting web page to a server, this paper compares conversion time of mobile contents between On-line and Off-line. Cache does not apply to the process of this experiment. Since it was difficult to experiment many real terminals logging on to web page, we used test client to simulate the identical process.

Table 1. Response Time

The number of Access	Off-Line	On-Line
100	290	625
200	567	1125
300	748	1688
400	1131	2187
500	1240	2688
600	1421	3375
700	1618	3687
800	1881	4312
900	2058	4750
1000	2309	5250

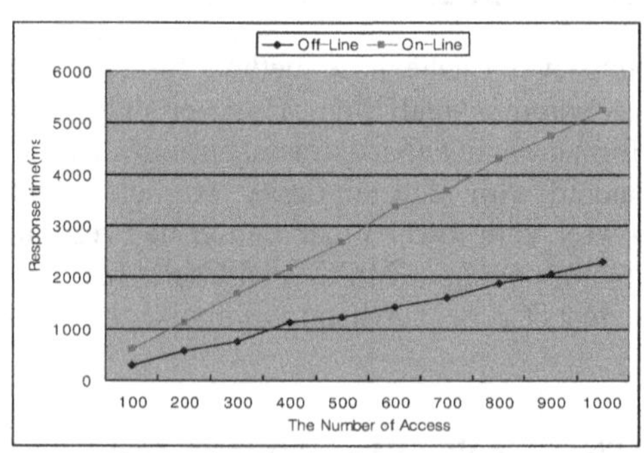

Fig. 5. The graph of response time

As depicted in [Fig 5], Off-line has faster response time than On-line because additional time for converting and creating contents is unnecessary by retaining pr-transcoded content prior to service. Mobile Gate System is showing slower response time than the case of contents converting previously. The reason is that even though resource conversion is taken in off-line beforehand, it is necessary to convert description so as to be fit characteristic of mobile device in case that there is no contents in Contents Cache. In addition, the System not only estimates specification of terminals but also generates mobile contents that suits these terminals in real-time.

7 Conclusion

In a ubiquitous environment, wireless internet has been constantly trying to provide users with scattered information whenever and wherever. However, providing internet service through personal communication system requires recreating contents to match its environment. This is not able to avoid additional cost for development and maintenance. To overcome such problem, this paper suggests Mobile Gate System based on MPEG-21 that enables PC Web contents to display in Cellular phones and PDAs. In addition, we have suggested DIDL to efficiently and flexibly provide various contents to mobile terminals, and proposed Multi-Level Caching to minimize the response time. The suggested system shows slower response time than the case of convergence in off-line, but it has an advantage of providing various and new terminals with flexibly created exact pages. Moreover, it is possible for existing pages to be served to wireless environment without re-creation of mobile contents newly. To make up for the weak point of longer response time, Multi-Level Caching Schema will be considered to be a good method.

References

1 I. Burnett et al.: MPEG-21: Goals and Achievements. IEEE Multimedia, vol. 10, no. 6, Oct-Dec. 2003, pp. 60-70.
2 Yong-hyun Whang, Changwoo Jung, et al. : WebAlchemist: A Web Transcoding System for Mobile WebAccess in Handheld Devices. Proc. SPIE Vol. 4534, p.37-47, 2001.
3. Maria Hong, DaeHyuck Park, YoungHwan Lim. : Transcoding Pattern Generation for Adaptation of Digital items Containing Multiple Media Streams in Ubiquitous Environment. LNCS 3583, pp.1036-1045.
4. Euisun Kang, Maria Hong, Younghwan Lim.: A Guided Search method for Real time Transcoding a MPEG2 P frame into H.263 P Frame in a Compressed Domain. LNCS 3581, pp 242-251.
5. MPEG MDS Group. : Information technology - Multimedia framework (MPEG-21) - Part 2: Digital Item Declaration. ISO/IEC TR 21000-1:2005, Final Draft.
6. MPEG MDS Group. : MPEG-21 Multimedia Framework, Part 7: Digital Item Adaptation (Final Committee Draft). ISO/MPEG N5845, July 2003.
7. Association in Korea. : Digital Multimedia Broadcasting. Telecommunications Technology, 2003SG05.02-046, 2003.
8. Munchurl Kim, Jeongyeon Lim, Kyungok Kang and Jinwoong Kim. : Agent-based intelligent Multimedia Broadcasting within MPEG-21 Multimedia Framework. ETRI Journal, April 2004.
9. Goodman, D.J. : The Wireless Internet: Promises and Challenges. *IEEE Computer*, pp. 36-41, Volume 33, No. 7, July 2000.
10. D. Lee, J. choi, J. Kim, S. Noh S. Min Y. Cho, and C. Kim. : LRFU replacement policy: a spectrum of block replacement policies. in IEEE Transactions on Computers, 1996.
11. Venkata N. Padmanabhan and Jeffrey C. Mogul, : Using Predictive Prefetching to Improve World Wide Web Latency. *Proceedings of SIGCOMM 96*, 1996.
12. http://www-306.ibm.com/software/websphere/
13. http://www.anybil.com
14. http://www.uniwis.com/

Table 2. DIDL in Mobile Gate

Element Name	Description	Element Name	Description
Mobile Content		**Image Attribute**	
<CONTAINER>	Definition of Content	<mbd:CLIP>	selecting specific part within the a image
<mbd:TITLE>	Title of Content		
<mbd:MAXOBJECT>	The number of Resouce in Content	<mbd:ROTATION>	Rotation angle
<mbd:BACKCOLOR>	Background Color	<mbd:BRIGHT>	Brightness
<mbd:BACKIMAGE>	Background Image		
<mbd:BACKAUDIO>	Backgound Sound	<mbd:CONTRAST >	Color contrast
Resource command Attribute		**Form Attribute**	
<ITEM>	Definition of Resouce	<mbd:NAME>	Retrive parameter
<RESOURCE>	URL of Resouce		
<mbd:OBJECTTYPE>	Resource Type		
<mbd:OBJECTSUBTYPE>	SubType of Resource	<mbd:CAPTION>	text to be displayed
<mbd:SIZE>	Potision and Size		
<mbd:Z_INDEX>	Priority given when several resources are overlapped		
<mbd:ALIGN>	Alignment of Resouce on a line	<mbd:MODE>	object of the form that mode will be applied
<mbd:HYPERLINK>	Hyperlink of Selected Resouce	<mbd: DEFVALUE>	default value to be shown in the screen
Text Attribute		**Video and Audio Attribute**	
<mbd:FAMILY>	Font	<mbd:VOLUME>	Volume
<mbd:FSIZE>	Font size		
<mbd:COLOR>	Color	<mbd:LOOP>	The number of loop for Playing
<mbd:ITALIC>	Italic		
<mbd:BOLD>	Bold	<mbd:MUTE>	Mute
<mbd:LINE>	Underline	<mbd:PLAYPOS>	Range of playing

Recommendation Framework for Online Social Networks

Przemysław Kazienko[1] and Katarzyna Musiał[1,2]

[1] Wrocław University of Technology, Wyb.Wyspiańskiego 27, 50–370 Wrocław, Poland kazienko@pwr.wroc.pl
[2] Blekinge Institute of Technology, S–372 25 Ronneby, Sweden weaving@wp.pl

The recommendation framework that supports the creation of new interpersonal relationships within the social networks is presented in the paper. It integrates many sources of data in order to generate the relevant personalized recommendations for network members. The unique social filtering techniques and measures of the activity and strength of relationship are encompassed by the framework.

1 Introduction

Recommender systems became an important part of recent web sites; the vast numbers of them are applied to e–commerce. They help people to make decisions what items to buy, which news to read [17] or which movie to watch. Recommender systems are especially useful in environments with information overload since they cope with selection of a small subset of items that appears to fit to the users preferences [2, 12, 15]. Furthermore, these systems enable to maintain the loyalty of the customers and increase the sales [10]. However, not only products or multimedia content can be suggested to users. The new area where recommender systems can be applied are various online communities called social networks that rapidly develop in the web and usually have thousands or even millions of members like Friendster or LinkedIn. The main goal of a recommendation system, in this case, is to help the user to establish new relationships and in consequence to expand the human community.

In general, there are three main approaches to recommendation: collaborative filtering, content–based filtering, and hybrid recommendations [2]. The collaborative filtering technique relies on opinions about items delivered by users. The system recommends products or people that have been positively evaluated by other people, whose ratings and tastes are similar to the preferences of the user who will receive recommendation [2, 5, 17]. There are two main variants of collaborative filtering. The first one is the k–nearest neighbour and the second one is the nearest neighbourhood. In the content–based

Mark Last et al. (Eds.): Advances in Web Intelligence and Data Mining (SCI) **23**, 111-120 (2006)
www.springerlink.com

filtering the items that are recommended to the user are similar to the items that the user had liked previously [13]. The hybrid method combines two previously enumerated approaches [8, 10, 17].

The proposed recommendation framework is supposed to be applied to the social networks that have recently become more and more important element of information society [1, 6]. A social network is the set of the actors (a single person is the node of the network) and ties, called also relationships, that link the nodes [1, 4]. The evolution of the social network depends on the mutual experience, knowledge, relative interpersonal interests, and trust of human beings [3, 14]. The measurements can be collected to investigate the number and the quality of the relationships within the network.

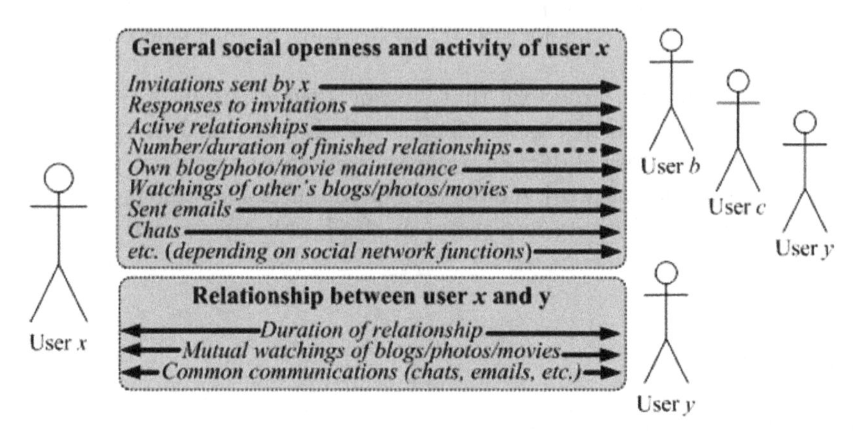

Fig. 1. Social features of relationships in a social network.

2 Problem Description

Recommender systems for social networks differ from typical kinds of recommendation solutions since they suggest rational human beings to other ones rather than inanimate goods. After recommendation selection, one person initializes the relationship with another one, and the latter can respond positively or negatively to the invitation. Such an interaction is impossible with products or content. Moreover, the bond between people is bidirectional in opposite to the relationship between a person and an object. Thus, we would like to find for the current user another person who would also like to react in a favourable way. Things possess no free will and cannot refuse to be sold. For that reason, the recommender systems for social networks need to respect preferences and human limitations of people who would be suggested. In conclusion, the recommendation system should suggest to user x only those network members, who would be potentially good friends or co–workers for user x. In consequence, new relationships of user x can appear in the network.

On the other hand, the owners of the online social network system can execute their own policy. Hence, their first goal can be to build the widest social network, with as many connections between individuals as possible. Yet another possible purpose would be to create the network in which the relationships reflect the strong similarity between people and in consequence, the network consists of many close groups. This evolution of the social network can be stimulated by different kinds of recommendation. In both cases, the aim is to achieve the community in which the connections between human beings are permanent. The common purpose of the framework proposed below is to enable the adjustment of recommendations to the profile of the particular user as well as to the general policy of the network.

3 Recommendation Framework for Social Network

3.1 Relationships between People

Before building the recommendation framework, let us point the main sources of data useful for the recommendation. In the social networks, not only the typical information about a particular user like their interest, demographic data, etc. can be considered, but also their activities and some measures of relationships with other users, especially those related to the process of initialization of bounds and some further, successive activities. The social statement of the user x in the network consists of two data sets: general, aggregated openness and activities features of this user in relation to all others, also in the past, and measures of the relationship between user x and other members of the network (Fig. 2). The most significant elements in the first set are the user's willingness for initialization of relationships and responding to invitations from others. The most important issue within the second set is frequency and intensity with which all relationships are maintained.

3.2 User Profile

All the data related to the characteristic of the individual user is called a user profile and consists of two main parts: static and dynamic. The users themselves deliver the former by filling in the special forms while the latter is monitored and gathered by the system. Based on the analysis of the existing online communities we split the user profile into a set of components which can be easily extended in the future. Each of the components consists of several separate attributes (Fig. 3). The *preferred* component is the direct hint for the system about people sought by the user. The *search* component includes the data about all searches made by the user within the network. The two most complex components are *activity* and *relationship*. The first one measures the activity of the user within the community. The relationship profile describes the number and duration of the users relationships and some other features that characterize them.

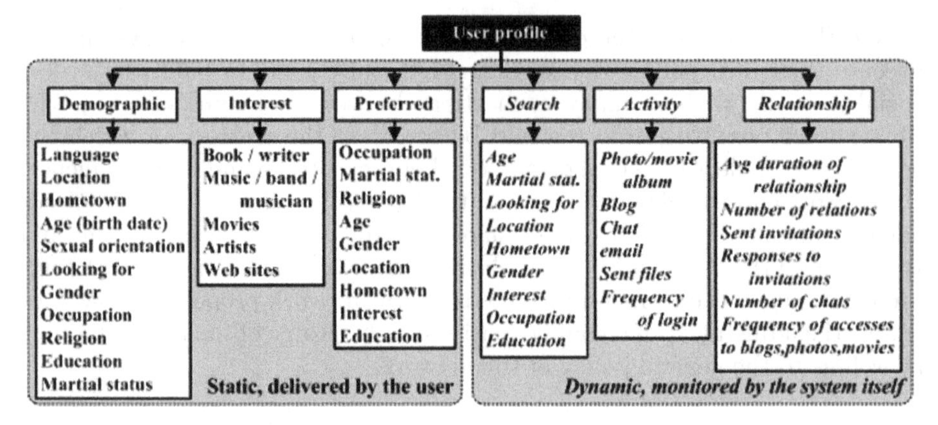

Fig. 2. User profile components. *Location* - the place of residence, *Hometown* - the place of origin, *Looking for* - the purpose of inviting new friends, e.g. friendship.

3.3 Recommendation Process

Based on the gathered information we have built the recommendation framework that supports the creation of the recommendation for the social network. The main goal of the system is to provide the most relevant recommendations to users. Moreover, by combining the several, different sources of data, the method facilitates a new user to join the network and satisfy their expectations.

The overall view of the recommendation process for the social network is presented in Fig. 4. Firstly, knowledge about the users is gathered. Next, the system calculates final similarity function $r(x \Rightarrow y)$ for each pair of users (see below). The final similarity consists of four elements: the direct similarity derived directly from users static attributes, the complementary of relationship initialization, the general activity measure of a candidate for recommendation (user y), and the strength of relationships maintained by user y. The complementary of relationship matches the will of initiation of relationships for user x with the willingness of user y to respond the invitation. Calculation of $r(x \Rightarrow y)$ is periodically repeated due to possible changes in source data according to the network policy. In the next step, static list L of the users that match to particular person x is created when user x logs into the system. Based on user profiles, especially their static components, all users are clustered in separate groups using any of clustering algorithms [16]. Next, depending on the strategy of the evolution and aim of the network, the connectivity social filtering is utilized to promote the creation of connections either within the same group or between the groups. The people from the same group will be recommended if the strong groups are supported, while cross–group recommendations are created to flatten the social network.

In addition, some research revealed that the number of stable relationships that one human being can maintain is about 150. This number called Dunbar

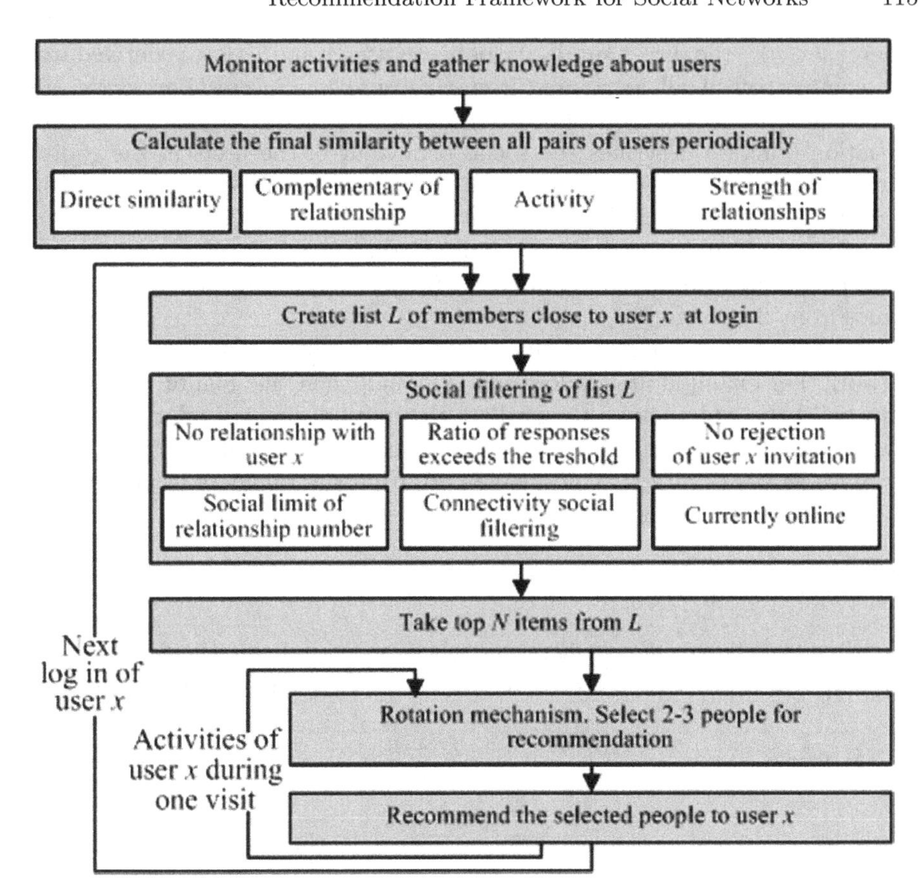

Fig. 3. The process of recommendation for user x in a social network

Number [7] defines the social limit of relationships number within the proposed network. For that reason, the system should not recommend members who have already had more than 150 relationships.

Additionally, the system, within its social filter, promotes people who are currently online to enable the possibly quick response to the invitation. Due to efficiency limits, only top N most suitable users are fixed for recommendation to user x during their stay in the network. Finally, only 2–3 selected people are suggested to person x. The rotary mechanism is used, to prevent the same people to be recommended to user x all the time.

3.4 Final Similarity between Users

The goal of the recommendation process is to find out whether person y ought to be recommended to person x. It is achieved by using the final similarity function $r(x \Rightarrow y)$:

$$r(x \Rightarrow y) = \alpha \cdot s(x, y) + \beta \cdot c(x, y) + \gamma \cdot a(y) + \delta \cdot sr(y) \qquad (1)$$

where: $s(x, y)$ – the direct similarity between user x and y that is derived from the comparison of all attributes from *demographic*, *interest*, *preferred*, and *search* user profile components; $c(x, y)$ – the complementary of relationship initiation function describes the social behaviour of the users in the context of established relationships; $a(y)$ – the activity of user y is calculated basing on the information included in *activity* component; $sr(y)$ – the strength of the relationship between person y and all other members; it is derived from *relationship* user profile component; $\alpha, \beta, \gamma, \delta$ – importance coefficients with values from the range $[0, 1]$.

Coefficients are used to simulate and adjust the evolution of the social network. For example, if α is low and β is high then the goal of the network is to build the wide network regardless the mutual direct similarity of users. The values of these factors tightly depend on the social policy.

Since values of all four components are from the range $[0, 1]$, the value of final similarity function $r(x \Rightarrow y)$ belongs to the range $[0, 4]$.

3.5 Direct Similarity

The direct similarity $s(x, y)$ between two users x and y is the function that compares the set of attributes that characterize these members of the network:

$$s(x,y) = \frac{w_1 \cdot f_1(x,y) + w_2 \cdot f_2(x,y) + \ldots + w_n \cdot f_n(x,y)}{w_1^{\max} \cdot f_1^{\max}(x,y) + w_2^{\max} \cdot f_2^{\max}(x,y) + \ldots + w_n^{\max} \cdot f_n^{\max}(x,y)} \tag{2}$$

where: n – the number of attributes; $w_1, \ldots, w_n$ — the weights assigned to the attributes; $f_1(x, y), \ldots, f_n(x, y)$ — the similarity functions between x and y with respect to attribute $1, \ldots, n$, respectively; $w^{\max}$ – the maximum weight that can be assigned to the attribute; $f_1^{\max}(x, y), \ldots, f_n^{\max}(x, y)$ – the maximum values of the functions $f_1(x, y), \ldots, f_n(x, y)$, respectively.

All attributes can be divided into several classes such as categorical, which are ordered or disordered, and continuous ones. The former ones can be either single-valued like *Gender*, *Education*, *Location* or multiple-valued, e.g. *Interest* or *Language*. The values of these attributes can be one or several from the set of potential values. For categorical attributes their similarity function is usually a binary value that denotes whether any of attribute values is the same for both users, e.g. it is enough that users have at least one language in common. However, for some other categorical attributes like *Interest* we can assume that the more values two users share, the better. For continuous attributes e.g. *Age*, the simple, normalized inverse of difference may be utilized: $f_{age}(10, 10) = 1$, and $f_{age}(1, 100) = 0$, where the max. $age = 100$ and min. $age = 1$, see also the concept in formula (5).

To each of the attribute the appropriate weight, which expresses the usefulness of this attribute for recommendation, is assigned. These weights are adjusted to the specific network using the set of special, non-overlapping rules. A rule can concern the whole attributes and/or only their specific values. Each weight w_i for attribute i (or only for its specific values) is derived from two elements: the static system weight and the rule depended weight:

$$w_i = w_i^{rule} \cdot w_i^{sys} \tag{3}$$

where: w_i – the weight assigned to attribute i that results from the rule set; w_i^{sys} – the system static weight. If no rule related to attribute i exists, then $w_i^{rule}(x,y) = 1$. In the particular case, the whole attribute can be excluded by assigning $w_i^{rule} = 0$. Let us consider the following rule: if the person searches someone for business purpose, then the gender is not an important attribute and $w_{gender}^{rule} = 0$. However, if the person looks for serious relationship, the gender probably becomes crucial and then $w_{gender}^{rule} = 1$.

Additionally, both personal and system weights can be adjusted basing on the user's feedback, i.e. the contribution of the specific feature in the recommendation that user has selected increases the weights for this feature while all the others are slightly reduced [10].

3.6 Complementary of Relationship Initiation

The complementary of relationship initiation matches the willingness of initiation of relationships for user x with the will of user y to respond the invitation.

It enables to compare the frequency of sending the invitation with the probability of response to invitations. All network members are classified into three classes (levels) of initiators: low–medium–high, which are equivalent to $0 - 0.5 - 1$, using the appropriate thresholds applied to the average number of set invitations per month within last year. 0 means that a user sent invitations to others rather rarely or not at all. Value of 1 denotes a very open–minded person who wants to extent their contacts very quickly. In the same way the members are divided into three similar classes of responders: $0 - 0.5 - 1$, based on the percentage of positive replays to requests from others.

Thus, every user has two measures assigned: initialization and response level. The complementary of relationship initiation $c(x,y)$ is the function that combines both these measures and it is calculated as the absolute difference between both of them, e.g. if user x belongs to class 1 as initiator and user y to class 0 as responder then $c(x,y) = |1 - 0| = 1$. It means that we would try to initialize some relationships with resistant members (y) with the help of very open–minded people (x). Similarly, if user x is the bad initiator (0) and user y is the very good responder (1) then $c(x,y) = |0 - 1| = 1$, and we prevent these rare invitation of x from being rejected by recommended y.

During estimating the response level, time is an important factor. Hence, we introduced the "grey period". Invitations sent within this period are not included in calculation of percentage of responses. Moreover, the frequency of invitation takes into account the month: last month has more influence on the final class than the old ones.

3.7 Activity of the User

To define the activity of a user, e.g. their updates of photo/movies/blogs and the assessment whether their behaviours are frequent while some others are not ought to be done. Moreover, we need to respect the general activity context in the network, i.e. all other members who are the best and the worst in their update activities. In addition, the user can be more active in one period and less in another one and the oldest periods should have the least influence on the final measure of an activity. The activity function $a(y)$ of user y is defined as follows:

$$a(y) = \frac{a_1(y) + a_2(y) + \ldots + a_n(y)}{a_1^{\max}(y) + a_2^{\max}(y) + \ldots + a_n^{\max}(y)} \tag{4}$$

where: $a_1(y), \ldots, a_n(y)$ – component activity functions that describe the frequency of the particular activity (e.g. video updates, maintenance of the photo album or frequency of login into the system) in the context of the rest members of the community; $a_1^{max}(y), \ldots, a_n^{max}(y)$ – the maximum values of the functions $a_1(y), \ldots, a_n(y)$; n – the number of attributes.

Each of the functions $a_1(y), \ldots, a_n(y)$ is calculated in the same way. Let us consider for example the frequency of updating the photo album. We find the people in the community who update their photo albums most frequently and most rarely. Additionally, we introduce a time factor. Thus, the component activity function $a_1(y)$ related to photo updates, for user y is:

$$a_1(y) = \frac{1}{1} \cdot \frac{y_1 - l_1}{m_1 - l_1} + \frac{1}{2} \cdot \frac{y_2 - l_2}{m_2 - l_2} + \ldots + \frac{1}{n} \cdot \frac{y_n - l_n}{m_n - l_n} \tag{5}$$

where: n – the fixed number of months that we consider; $y_1, \ldots, y_n$ – the number of updates of a photo album made by person y in the first,..., n–th period (usually month), respectively; $l_1, \ldots, l_n$ – the number of updates of a photo album in the first,..., n–th period, , respectively, made by the least active person in this area; $m_1, \ldots, m_n$ – the number of maximum updates of a photo album in the first,..., n–th period, respectively. If for any period n, $m_n = l_n$, we assume that $\frac{1}{1} \cdot \frac{y_1 - l_1}{m_1 - l_1} = 1$.

This function counts how many times the member updates a photo album in a given period in comparison to the most and least active users within this area. Moreover, the last period has the highest influence on the final value of $a_1(y)$ due to 1/1 factor, while the last period is the least significant.

Note that the value of $a_1(y)$ can exceed 1 and for that reason we need to use maximum values in the denominator of (4).

All elements that are in the activity profile are dynamic because they change over the time. The calculation of activities of users should be repeated for all network members regularly, i.e. after each period (month).

3.8 Strength of the Relationship

The strength of relationship $sr(y)$ is the measure that indicates how firm the relationships between user y and all the others are in the social network:

$$sr(y) = \frac{sr_1(z_1, y) + sr_2(z_2, y) + \ldots + sr_n(z_n, y)}{max_{sum}} \tag{6}$$

where: n – the number of the relationships maintained by user y; $sr_1(z_1, y), \ldots, sr_n(z_n, y)$ – the partial function that denotes the strength of the relationship between user y and user $z_1, \ldots, z_n$, respectively; max_{sum} - the highest sum of $sr_1(z_1, y) + \ldots + sr_n(z_n, y)$ among all users y_j, used for normalization.

$sr_i(z_i, y)$ is calculated based on the number of sent emails to user z_i by user y, the number of received emails by user y from the user z_i, the number of the mutual readings and comments on their blogs, the number of common chats in specified time, e.g. per month, the frequency of mutual accesses to photo albums, movies, etc. This range of elements can vary between systems and tightly depends on the functionality that the specific system provides.

The i–th partial function $sr_1(z_1, y)$ between person x and y_i is:

$$sr_i(z_i, y) = w_1 \cdot h_1(z_i, y) + w_2 \cdot h_2(z_i, y) + \ldots + w_n \cdot h_n(z_i, y) \tag{7}$$

where: n – the number of criteria: readings, emails, chats, etc.; $w_1, \ldots, w_n$ – the weight assigned to criterion 1,..., n, respectively; $h_1(z_i, y), \ldots, h_n(z_i, y)$ – the functions that describe the specific criteria, e.g. a total number of mutual emails, a number of chats, readings or comments of blogs, etc.

Due to many changes in source data - changing $h_n(z_i, y)$, similarly to activity $a(y)$, also values of functions $sr(y)$ have to be recalculated periodically.

4 Conclusions and Future Work

The proposed framework facilitates the creation of the recommendation within a new application domain: in social networks, where one person is suggested to another one. Besides the typical demographic matching of network members, the unique social elements were introduced. They make use of behaviours and activities of users as well as their common interaction and relationship quality. The social filtering, which is one of the parts of the recommendation process, includes some social elements of the network like the limit of the stable relationships that one person can maintain or connectivity social filtering mechanism that enables to create either close or distributed human communities. The interpersonal similarity encompasses demographic, activity and the strength of relationship components and ensures the appropriate balance between people who are able to initialize new relationships and those who willingly respond to invitations.

The future work will focus on the extension of recommendation by means of the new mechanism that supports the renewal of the declined relationships, which previously were strong but now the information flow between its members is infinitesimal. Besides, the feedback from the used recommendation will be also considered [10]. as well as the usage of the concept of social capital [11]. The monitoring of the user feedback would also help to evaluate the effectiveness of the system.

References

1. Adamic L, Adar E (2003) Friends and Neighbors on the Web. Social Networks 25(3) 211-230
2. Adomavicius G, Tuzhilin A (2005) Toward the Next Generation of Recommender Systems: A Survey of the State-of-the-Art and Possible Extensions, IEEE Transactions on Knowledge and Data Engineering 17(6) 734-749
3. Golbeck J (2005) Computing and Applying Trust in Web-Based Social Networks. Ph.D. Thesis, University of Maryland http://www.cafepress.com/trustnet.20473616
4. Golbeck J, Hendler J (2004) Accuracy of Metrics for Inferring Trust and Reputation in Semantic Web-Based Social Networks. EKAW 2004, LNCS 3257 116-131
5. Ha S (2002) Helping Online Customers Decide through Web Personalization, IEEE Intelligent Systems 17(6) 34-43
6. Hanneman R, Riddle M (2005) Introduction to social network methods. Online textbook, http://faculty.ucr.edu/ hanneman/nettext/
7. Hill R, Dunbar R (2003) Social Network Size In Humans, Human Nature 14(1) 53-72
8. Kazienko P, Kiewra M (2004) Personalized Recommendation of Web Pages. Chapter 10, Nguyen T (ed.) Intelligent Technologies for Inconsistent Knowledge Processing. Advanced Knowledge International, Adelaide, Australia 163-183
9. Kazienko P, Kiewra M (2003) ROSA - Multi-agent System for Web Services Personalization. AWIC'03, LNAI 2663, Springer Verlag 297-306
10. Kazienko P, Kołodziejski P (2005) WindOwls Adaptive System for the Integration of Recommendation Methods in E-commerce, AWIC'05, LNAI 3528, Springer Verlag, 218-224
11. Kazienko P, Musiał K (2006) Social Capital in Online Social Networks
12. McDonald D (2003) Ubiquitous recommendation systems, IEEE Computer 36(10) 111-112
13. Mooney R, Roy L (2000) Content-based book recommending using learning for text categorization, ACM Conf. on Digital Libraries 195-204
14. Palau J, Montaner M, Lpez B, de la Rosa J (2004) Collaboration Analysis in Recommender Systems Using Social Networks, CIA 2004, LNCS 3191 137-151
15. Perguini S, Goncalves M, Fox E (2004) Recommender systems research: A Connection-Centric Survey, J. of Intelligent Information Systems 23(2) 107-143
16. Tan P, Steinbach M, Kumar V (2006) Introduction to Data Mining, Chapter 8 and 9. Addison-Wesley, 487-649
17. Terveen L, Hill W, Amento B, McDonald D, Creter J (1997) PHOAKS: A system for sharing recommendations, Communications of the ACM 40(3) 59-62

Packet Manipulating Based on Zipf's Distribution to Protect from Attack in P2P Information Retrieval

Byung-Ryong Kim[1] and Ki-Chang Kim of Author[2]

[1] School of Computer Science and Engineering, Inha Univ., 253, YongHyun-Dong, Nam-Ku, Incheon, 402-751, Korea doolyn@inha.ac.kr
[2] School of Information and Communication Engineering, Inha Univ., 253, YongHyun-Dong, Nam-Ku, Incheon, 402-751, Korea kchang@inha.ac.kr

1 Introduction

P2P is a resource's share among directly or indirectly connected computers. With the arrival of P2P system, PCs connected to Internet, whose roles are only as client in central server/client model, are moving its stage to internet itself. P2P made it possible for PCs placed in remote area to share storage, CPU power, contents and so forth using dynamic routing. Therefore PCs, which were responsible only for client in the past, became obtain computing power similar to costly workstation and mainframe level. In traditional centralized method, it takes a lot to obtain storage and space for main server and administrator to manage it are required whereas in P2P since general user's computers are used it does not require any space or separate administrator to maintain mass server and even with overlapped data it wastes storage less than in the traditional one.

Under server/client environment in order to retrieve desired data users request retrieval to server. At this moment since retrieval requests of all users are transferred to main server and large amount of retrieval request should be simultaneously transacted overload is incurred and it leads to degraded service quality to be offered to client by server or it costs much to improve server's processing capability. But in pure P2P network retrieval does not achieved in one PC or main server but decentralized retrieval is performed that it cost less to retrieve and update as well.

This P2P system can have many advantages, getting out from server/client architecture but there are some problems to consider. Some unsolved problems include speed, extension with other systems, irresponsibility on resource share and difficulty to manage and measure how much it can be is responsible. Although it can know that each host communicates which host, it is unable to know about correction information on which the host is. In terms of

Mark Last et al. (Eds.): Advances in Web Intelligence and Data Mining (SCI) **23**, 121-130 (2006)
www.springerlink.com

security P2P systems are too free and irresponsible to maintain stable reliability as much as displayed in server/client environment. Nevertheless perfect management cannot be performed considering P2P's characteristic and host should manage itself with independent authority so each host should hold anonymity. With Gnutella Protocol anonymity is basically offered to broadcasting Query[1, 6], Ping packet[1] but when downloading and uploading[7], which are final objective, it is not routed so anonymity is not secured this time. Accordingly the exposed host can have chance of attack such as service refusal attack and storage flood even if it is not malicious.

The P2P protocols like Gnutella follow message forwarding mechanisms which make the search completely anonymous. Users can search for any kind of data without exposing their identity. However, the anonymity is not so easy to maintain on the publisher/service provider side. Most of packets transferred from node to node do not contain identity information on node that sent packet. And these packets are transmitted to the destination through the routing systems dynamically composed of intermediate nodes. Therefore it is impossible to know who transmitted it for the first and who the designated recipient is. But since downloading and uploading host's IP address is exposed it does not provide anonymity. This study introduces techniques to provide anonymity for protecting identification of users and resource providers by packet manipulating based on Zipf's distribution QueryHit(Query-reply) packets in systems where anonymity can cause trouble.

This study proposes QueryHit packet[1] generation technique which can provide anonymity even when uploading and downloading among hosts. Where there is retrieved data correspondent to host receiving Query packet, QueryHit Packet is to be returned, this time host which received QueryHit packet transfers it to host which requested retrieval through dynamic routing[1, 2] however in this study host which finds QueryHit for the first transfers IP address and port(IP and Port whose host transfers the relevant QueryHit for the first) field of QueryHit after replacing it with its own IP and Port. When requested to download, it functions like Proxy. In order to complement the problems with simple caching technique, it is based on calculated statistics through Query and QueryHit Packet. This leads to providing anonymity between downloading host and uploading host.

2 Related Researches

2.1 P2P Networking

Gnutella protocol is pure P2P protocol. Gnutella has no administrator composing of network and managing intermediate transaction. Once it informs of its share to hosts connected to the network, without any help from intermediate broker it can receive desired file. In order to participate in Gnutella network it only has to connect to one host currently participating in Gnutella network. Transferring Ping packet after connecting to connectible host and host which received Ping packet broadcasts packets to adjacent host connected to it and at the same time Ping packet is to be returned as respond value. These broadcasted Ping packets receive Pong packets through hosts and returns to the final destination by dynamic routing of intermediate hosts. Finally if the first host which sent Ping packet knows only the connection information of the first one connectible host it can have information on IP address and Port number of lots of hosts participating in Gnutella network through Pong packet.

In case of retrieval request Query packets are transmitted to all the hosts obtained from Ping/Pong process. Where there are contents equivalent to user's retrieval, host which receives Query packet returns QueryHit packet, and this QueryHit packets are also transmitted to the final destination through dynamic routing of intermediate hosts. QueryHit packets include the number of contents equivalent to retrieval, IP address, port number, bandwidth (speed), result set, and so forth. In addition it can be connectible to hosts which do not support incoming connection such as firewall by transmitting Push packets[1].

2.2 Dynamic Routing

Gnutella network does not require continuous (lasting) connection between two hosts. It is sure that the two hosts are participating in network but it does not necessarily mean that they should be directly connected. While relaying messages hosts are connected to each other in spite of itself. Like Query and Ping packets they broadcast message each other, and relay them so they should return packets to the first sender. Therefore Gnutella network assigns UUID(GUID) or 128-bit personal identifier[1] to each host and whenever these UUID values get to each host, it is stored to the affected host and it is used later when it determines which host sent the packet. By these broadcasted packets dynamic routing is made. So packets are not distinguished by IP address and port information but it makes identify them by UUID. Therefore without UUID there is no way to take the packet back through routing.

2.3 Gnutella Disadvantages

Gnutella has several disadvantages and here two of them will be explained. First, in regard to broadcasted packet disappearance the extent to be transferred is defined as TTL and it is basically 7 hop. Which means that one packet is propagated in great amount of packets. Typical Gnutella Query packets occupy 1/4 of all the packets and the half of the rest is occupied by Ping messages. Therefore where periodical transmission of packets is performed hosts located in the middle shall cover big amount. The second, the other disadvantage is from the security perspective. It is difficult to deal with in traditional network, and quite difficult also in P2P but what it counts is the matter of anonymity to hide host identity. It is because relatively weak host information in terms of security obtained when downloading and the information that certain contents has been downloaded may result in malicious use.

2.4 Zipf's Law and Zipf Distribution

Zipf's Law is based on linguistic experimental fact computed in relation to rank and frequency in natural language and the law can be represented as follow.

$$\mathbf{R} \times \mathbf{F_r} = \mathbf{Constant} \tag{1}$$

This indicates that the relation between rank and frequency is represented as effective and correct general empirical law to any inclusive text. That is, product of generated word frequency($\mathbf{F_r}$) and ranking value($\mathbf{R}$) with the frequency is computed approximately as certain constant value. This is seized distribution of the different word each other within a certain document and according to this rule, the frequency of ith word in the sequenceof occurrence frequency is $\frac{1}{i^{\o}}$ (1.5< ø <2.0) times to the highest frequent word. This means that ith frequent word in a text composed of vocabulary number, V and number of word, n occurs $\frac{n}{i^{\o}(H_v(\o))}$ times. Here $H_v(\o)$ is V's $\o^{th}$ harmonic number defining the total occurrence frequency is n. It is important that the relation of connection frequency and hit number with the ranking of that under WWW where numerous retrieval processes are performed also follows this law. Therefore it can be said that web requests follow the distribution of Zipf. The relation of information hits with the information size and of server's connection frequency with ranking made by this produces similar results.

3 Broadcast in Gnutella

3.1 Broadcasted Packets

In Gnutella protocol broadcast is powerful but it may be considerable risks. Gnutella performs broadcasting above TCP. Broadcasted packets include Ping and Query and Request packet also can be broadcasted. Due to broadcast there are problems with communication amount and many efforts has been made to correct this problem. With UUID value embedded in Gnutella itself loop which can be caused in network can be removed and broadcast was diminished in many ways such as host cache and un-relaying Ping packet. What is most important is that routing is made without broadcasting and makes sure that the network is not congested.

3.2 Distribution of Broadcasted Packets

Ping and Query packets are broadcasted periodically or on request. Fig. 1 shows the distribution of packets with TTL value. TTL value on simulation is 7 and the maximum connection was 7. It is the result of broadcasting one packet once for one host and based on Newtella Gnutella Clone[11, 12] on Gnutella protocol basis. Where TTL value is 7 and the maximum connection number is 7, 7^7=823,543 hosts can be retrieved. But it is not reliable numbers. Assuming that 1/10 of these share files, approximately 82,000 hosts can be retrieved. Incidentally http://ed2kx4u.de/list.html administering server lists of edonkey200 showed the similar results in the average connectors of the top 20 servers.

The rate of TTL value 4, 5, 6, 7 was omitted for they are relatively very small quota. Packets transmitted as Fig. 1 are in practical removed by UUID value to prevent loop. Therefore non-discarded packets respond at each host in the end. Fig. 2 indicates distribution of effective packets among the overall packets.

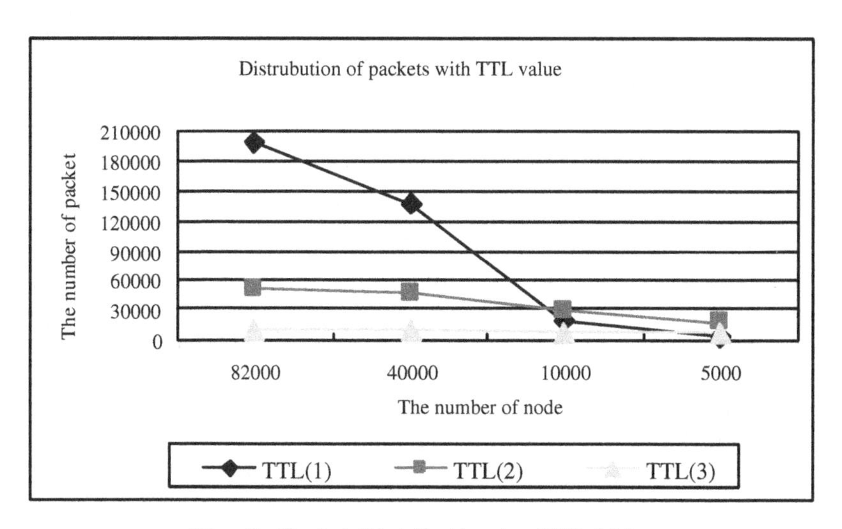

Fig. 1. Packet Distribution by TTL Value

When TTL is 1, the more the number of nodes decreases the more rapidly the effective packet amount decreases and the effective packet amount is quite small to the number of node. With TTL 2 and 3 assuming that the node is not many it does not show any rapid change. In the end assuming that node is not considerably many, with TTL value 2 and 3 effective packets from one broadcasted Query packets by one host showed the most and it can be seen that TTL value 2 produced the more. More retrieval results (QueryHit) will be transmitted by host which receives packets of TTL value 2. Therefore it can be said that it is very likely to download contents of these hosts. QueryHit packet generating technique proposed at this study uses QueryHIt packet responded by host receiving packets of TTL value 2.

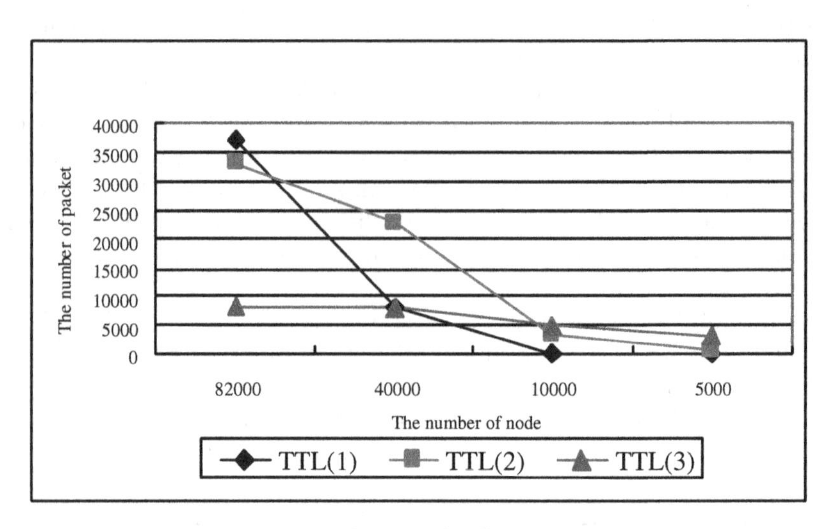

Fig. 2. Distribution of Effective Packets

4 Message Manipulation

Anonymity basically offered by Gnutella network has limitation by exposing user identity on downloading and uploading. And this can cause unexpected situation. For example after sharing malicious file (e.g. remote controller) using information of host downloading this file various attacks can be brought and by maliciously using exposed information denial of service can be performed. The technique of providing anonymity proposed in this study is that IP address, port number and UUID value are sent to the host having sent Query packet for the first after manipulating them. And when receiving download request it behaves like Proxy server. Fig. 3 shows the QueryHit packet transfer with the example of Newtella Clone. Node 6 transmits Query packet received from Node1 to Node 7, and updates SearchRoute routing table having key of UUID of Node 1. It is common that when QueryHit message arrived from node 7 routing table is retrieved with UUID value and it is transferred to Node 5. Ultimately it is transmitted to Node 1 and Node 1 requests contents download using Node 7 information. But this makes lose anonymity due to the information exposure of both Node 1 and Node 7.

But there still exists problem. Manipulating Node 2, 3, 4, 5 as Node 6 will incur great deal of network traffic. Furthermore if Node 1 is connected to Node 6 in a very low bandwidth and connected to Node 7 in very high bandwidth, it takes burdens too much to provide anonymity. In addition where there are

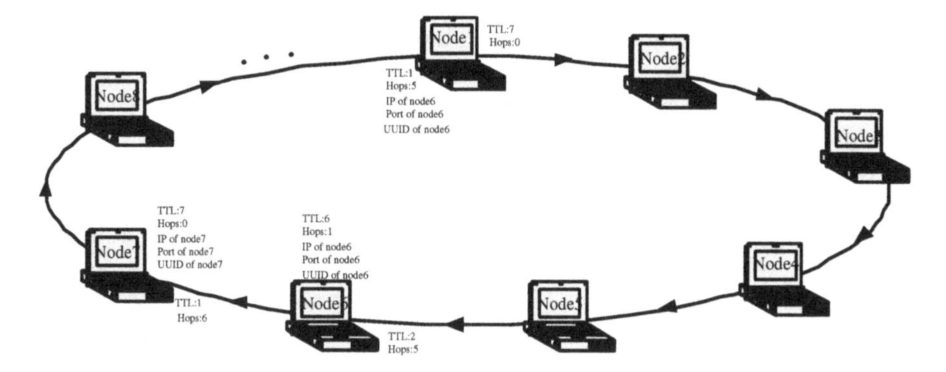

Fig. 3. Transmission of QueryHit Packets

very popular contents in Node 7 non-powerful Node 6 may not handle all request.

Therefore this study proposes two methods to settle the problems. Fist, in Fig. 3 on Node 7's creating QueryHit packet it transmits the packet making TTL=hops(of Query Packet)+2, Hops=0[11, 12]. Therefore Node 6 is Hops=0 and it needs only to manipulate QueryHit packets having Hops=0 and greater than TTL value 2 then other Nodes will transmit packets as it were without manipulating them. The second, Fig 1,2 in Sec. 3.2 shows that it has effective packets most with TTL value 2. Since Node 6 is the sender which sent packets with TTL value 2 that ultimately the most QueryHit packets get routed. During the routing Node 6 can check what contents is requested most frequently and obtain the statistics. Using the value obtained from statistics up to the storing capacity it caches contents which are requested frequently and may be requested frequently later on. And if Node 6 has many requests it can provide service without communicating with Node 7.

5 Test and Performance Evaluation

We simulate that proposed two methods in sec.4 based on Newtella Gnutella Clone on Gnutella protocol basis. First method had no problem on our simulation results. Question of Proposed method is the second method that using the value obtained from statistics. We used Zipf's law to solve problem. Zipf's Law is based on linguistic experimental fact computed in relation to rank and frequency in natural language and the law can be represented as Rank×Frequency=Constant[3, 8]. This indicates that the relation between rank and frequency is represented as effective and correct general empirical law to any inclusive text. That is, product of generated word frequency and ranking value with the frequency is computed approximately as certain constant value. It is important that the relation of connection frequency and hit number with the ranking of that under WWW where numerous retrieval processes are performed also follows this law[3, 4, 5]. Therefore it can be said that web requests follow the distribution of Zipf's[3, 4, 5]. As shown on Fig.1,2 a great deal of messages are discarded at intermediate peers. And the most appropriate Query messages are those with TTL value 2 to apply Zipf's law. Therefore after inducing approximate constant value, which is product value of frequency, and ranking according to the law, QueryHit message, which is relevant high ranking value and retrieval Query, is updated periodically, or whenever relevant QueryHit message arrives. By this QueryHit message related to high ranking is updated always as new value highly possible for connection. If approximate constant value is induced then cache the QueryHit message and shared file which is relevant retrieval Query in the high 5% range. This is because about 50% of the total demand falls in the high 5% of demand frequency[4, 5]. This means that around 50% of many retrieval messages demand similar messages. So if retrieval Query message of TTL value 2 is ranked high in each peer, that cached QueryHit messages are returned without relaying them. In the same way cached files are serviced.

Result from simulation indicates that with TTL value 1 many Query messages amounting to about 49.83% were reduced and about 17% was reduced in total because without relaying. Fig. 4 show that the result of broadcasting one packet once for one host and based on Newtella Gnutella Clone[11, 12] on Gnutella protocol basis. Fig. 4 simulated message throughput rate by number of peer.

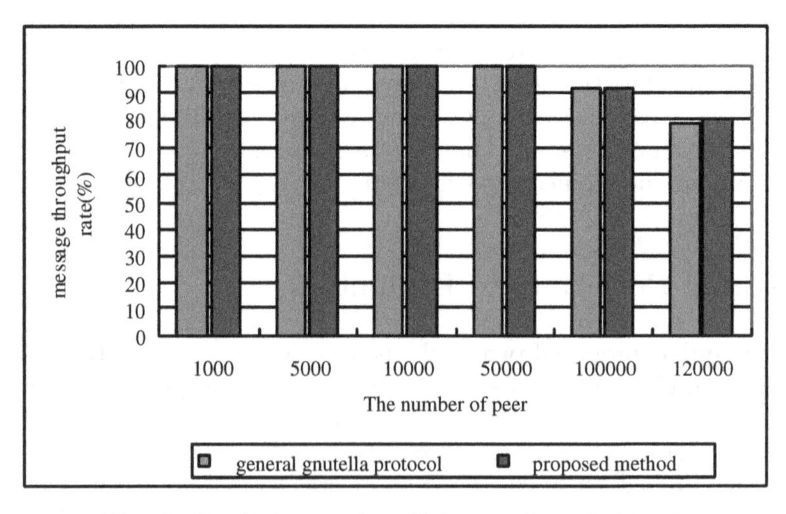

Fig. 4. Simulation results of Message throughput rate

The result shows that message throughput rate which indicates all the Query messages sent by non-specific peer were transferred to all the currently participating peers was constantly remained. Although large amount of broadcasted messages were reduced, messages were transferred to all peers. It means that request sent by peer is able to retrieve every matching data among all peers participating network. Therefore retrieval result which can be obtained by user is constant. And the message throughput rate was not changed to the result applied caching technique proposed at this study whereas message amount was considerably reduced instead. This shows that despite the reduction of message amount it hardly affect the retrieval result.

6 Conclusion

On composing the P2P systems the most important point is the possibility that users exchange information and contents under anonymous condition having exclusive right. Most of packets transferred from node to node do not include sender's IP address and these packets are transmitted through dynamic routing carried out by intermediate hosts. In addition this may be temporary since dynamic routing is renewed (updated) periodically so it is impossible to know which host transfers packet for the first and which the designated recipient host is. Therefore basically it provides anonymity. However when contents upload and download is made between user and provider information on both sides is exposed and this attenuates the possibility mentioned first. In order to settle this problem this study calculates packet distribution within the whole network of Query and QueryHit, which is different from downloading cached information to protect identity of user and provider, manipulates QueryHit on the basis of the calculation and transfers contents after caching them. This provides secured anonymity to the intermediate node performing Proxy role between user and provider.

Acknowledgements

This work was supported by INHA UNIVERSITY Research Grant.

References

1. The Gnutella Developer Forum(GDF)(2003). The Annotated Gnutella Protocol Specification v0.4(1), http://dlaikar.de/gdf_files/Developement/GnutellaProtocol-v0.4-r1.8.html
2. Dynamic Peer-2-Peer Source Routing, http://dynamo.ecn.purdue.edu/ ychu/projects/dpsr/
3. Lee Breslau,Pei Cao,Li Fan,Graham Phillips, Scott Shenker (1999) Web Caching and Zipf-like Distributions, IEEE INFOCOM
4. Kunwadee Sripanidkulchai (2004) The popularity of Gnutella queries and its implications on scalability, http://www-2.cs.cmu.edu/~ kunwadee/research/p2p/paper.html

5. V.Almeida, A.Bestavros, M.Crovella and A.de Oliveira (1996) Characterizing reference locality in the WWW, International Conference on parallel and Distributed
6. Andy Oram (2001) PEER-TO-PEER, O'Reilly and Associates Sebastopol:117–119
7. Andy Oram (2001) PEER-TO-PEER, O'Reilly and Associates Sebastopol:119–120
8. G.K Zipf (1949), Human Behavior and the Principles of Least Effort, ddison-Wesley, Cambridge
9. L. Fan et al. (1998), Summary Cache: a scalable wide-area web cache sharing protocol. in: Proc. Of ACM SIGCOMM 98:254–165
10. J. Pitkow (1998), Summary of WWW characterizations. in: Proc. WWW
11. Newtella- P2P Fliesharing, http://p2p.at-web.de/newtella.htm
12. Introduction to Newtella, http://www.schnarff.com/gnutelladev/source/newtella/

Investigation of the Fuzzy System for the Assessment of Cadastre Operators' Work

Dariusz Król[1], Grzegorz Kukla[1], Tadeusz Lasota[2], and Bogdan Trawiński[1]

[1] Institute of Applied Informatics, Wrocław University of Technology, Poland
dariusz.krol@pwr.wroc.pl grzegorz_kukla@o2.pl trawinski@pwr.wroc.pl
[2] Faculty of Environmental Engineering and Geodesy, Agricultural University of Wrocław, Poland tadeusz.lasota@wp.pl

1 Introduction

Cadastre systems are mission critical systems designed for the registration of parcels, buildings and apartments as well as their owners and users. Those systems have complex data structures and sophisticated procedures of data processing. They are constructed in client-server architecture for LAN as well as in Web technology to be used in intranets and extranets. There are above 400 information centres located in district local self-governments as well as in the municipalities of bigger towns. Managers of information centres often complain they have no adequate tools for the assessment of work of cadastre system operators. The fuzzy model for the assessment of operators' work of a cadastre information system was proposed in [6]. According to centre managers' suggestions the following four input criteria were designed: i.e. productivity, complexity, time and quality. Productivity was expressed by the number of changes input into the cadastre database within a given period, complexity of changes was specified as the mean number of objects which were modified in the database falling per one change, time was determined by average time of inputting one change and the quality of work defined as the percentage of changes without any corrections.

The results of the investigation of the fuzzy model are discussed in the present paper. The tests have been carried out using real data taken from one cadastre centre. The data comprised all change records input into cadastre database during the period of one year from October 2004 till September 2005.

In general numerous methods are used for the determining optimal parameters of fuzzy models [7] including such approaches as neuro-fuzzy systems [1], genetic algorithms [4] and fuzzy clustering techniques [3]. Fuzzy models are also evaluated using specific analyses like interpretability [9], sensitivity [8] or regression [2]. In our approach descriptive statistics, correlation, multiple

Mark Last et al. (Eds.): Advances in Web Intelligence and Data Mining (SCI) **23**, 131-140 (2006)
www.springerlink.com

regression and the distance between rankings have been used in the analysis of the test results.

2 The Structure of the Fuzzy System

2.1 General description of the system

The fuzzy Mamdani's model with Larsen's implication proposed in [6] constitutes the basis of the fuzzy system which is intended to rationalize the management of information centres, to improve the organization of work and to determine wages of part-time employees. General architecture of the fuzzy system is shown in Fig. 1. It comprises five main modules of operators' work statistics, fuzzification, inference, defuzzification and visualization.

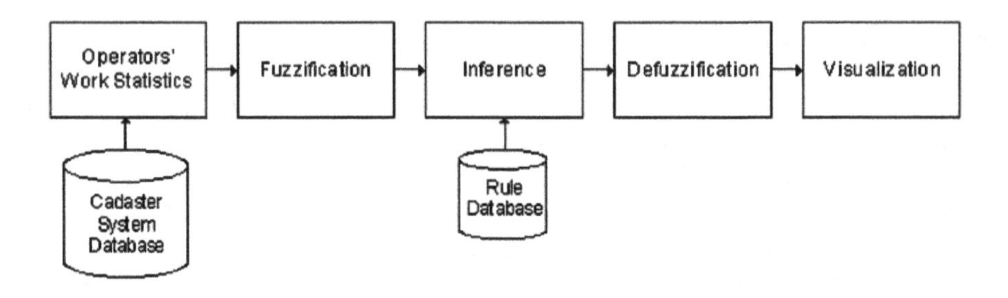

Fig. 1. Architecture of the fuzzy system for the assessment of operators' work

For each input criterion i.e. productivity (P), complexity (C), time (T) and quality (Q) as well as for output assessment triangular and trapezoidal membership functions have been defined. The statistics module provides initial parameters of the model and values of input criteria. The idea of obtaining the final assessment consists in calculating the average value of P, C, Q and T criteria taking into account the change records saved in cadastre database for all operators and for long period of time, e.g. a year or a half of year. These average values are used as the reference values of 100% for calculating what percentage of corresponding average value a given operator achieved within an assessment period. The domain for Q, P and T has been set up from 0% to 200% that means, if an operator achieves better results than 200% of a mean during the assessment period, his result will be trimmed to 200%. Data for quality variable are used directly, because this criterion is expressed in percents. Standard deviations, calculated for each criterion separately, determine the initial width of the basement of triangle and trapezoid. The domain for output is an arbitrary assessment scale from 0 to 200; with 0 being the lowest and 200 the highest mark.

2.2 Characteristics of the inputs and the output of the system

The first step of data analyses was to examine significant relations between input criteria. Work statistics for the period of 12 months for 10 operators were calculated and the values of input criteria were obtained. Criteria with zero values, for months where a given operator did not work and criteria with values trimmed to 200, for months when a given operator achieved results more than 200 percent better than average, were neglected, so finally the correlation matrix was calculated for 97 quadruplets of input values (see Table 1). Correlation coefficients between T and C as well as between T and P turned to be significant. This result has led to the decision to remove the input variable of Time from the fuzzy model.

Table 1. Correlation matrix for input criteria

	Complexity	Productivity	Quality	Time
Complexity	1.0			
Productivity	−0.04	1.0		
Quality	0.05	0.11	1.0	
Time	**-0.45**	**0.27**	0.01	1.00

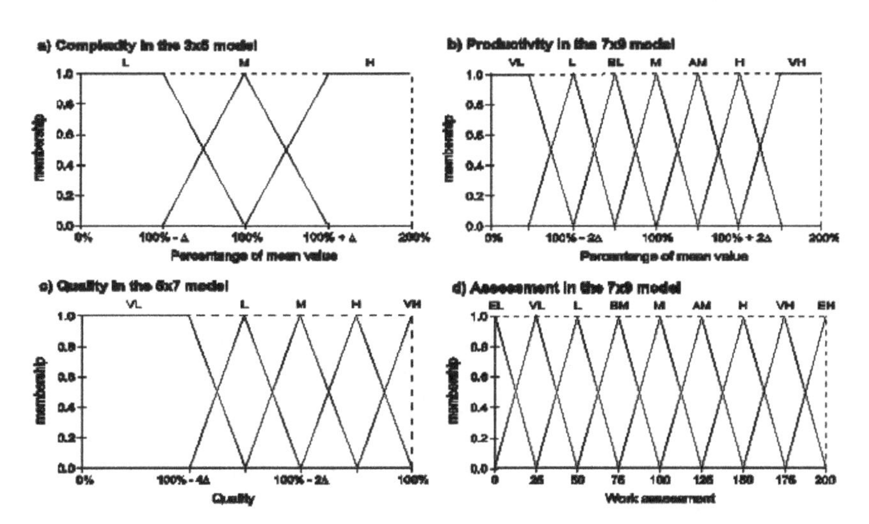

Fig. 2. Examples of membership functions of input and output variables

Three main models of input and output variables have been programmed and named 3x5, 5x7 and 7x9. In the 3x5 model 3 fuzzy sets determine each input criteria and the output is defined by 5 fuzzy sets. In the 5x7 and 7x9 models there are 5 and 7 fuzzy sets for each input as well as 7 and 9 for each output respectively. Examples of fuzzy membership functions used to define

C, P and Q criteria and an output are shown in Fig. 2, where EL, VL, L, BM, M, AM, H, VH, EH denote Extremely Low, Very Low, Low, Below Medium, Medium, Above Medium, High, Very High and Extremely High respectively. Delta (Δ) is a parameter which determines the width of the basement of a triangular membership function. Initial value of Δ is set up with standard deviation (σ), calculated for each criterion separately. During tests the value of Δ was changed from 1.0σ to 0.2σ.

2.3 Rule base and inference process

The rule base for each model contains simple IF-THEN rules where the condition consists of only two input variables combined by AND operator and the conclusion is built by one variable. An example of a rule is as follows:
IF Complexity is low AND Productivity is medium THEN Assessment is low.

a) Rules for 3x5 model

(X_1, X_2)	L	M	H
L	VL	L	M
M	L	M	H
H	M	H	VH

b) Rules for 5x7 model

(X_1, X_2)	VL	L	M	H	VH
VL	VL	L	L	BM	M
L	L	L	BM	M	AM
M	L	BM	M	AM	H
H	BM	M	AM	H	H
VH	M	AM	H	H	VH

Fig. 3. Representation of rule base in matrix form for 3x5 and 5x7 models

Thus the rules for one pair of input criteria can be given in the form of a matrix shown in Fig. 3 and 4. Three matrices for each pair of input criteria i.e. (C,P), (C,Q) and (P,Q) have been designed and they comprise 9, 25 and 49 rules for the 3x5, 5x7 and 7x9 models respectively. In order to express the strength of rules belonging to particular combination, rule weights can be assigned to each combination as the multipliers of rule conditions in aggregation step in the inference module, for example $w_{(C,P)} = 0.60$, $w_{(C,Q)} = 0.20$ and $w_{(P,Q)} = 0.20$.

In order to assure that each input value and each rule will have an impact on the final assessment following operators has been used: PROD for aggregation of rule conditions, PROD for activation of rule conclusions and ASUM for accumulation of output membership functions, where PROD means algebraic product and ASUM denotes algebraic sum [5]. In the defuzzification step the centre of gravity method has been used.

c) Rules for 7x9 model

(X_1, X_2)	EL	VL	L	M	H	VH	EH
EL	EL	EL	VL	L	L	BM	M
VL	EL	VL	L	L	BM	M	AM
L	VL	L	L	BM	M	AM	H
M	L	L	BM	M	AM	H	H
H	L	BM	M	AM	H	H	VH
VH	BM	M	AM	H	H	VH	EH
EH	M	AM	H	H	VH	EH	EH

EL - Extremly Low
VL - Very Low
L - Low
BM - Below Medium
M - Medium
AM - Above Medium
H - High
VH - Very High
EH - Extremly High

Fig. 4. Representation of rule base in matrix form for 7x9 models

3 Plan of the Investigation

The experiment has been carried out using cadastre database taken from one information centre and change records added by 10 operators into the database during the period of one year from October 2004 till September 2005. The fuzzy system has been treated as a black box, that means only input values of Complexity, Productivity and Quality and corresponding output assessments have been taken into account in the study. Multiple regression, descriptive statistics, and the distance between rankings have been used in the analysis of the test results. In order to examine how the output assessments change for different parameters of the system 180 variants of fuzzy model have been constructed by a simulation program and tested. The variants covered all possible combinations of three basic 3x5, 5x7 and 7x9 models, five values of Δ parameter determining the widths of triangle basements, three sets of rules and four sets of rule weights. Each variant of the model has been coded according to the method shown in Table 2 where (1), (2), (3) and (4) by a code caption indicate the position of a digit in the code. For example 7413 denotes the 7x9 model with Δ=0.4σ using the rule set of (C,P) with the weight equal 0.8, (C,Q) with the weight equal 0.4 and (P,Q) with the weight equal 0.1 In turn 3134 denotes the 3x5 model with Δ=1.0σ using the rule matrix of (C,P) with the weight equal 0.6, and (P,Q) with the weight equal 0.2.

The purpose of the experiment was to examine how input values influence the output of the system, how the assessments produced by the system make it possible to differentiate the results of operators' work and how close are the system assessments to subjective manager's judgments.

Table 2. Coding method of variants tested

C(1)	Model	C(2)	Delta	C(3)	Rule sets	C(4)	Weights
3	$3x5$	1	1.0σ	1	$(C,P),(C,Q),(P,Q)$	1	$1.0,1.0,1.0$
5	$5x7$	2	0.8σ	2	$(C,P),(C,Q)$	2	$0.9,0.6,0.3$
7	$7x9$	3	0.6σ	3	$(C,P),(P,Q)$	3	$0.8,0.4,0.1$
		4	0.4σ			4	$0.6,0.2,0.2$
		5	0.2σ				

4 Results of the Investigation

Data for the analysis of descriptive statistics and multiple linear regression have been prepared in analogous way as data which were used during correlation study. Input criteria with zero values, for months where a given operator did not work and criteria with values trimmed to 200, for months when a given operator achieved results more than 200 percent better than average, were neglected. However the comparison of the assessments produced by the fuzzy system with subjective judgments of information centre manager has been conducted using statistical data of changes added by operators into the cadastre system during September 2005.

4.1 Multiple Linear Regression Analysis

The multiple linear regression with no intercept has been calculated for all 180 models. In each case the coefficient R was greater than 0.9 (minimum value equal 0.935 and maximum value equal 0.997), F-value scaled between 219.473 and 4760.762 and p-value very close to zero. This indicates that input criteria are strongly related to the output assessments. The analysis of β coefficients has revealed that p-value for β_Q coefficient by Quality variable was greater than 0.05 in 37 cases, i.e. 20.6%. Moreover the value of β_Q coefficient was negative in 156 cases what may be interpreted that operators achieved better complexity or productivity at the cost of decreasing quality. The results of the regression analysis of 9 selected models for which all β_C, β_P and β_Q coefficients were significant are shown in Table 3.

4.2 Analysis of descriptive statistics

General question of the analysis of descriptive statistics was how the output generated by the system made it possible to differentiate the results of operators' work. So two measures have been taken into account namely the variability coefficient which is expressed by the standard deviation divided by the mean and the range which equals the difference between maximum and minimum assessments. It may be expected that if the fuzzy system provides more differentiated results then it will better assist managers' in assessing their workers. The variability coefficient calculated for 180 models has had

Table 3. Results of multiple linear regression analysis for 9 selected models

Model type	Model code	Multiple R	F-value	β_C	β_P	β_Q
$3x5$	3111	0.984	933.819	0.326	0.411	0.201
$3x5$	3223	0.981	807.087	0.626	0.439	−0.200
$3x5$	3434	0.969	475.597	0.538	0.724	−0.453
$5x7$	5131	0.993	2264.987	0.256	0.543	0.289
$5x7$	5313	0.984	973.042	0.617	0.447	−0.137
$5x7$	5422	0.971	507.758	0.706	0.402	−0.285
$7x9$	7312	0.965	426.105	0.555	0.516	−0.220
$7x9$	7424	0.956	331.688	0.759	0.627	−0.613
$7x9$	7531	0.948	275.480	0.392	0.791	−0.498

the values between 0.283 and 0.773 and the range between 128 and 182. The values of variability coefficient presented in Fig. 5 are greater for 7x9 models than for 3x5 and 5x7 models.

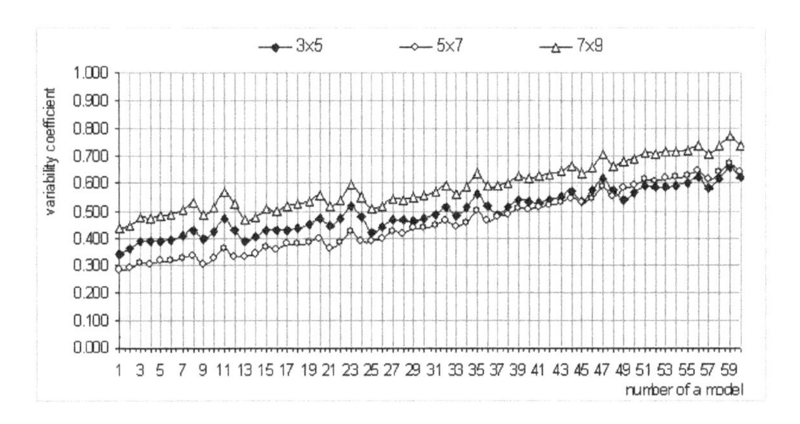

Fig. 5. Values of variability coefficient for 180 models tested

You can draw similar conclusions when you examine the fuzzy system output surface. The plots generated by Matlab Surface Viewer has shown that the 7x9 model assures more distinguishable assessments than the assessments produced by the 3x5 model (see Fig. 6).

In Fig. 7a range values for different Δ sizes for corresponding 3x5, 5x7 and 7x9 models (the same rule sets: (C,P), (C,Q), (P,Q) and the same weight variant: 0.6, 0.2, 0.2) are presented, where 1, 2, 3, 4, 5 on X axis denote Δ equal 1.0σ, 0.8σ, 0.6σ, 0.4σ, 0.2σ respectively. In Fig. 7b range values for different rule weight variants for corresponding 3x5, 5x7 and 7x9 models (the same rule sets: (C,P), (C,Q), (P,Q) and the same Δ=0.6σ) are presented, where 1, 2, 3, 4 on X axis denote 1.0, 1.0, 1.0 and 0.9, 0.6, 0.3 and 0.8, 0.4, 0.1 and 0.6, 0.2, 0.2 variants respectively. In both Fig. 7a and 7b it is clearly seen that 7x9 models provide more distinguishable results than other models.

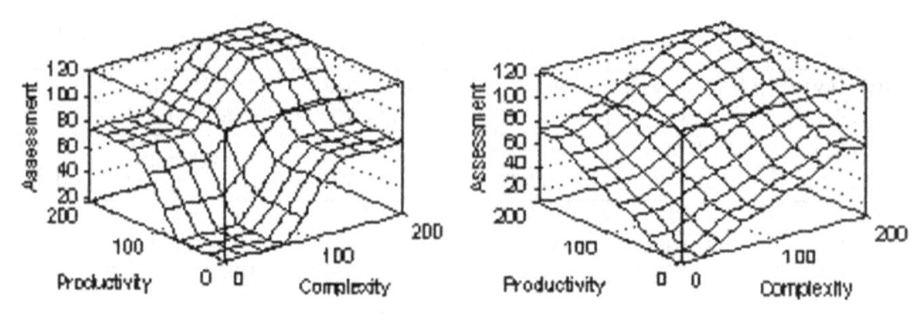

Fig. 6. Assessment surface versus C and P criteria for 3x5 and 7x9 models

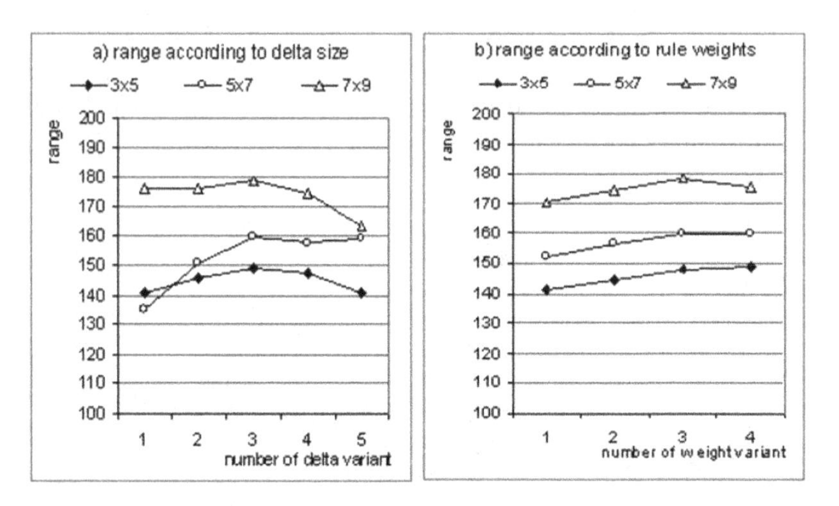

Fig. 7. Value of a range for a) different variants of Δ and for b) different variants of rule weights

4.3 Comparison of the assessments assigned by the system and by a centre manager

In order to evaluate how the output of the system is related to a centre manager's judgment one information centre manager was asked to appraise his operators' work in September 2005. He was not informed how fuzzy system worked and he did not see the results of statistics module so that his judgments were entirely subjective. The manager was able to give relatively rough assessments expressed in percents: 150%, 120%, 120%, 120%, 110%, 100%, 80%, 80%, and 70% for successive operators. It could be easily seen that the manager had difficulties in differentiating individual operators. Nevertheless in the case of equal assessment he was asked to rank the operators. So we were able to compare the rankings determined by the manager with produced by the fuzzy system. The tenth operator was not classified by the manager who stated that operator fulfilled different tasks and added changes to the cadastre database sporadically and therefore was assigned last position in the manager's ranking. We used following measure of the distance between these two rankings:

$$DRank = \sum_{i=1}^{10} |r_{mi} - r_{si}| \tag{1}$$

where r_{mi} denotes the position of i-th operator in the manager's ranking and r_{si} the position of i-th operator in the ranking produced by the fuzzy system. The DRank measure was calculated for each of 180 models tested and its value was between 18 and 26.

Rank positions of individual operators produced by the system for three selected 3x5, 5x7 and 7x9 models and positions assigned by the centre manager are presented in Fig. 8.

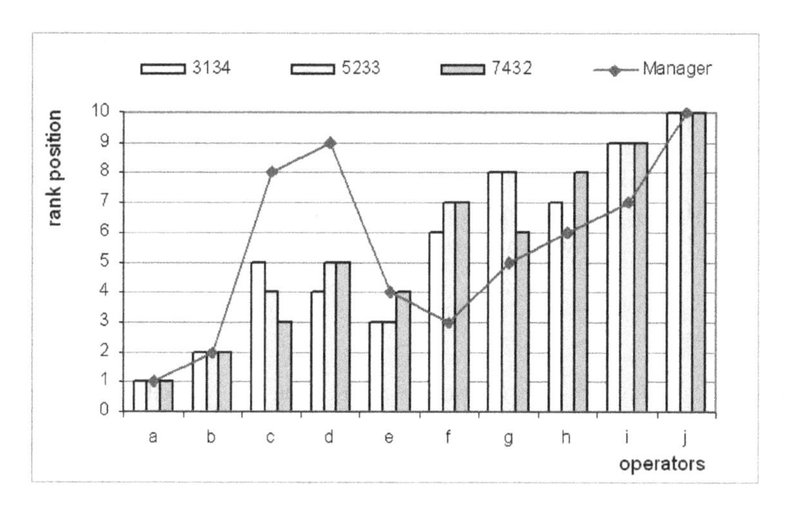

Fig. 8. Rank positions assigned to individual operators by the manager and the system

It can be seen that the manager and the system equally recognized the best and the worst operators. However the manager clearly underestimated operator c and operator d. Maybe it has been caused by manager's subjective approach, which for example when assessing the d operator's work for 70% stated that operator admittedly was a very experienced person but she tended to work slowly. It is also possible that there are other criteria of operators' work assessment, maybe even immeasurable, which the manager took into consideration.

5 Conclusions and Future Works

The fuzzy system for the multi-criteria assessment of information system operators' work has been evaluated using real data taken from one cadastre centre. Input data generated by the statistical module have been processed using automatically created 180 variants of fuzzy models. The variants covered all possible combinations of three basic 3x5, 5x7 and 7x9 models, five values of parameter determining the widths of triangle basements, three sets of rules and four sets of rule weights. Multiple linear regression, descriptive statistics, correlation and the distance between operator rankings have been used in the analysis of the test results.

The experiment allowed us to investigate the properties of the fuzzy system. In 79% all input variables influenced the output significantly. The assessments generated by the models differed in the value of variability coefficient and the range. The 7x9 models assured better differentiation of the results. It is not possible to determine definitely which model is optimal, nevertheless the study proved usefulness of the model. It is planned to carry out further evaluation experiments with the active participation of the centre managers. This time the centre managers will be instructed how the fuzzy system operates and will be got familiar with the statistics of operators' work within a given time. Moreover they will be able to determine the weights of the rules in order to adjust the system to their preferences.

References

1. Ajith A (2001) Neuro-Fuzzy Systems: Sate-of-the-Art Modelling Techniques. In: Proceedings of the 6th International Conference on Neural Networks 269–276
2. Cheung W, Pitcher T, Pauly D (2004) A Fuzzy Logic Expert System for Estimating the Intrinsic Extinction Vulnerabilities of Seamount Fishes to Fishing. Fisheries Centre Research Reports 12(5):33–50
3. Gomez A, Delgado M, Vila M (1999) About the use of fuzzy clustering techniques for fuzzy model identification. Fuzzy Sets and Systems 106(2):179–188
4. Herrera F (2005) Genetic Fuzzy Systems: Status, Critical Considerations and Future Directions. Journal of Computational Intelligence Research 1(1):59–67
5. IEC 1131 - Programmable Controllers (1997) Part 7 - Fuzzy Control Programming. Committee Draft CD 1.0 (Rel. 19 Jan 97)
6. Król D, Kukla G S, Lasota T, Trawiński B (2006) Fuzzy Model for the Assessment of Operators' Work in a Cadastre Information System (to be published
7. Piegat A (2003) Fuzzy Modelling and Control (in Polish). Akademicka Oficyna Wydawnicza EXIT Warszawa
8. Saez D, Cipriano A (2001) A new method for structure identification of fuzzy models and its application to a combined cycle power plant. Engineering Intelligent Systems 9(2):101–107
9. Xing Zong-Yi, Jia Li-Min, Zhang Yong, Hu Wei-Li, Qin Yong (2005) A Case Study of Data-driven Interpretable Fuzzy Modeling. Acta Automatica Sinica 31(6):815–824

DLAIP: A Description Logic Based Approach for Dynamic Semantic Web Services Composition *

Yingjie Li, Li Wang, Xueli Yu, Wen Li, Yu Xing

College of Computer and Software, Taiyuan Univ. of Tech.,
Taiyuan, Shanxi, P. R. China, 030024, Phone (86) 351-6111253
{lyj_613@hotmail.com}

Abstract. The Description Logic, which possesses strong knowledge representation and reasoning capabilities, is the logic basis of the Semantic Web ontology languages such as OWL and OWL-S. The AI planning, which provides an effective method for solving the planning problem and task decomposition in AI, possesses better modeling capability of the action state transformation. Based on the merits of the Description Logic and the AI planning above, this paper proposes a service composition mechanism $DLAIP$ and testifies its feasibility in Description Logic. The results show that this composition mechanism can not only be feasible but also be helpful for the semantic modeling of the service composite process in the Semantic Web.

Keywords: Semantic Web, Service Composition, Description Logic, AI planning, DLAIP

1 Introduction

The DL (Description Logic) [1], whose basic components are Concepts, Roles and Individuals, is the logic basis of the Semantic Web. Based on DL, the semantic modeling of some specific domain and service can be successfully achieved, but the semantic description is deficient in the semantic modeling of the dynamic services composition. The AI planning [2] is feasible to realize the dynamic web services composition through introducing the action notion and the reasoning about these actions. But from the current researches about AI planning [3][4], the services composition mechanisms proposed by them either lack the utility of semantic or do not consider the logic reasoning of the composition mechanism. Considering the deficiency above and based on the merits of the DL and the AI planning, this paper proposes a service composition mechanism $DLAIP$ (AI Planning based on DL), which combines the strong knowledge representation and reasoning capabilities of the DL with the modeling capability of the action state transformation of the AI planning, and finally testifies its availability in Description Logic. In Section 2, the Description Logic is presented. Then section

* **Sponsored by** National Science Foundation of China (No. 60472093)

Mark Last et al. (Eds.): Advances in Web Intelligence and Data Mining (SCI) **23**, 141-150 (2006)
www.springerlink.com

3 shows the service composition mechanism $DLAIP$, testifies its availability in DL and analyzes its complexity in detail. Section 4 briefly discusses the application of the service composition mechanism through an example. Finally in Section 5, the conclusion and our future work are presented.

2 Description Logics

The DL, which inherits the principles of the $KL-ONE$ [5] system, is a decidable subset of the First-Order Logic and supports the decidable reasoning. The DL sublanguage $ALCQIO$ forms the core of the OWL-DL with the additional OWL-DL constructor Q, O, and I based on the DL sublanguage ALC. So in this paper, we employ the DL $ALCQIO$ to prove the service composition feasibility. Table 1 shows the basic minimal set of constructors of the $ALCQIO$. In $ALCQIO$, we assume that N_c stands for the set of defined concepts, N_r stands for the set of defined roles and N_I stands for the set of asserted individuals.

Table 1. Syntax and Semantics of $ALCQIO$

Name	Syntax	Semantics
Inverse role	s^-	$\{(y, x)\|(x, y)\in s^I\}$
conjunction	$C \cap D$	$C^I \cap D^I$
negation	$\neg C$	$\Delta^I \backslash C^I$
At-least number restriction	$\geq nrC$	$\{x \mid card(\{y \mid (x,y) \in r^I \wedge y \in C^I\}) \geq n\}$
nominal	$\{a\}$	$\{a^I\}$

The reasoning functions of DL mainly include concept satisfiability reasoning, subsumption reasoning, consistency checking reasoning and instance checking reasoning. For the purpose of this paper, it suffices to introduce concept satisfiability reasoning and consistency checking reasoning. As follows:

(1) The concept C w.r.t TBox T is satisfiable iff there exists a interpretation model I of T such that $C^I \neq \emptyset$;

(2) The ABox A is consistent w.r.t the TBox T iff there exists an interpretation I that is a model of both T and A.

3 The Service Composition Mechanism: *DLAIP*

The service composition, which can shield the service transition at the low level and realize the service composite application at the high level, is achieved through combining the existing and autonomic services. This part mainly focuses on the reasoning about the Semantic Web service composition mechanism $DLAIP$. For simplicity, we concentrate on the services with no parameters or the ground services. Parametric services, which contain variables in place of individual names, should be viewed as a compact representation of all its ground instances. We may restrict ourselves to ground services since all the reasoning tasks considered in this paper presuppose that parametric services have already been instantiated.

3.1 Relative Definitions in *DLAIP*

According to the overview of the features of the DL and AI planning in part 1, we propose a kind of semantic web services composition mechanism *DLAIP* in order to realize the semantic modeling and reasoning of the services composition.

Definition 1. *(Atomic Service)Let T be an acyclic TBox in an ALCQIO knowledge base KB={TBox, ABox}. Assuming that $N_c$stands for the set of defined concepts, N_r stands for the set of defined roles and $N_I$stands for the set of asserted individuals. An atomic service for an acyclic TBox is defined as:*

S= (pre, sname, inputs, outputs, post)

where S consists of:

(1) pre: a finite assertion set based on the ABox. This component describes the dynamic and static precondition information of the atomic service;

(2) sname: name identification of the atomic service;

(3) inputs: the domain classes related with the service inputs. The forms in the inputs include $A(a)$or $r(a,b)$, where A is a primitive concept w.r.t T, $A \in N_c$, and r is a role, a, b$\in N_I$;

(4) outputs: the domain classes related with the service outputs. The form definition is similar with that in the "inputs";

(5) post: a finite set of post conditions of the service execution, which includes the dynamic and static information. The "post" form is expressed as P/E, where P is an ABox assertion, E is a primitive concept w.r.t the TBox T, i.e., $A(a)$, $\neg A(a)$, $r(a,b)$ or $\neg r(a,b)$. The semantics of P/E is that if P is true before executing the service, then E should be true afterwards.

Definition 2. *(Composite Service) Let T be an acyclic TBox in an ALCQIO knowledge base KB={TBox, ABox}. A composite service for an acyclic TBox is a finite ordered services flow, which is composed of a series of proper services (including atomic and composite services) S_1, S_2,..., and S_k, and can be expressed as S={SL, CF}, where SL (Services List) is the list of the services used to construct the composite service, and CF (Control Flow) is designed to control the executing sequence of the services in the services list SL, $CF \in \{sequentialpattern, parallelpattern, otherpatterns\}$[6]. The relation between the SL and CF can be expressed as SL=CF $\{S_1, S_2,, S_k\}$.*

According to Definition 1 and Definition 2, a composite service is a finite ordered services flow of a series of proper services (including atomic and composite services). During the composite process, the state of every service can be changed because of the execution of the different services. As a result, the interpretations model *I* of the different composite services or the different phrases in only one composite service can also be different. In order to express the relations of these interpretation models, the binary relation '<' among the interpretations is defined as follows (Definition 3).

Definition 3. *(Binary Interpretation Relation) Let T be an acyclic TBox in an ALCQIO knowledge base KB={TBox, ABox}, S= (pre, sname, inputs, outputs, post) be an atomic service w.r.t T and I be an interpretations model w.r.t T. We define the binary relation '<'on models of T by setting $I^{'} < I^{''}$ iff*

(1) $((A^I \oplus A^{I'})\backslash\{a^I \mid A(a) \in \{sname, inputs, outputs\}\}) \sqsubseteq A^I \oplus A^{I''}$

(2) $((r^I \oplus r^{I'})\backslash\{(a^I, b^I) | r(a,b) \in \{sname, inputs, outputs\}\}) \sqsubseteq r^I \oplus r^{I''}$

where $\oplus$ denotes the symmetric difference, I'' is called the successive model of I'.

According to Definition 3, when a service $S = (pre, sname, inputs, outputs, post)$ is executed, and there exists two interpretation models I, I' satisfying that $I' \models post$, $I<I'$ and there does not exist an interpretation I'' satisfying $\{I < I' < I'' \cap I'' \models post\}$, we can conclude that through executing this service, the state of this service is transformed from I to I', namely if $post = P/E$, then $\{I, I' \models post\} \iff$ {if $I \models P$ then $I' \models E$}. The definition of service execution is as Definition 5. Before being executed, the service must be compatible with the knowledge base. So, we first present the service consistency definition (Definition 4).

Definition 4. *(Service Consistency) Let T be an acyclic TBox in an ALCQIO knowledge base KB={TBox, ABox}, S= (pre, sname, inputs, outputs, post) be an atomic service w.r.t T and I be an interpretations model w.r.t T. If there dose not exist an interpretation model I' satisfying Definition 3, namely there is not a successive interpretation model I' of I, we can conclude that the service S is inconsistent with the interpretation model I w.r.t T, otherwise, the service S is consistent with the interpretation model I w.r.t T.*

Definition 5. *(Service Execution) Let T be an acyclic TBox in an ALCQIO knowledge base KB={TBox, ABox}, S= (pre, sname, inputs, outputs, post) be an atomic service w.r.t T and I, I' be the interpretations models w.r.t T sharing the same domain. After the service execution, the service S may transform I to I' ($I \Longrightarrow I'$) iff*

(1) I, $I' \models post$, and

(2) there does not exist an interpretation model J satisfying I, $J \models post$, $j \neq I'$ and $J < I'$.

Based on Definition 5, when the consistent composite service S={SL, CF},SL=CF{S_1, S_2,..., S_k} is executed, there must exist a series of interpretation models I_0, I_1,...,I_k, which take turns along with the service S execution. So we can get the formalized definition of composite service execution (Definition 6).

Definition 6. *(Composite Service Execution) Let T be an acyclic TBox in an ALCQIO knowledge base KB={TBox, ABox}, S={SL, CF}, SL = CF {S_1,*

$S_2,\ldots, S_k\}$ be an composite service w.r.t T. The composite service S may transform I to I′, iff there exists interpretation models $I_0, I_1,\ldots I_k$ with $I=I_0, I'=I_k$ and $I_{i-1} \Longrightarrow I_i (1 \le i \le k)$.

3.2 Logic Reasoning in *DLAIP*

Assuming a knowledge base KB={TBox, ABox}, where TBox is acyclic. Let S={SL, CF},SL=CF$\{S_1, S_2,\ldots, S_k\}$ be a consistent composite service. Intuitively, if the composite service S is executable, then the state transformation of every atomic service S_i $(1 \le i \le k)$ in S before and after the execution must be consistent with the current knowledge base. According to Definition 5, when an atomic service is executed, the service may transform the interpretation model I to I'. As a result, the concepts can be changed from A^I to $A^{I'}$ and the roles can be changed from r^I to $r^{I'}$. Here, we first present the lemma of calculating the $A^{I'}$ and $r^{I'}$ (Lemma 1).

Lemma 1 (Calculating the $A^{I'}$ and $r^{I'}$) Let T be an acyclic TBox in an *ALCQIO* knowledge base KB=$\{TBox, ABox\}$, S= {pre, sname, inputs, outputs, post} be an atomic service w.r.t T, and I, I' be the interpretations models w.r.t T, $I \Longrightarrow I'.I, I'$ are consistent with the service S. If A is a primitive concept and r a role name, then:

$A^{I'}=(A^I \cup \{a^I \mid P/A(b) \in post \cap I \Longrightarrow P\}) \cup \{a^I \mid P/\neg A(b) \in post \cap I \Longrightarrow P\}$,

$r^{I'}=(r^I \cup \{(a^I,b^I) | P/r(a,b) \in post \cap I \Longrightarrow P\}) \cup \{(a^I,b^I) | P/\neg r(a,b) \in post \cap I \Longrightarrow P\}$.

When every atomic service S$_i$ $(1 \le i \le k)$ of a composite service is executed, we must first judge that the pre_i component of the atomic service S_i is satisfied. Only if the pre_i is satisfied, this atomic service can be executed. This feature of the service is called "***Executable***". If the atomic service S_i is executable, the current interpretation model of S_i is I, the subsequent interpretation model of S_i is I', and the executive effects of the atomic service S_i is $post_i$, then S_i must satisfy the form $I' \models post_i$, namely the effects of service must be valid. This feature of the service is called "***Executive Effects Validity***". If the atomic service S_i is executable and executive effects of S_i is valid, then the $outputs_i$ of the former atomic service $S_i$must be matched with the $inputs_{i+1}$ of the subsequent atomic service S_{i+1} during the execution of the composite service S. Namely when the matching degree between the outputs of S_i and the inputs of S_{i+1} is in some value range, $S_i$and S_{i+1} can be composed. This feature of the service is called "***Matching Between Inputs and Outputs***". All these features are shown in Theorem 1.

Theorem 1 (Service Composition Validation) Let T be an acyclic TBox in an *ALCQIO* knowledge base KB={TBox, ABox}, and S={SL, CF}, SL=CF $\{S_1, S_2,\ldots, S_k\}$ be an composite service w.r.t T, where S_i= (pre_i, $sname_i$, $inputs_i$, $outputs_i$, $post_i$), $(1 \le i \le k-1)$. The composite service is valid iff

(1) ***Executable:*** S_1, S_2,..., S_k is executable in A w.r.t T iff the following conditions are true in all models I of A and T:

$I \models pre_i$,

For all i with $1 \leq i \leq k$ and all interpretations I_i, and $I \Longrightarrow I_i$, then $I_i \models pre_{i+1}$.

(2) ***Executive Effects Validity:*** Assuming *post* is the executive effect of the composite service S, iff, for all i in $1 \leq i \leq k$ and all the interpretation models I_i, and $I \Longrightarrow I_i$, then $I_i \models post$.

(3) ***Matching Between Inputs and Outputs:*** In the composite service S={SL, CF}, the outputs of the service S_{i-1} and the inputs of the service S_i is matching iff the matching degree must be in some range (θ, 1), ($1 \leq i \leq k$). Namely $Matching\_Degree(outputs\_S_{i-1}, inputs\_S_i) \in (\theta , 1)$, where θ is a specific value.

Lemma 2 In the execution of the composite service S={SL, CF}, SL=CF {S$_1$, S$_2$,..., S$_k$}, the "***Executable***" and "***Executive Effects Validity***" can be reduced to each other.

Proof:

(1) "***Executable***" can be reduced to "***Executive Effects Validity***": In the first atomic service S_1 of the composite service S, "pre_1" must be satisfied at the beginning of the execution of S. This condition can also be seen as an "***Executive Effects Validity***" problem for the empty service S=($\Phi, \Phi, \Phi, \Phi, \Phi$). Namely $I \models pre_1 \Longleftrightarrow S = (\Phi, \Phi, \Phi, \Phi, \Phi)$ is executed.

(2) "***Executive Effects Validity***" can be reduced to "***Executable***": Assuming the effects of the composite service execution is "P". We can consider the new composite service S_1, S_2,..., S_i, S', where S'=($\Phi, \Phi, \Phi, \Phi, \{P\}$). Then P is a consequence of applying S={SL, CF}, SL=CF {S$_1$, S$_2$,..., S$_k$} in A w.r.t T iff S={SL, CF}, $SL = CF\{S_1$, S$_2$,..., S$_k$, $S'\}$ is executable. END.

According to Lemma 2, the "***Executable***" and "***Executive Effects Validity***" can be reduced to each other. So we only consider how to transform "***Executable***" or "***Executive Effects Validity***" to the consistency reasoning in the DL (Theorem 2). For the attestation of the "***Matching Between Inputs and Outputs***", we can present the matching degree formulas and then transform these formulas to the satisfiability reasoning in DL (Theorem 3).

Theorem 2 "***Executable***" and "***Executive Effects Validity***" can be reduced to the consistency reasoning of the TBox and ABox in DL.

Proof:

Based on every component in the service S= (pre, sname, inputs, outputs, post), we can construct the TBox T' and ABox A', which are corresponding to this service. Assuming that ABox A'' is constructed according to the TBox T', evidently A' is the subset of A'', namely $A' \subseteq A''$. If "***Executive Effects Validity***" is satisfied, then A' must be consistent with T', otherwise A' is inconsistent with T', and vice versa. So we only construct the TBox T' and ABox A', which reflect the state of the service after its execution, and then testify that

A' is consistent with T' through using the proper DL reasoner such as RACER to determine whether the "***Executive Effects Validity***" is satisfied.

Preparation work:

(1)The set that contains all concepts and their super or sub concepts in the inputs and outputs is denoted with C. For every $c \in C$, we introduce a concepts name T_c^i, which stands for the interpretation of c in the corresponding interpretation I_i.

(2)Every primitive concept in the inputs and outputs is denoted as A^i.

(3)Every role in the inputs and outputs is denoted as r_i.

(4)The set of individual names in the inputs and outputs is denoted with Ins. For every $a \in Ins$, we introduce an auxiliary role name r_a.

(5)The named elements of the interpretations in the inputs and outputs is denoted as N.

(6)The auxiliary individual names $a \in Ins$ introduced by r_a is denoted as a_n.

1.Constructing T':

(1)$T_1 = \{N = \bigcup_{a \in Ins} \{a\}\}$

(2)T_2={A formulas set F, which consists of the union, intersection and negation of T_c^i. F is as follows.}

F={ $T_a^i \equiv (N \cap A^i)$,$T_{\neg c}^i \equiv \neg\, T_c^i$,$T_{C \cap D}^i \equiv T_C^i \cap T_D^i$,$T_{C \cup D}^i \equiv T_C^i \cup T_D^i$;

$T_{\geq mrc}^i \equiv (\mathrm{N} \cap \bigcup_{0 \leq j \leq m} ((\geq jr^i(\mathrm{N} \cap T_C^i)) \cap (\geq (m-j)r^i(\neg N \cap T_C^i))))$;

$T_{\leq mrc}^i \equiv (\mathrm{N} \cap \bigcup_{0 \leq j \leq m} ((\leq jr^i(\mathrm{N} \cap T_C^i)) \cap (\leq (m-j)r^i(\neg N \cap T_C^i))))$}

(3)$T' = T_1 \cup T_2$

2. Constructing A':

(1)The set that contains the individuals introduced by the introduction of auxiliary role r_a is $A_1 = \{a : (\exists r_b\{b\} \cap \forall r_b\{b\}) | a \in Ins \cup a_n, b \in Ins\}$

(2) The set of the generated individuals after the execution of the service is $A_2 = \{\bigcup_{1 \leq i \leq k} post_i\}$

(3) The set of the unchanged concepts unaffected by the service execution is $A_3 = a : ((A^{i-1} \cap \neg \forall r_a . T_c^i \cap \neg \forall r_a . \exists r^i . \{b\} \cap \neg \forall r_a . \exists r^i . \{\neg b\}) \cap (A(a) \in post_i) \longrightarrow A^i)$, where A is a primitive concept.

(4) The set of the unchanged roles unaffected by the service execution is $A_4 = a : (\exists r^{i-1} . \{b\} \neg \forall r_a . T_c^i \cap \neg \forall r_a . \exists r^i . \{b\} \cap \neg \forall r_a . \exists r^i . \{\neg b\}) \cap (r(a,b) \in post_i) \longrightarrow \exists r^i . \{b\})$, where r is a role.

(5) The initial individuals set before the service execution is A_5.

(6) $A' = A_1 \cup A_2 \cup A_3 \cup A_4 \cup A_5$

3.Checking the consistency:

Checking the consistency of the A' and T': This task can be achieved by using the proper DL reasoning engine RACER[7]. END.

Theorem 3 The "***Matching Between Inputs and Outputs***" can be reduced to the satisfiability reasoning in the DL knowledge base.

Proof:

Assuming that D_r stands for the requested service, D_p stands for the published service, S_r stands for the requested service individuals and S_p stands for the published service individuals, the "***Matching Between Inputs and Outputs***" is satisfied iff any of the following three conditions is satisfied.

(1)Concepts Conjunction:

$KB \cup D_r \cup D_p \cup \{\exists x : S_r(x) \wedge S_p(x)\} \Longleftrightarrow KB \cup D_r \cup D_p \cup \{c : S_r \cap S_p\}$

can be reduced to the satisfiability of the formula $KB \cup D_r \cup D_p \cup \{c : S_r \cap S_p\}$. END.

(2)Concepts Subsumption:

$KB \cup D_r \cup D_p \models \forall x : S_r(x) \longrightarrow S_p(x) \Longleftrightarrow KB \cup D_r \cup D_p \cup \{c : S_r \cap \neg S_p\}$;

$KB \cup D_r \cup D_p \models \forall x : S_p(x) \longrightarrow S_r(x) \Longleftrightarrow KB \cup D_r \cup D_p \cup \{c : S_p \cap \neg S_r\}$;

can be reduced to the satisfiability of the formulas $KB \cup D_r \cup D_p \cup \{c : S_r \cap \neg S_p\}$ and $KB \cup D_r \cup D_p \cup \{c : S_p \cap \neg S_r\}$respectively. END.

(3)Concepts Non-Disjointness:

$KB \cup D_r \cup D_p \models \exists x : S_r(x) \wedge S_p(x) \Longleftrightarrow KB \cup D_r \cup D_p \cup \{S_r \cap S_p \subseteq \perp\}$,

can be reduced to the satisfiability of the formula $KB \cup D_r \cup D_p \cup \{S_r \cap S_p \subseteq \perp\}$. END.

According to Lemma 2,Theorem 2 and Theorem 3, we can conclude that the three conditions for checking the validity of the service composition in Theorem 1 are decidable in DL.

3.3 Complexity Analysis of *DLAIP*

According to Theorem 1, the problem of checking the validity of *DLAIP* can be reduced to the satisfiabilty reasoning and consistency reasoning in the DL *ALCQIO*. So in this paper, we can employ the complexity calculation of the *ALCQIO* to calculate the complexity of the service composition. The complexity is as Theorem 4.

Theorem 4 (*DLAIP* Complexity Analysis) The service composition *DLAIP* complexity w.r.t the acyclic TBox is analyzed as follows:

(1) For the DL sublanguages *ALC*, *ALCO*, *ALCQ* and *ALCQO*, if numbers in number restrictions are coded in unary, the complexity of the service composition is a $PSPACE-complete$ problem.

(2) For the DL sublanguages *ALCI* and *ALCIO*, the complexity of the service composition is an $EXPTIME-complete$ problem.

(3) For the DL sublanguages *ALCQI* and *ALCQIO*, whether numbers in number restrictions are coded in unary or binary, the complexity of the service composition is a $co-NEXPTIME-complete$ problem.

4 Application Scenario

In this section, we consider a set of semantic web services for a new student Tom to borrow books from the campus library. In this application, there exists four different services: S_1, S_2, S_2 and S_4, where S_1 describes that the student has an enrolled notification, S_2 describes the service of applying a library card, S_3 describes the borrowing books service for the graduates and S_4 describes the borrowing books service for the undergraduates. The detail description of these services is as follows:

Service S_1:

$pre_1=\Phi$;

$sname_1$=hold_notice;

inputs_S_1={notice (b), new_student(a)};

output_S_1=holds (a, b);

post1={holds (a, b), notice (b)};

Service S_2:

pre_2={new_student (a), ∃ register.record (a)};

$sname_2$=apply_for_ library_card;

inputs_S_2={holds (a, b), notice(b)}; outputs_S_2={holds (a, c), library_card (c)};

$post_2$={holds (a, c),∃ holds.notice(b)/library_card(c),¬∃holds.notice(b) /library_no_card(c)};

Service S_3:

pre_3={graduate(a),∃ holds(a, b), library_card (b)};

$sname_3$= borrow_book; inputs_S_3={holds(a, b), library_card (b)};

outputs_S_3={ten_books _borrowed(a, c), ten_books (c)};

$post_3$={holds (a, c), ten_books (c)};

Service S_4:

pre_4={undergraduate(a),∃ holds(a, b), library_card (b)};

$sname_4$= borrow_book; inputs_S_4={holds(a, b), library_card (b)};

outputs_S_4={five_books_borrowed (a, c), five_books (c)};

$post_4$={holds (a, c), five_books (c)}

Based on the concepts in S_1, S_2, S_3, S_4 and the Tom's information in the problem description, we can construct the acyclic TBox T and ABox A in the knowledge base:

$T=\{student \equiv graguate \cup undergraduate \cup new\_student, book \equiv five\_book \cup ten\_book, holds(domain, range) \mid domain \in \{student\}, range \in \{librarycard, books\}\}$;

$A=\{new\_student(Tom), graduate(Tom), notice(graduate\_notice), \exists register.record(Tom)\}$

Based on the information in the ABox A and according to Theorem 1 in 3.2, we can conclude that the composite service S=CF {S_1, S_2, S_3} satisfies Tom's requirement, where CF stands for the sequential control flow of S.

5 Conclusion

Based on the merits of Description Logic and AI planning, this paper presents a service composition mechanism *DLAIP*, and testifies its feasibility in Description Logic. The results of the initial experiment show that this mechanism not only possesses strong knowledge representation and reasoning capabilities, and compensates the semantic modeling deficiency during the service composition, but also improves the service composition efficiency of directly using the OWL-S semantics. But in this paper, we mainly focus on the service composition w.r.t the acyclic TBox and do not consider the composition w.r.t the cyclic TBox. So it is our future work to process the service composition w.r.t the cyclic TBox and achieve its corresponding decidable reasoning.

6 Acknowledgement

Sponsored by the Natural Science Foundation of China (No. 60472093)

References

1. F. Baader, D. Calvanese, D. McGuinness, D. Nardi, and P.F. Patel-Schneider, editors. The Description Logic Handbook: Theory, Implementation, and Applications. Cambridge University Press, 2003.
2. R. Reiter. Knowledge in Action. MIT Press, 2001.
3. S. R. Ponnekanti and A. Fox. SWORD: A developer toolkit for Web service composition. In Proceedings of the 11th World Wide Web Conference, Honolulu, HI, USA, 2002.
4. E. Sirin, B. Parsia, D. Wu, J. Hendler, and D. Nau. HTN planning for web service composition using SHOP2. Journal of Web Semantics, 1(4): 377-396, 2004.
5. An Overview of the KL-ONE Knowledge Representation System, R.J. Brachman and J. Schmolze, Cognitive Sci 9(2), 1985
6. Yu Xing, Yingjie Li, Xueli Yu. Using Semantic Matching, Research on Semantic Web Services Composition. P.S. Szczepaniak et al. (Eds.): AWIC 2005, LNAI 3528, pp. 445-450, 2005. Springer-Verlag. 2005
7. Volker Haarslev and Ralf Moller Description of the RACER system and its applications. In Proc. of the 2001 Description Logic Workshop (DL 2001), pages 132-141. CEUR Electronic Workshop Proceedings, http://ceur-ws.org/Vol-49/, 2001

Adding Support to User Interaction in Egovernment Environment

Claude Moulin[1] and Fathia Bettahar[1] and Jean-Paul A. Barthès[1] and Marco Sbodio[2] and Nahum Korda[3]

[1] Compiègne University of Technology - UMR CNRS 6599, Heudiasyc, France
[2] Hewlett Packard, Italy Innovation Center
[3] Technion Institute of Technology, Israel

Abstract. The paper presents a way of pluging a dialog system on a platform dedicated to eGovernment services. All platform modules are compliant with a central multi-lingual ontology used to represent both the domain knowledge - social care domain -, the semantic descriptions of services and the semantic indexing of documents. The dialog system is built as a multi-agent system in charge of responding to various types of users' questions by querying the main modules of the platform and the ontology itself. The document shows some scenarios which illustrate some questions involving the discovery of services and the search of documents by respecting agents.

1 Introduction

For dialogging with a machine where applications are not integrated together, users must learn how to interact with a variety of applications (information servers, handhold devices, transaction servers, etc.). This requires knowing what vocabulary and syntax to use with each application, as well as having some way of ascertaining the capabilities and limitations of a given application [8].
A possible solution to this problem is to give an "intelligent" interface to the system, usually based on machine driven dialogs [6]. We followed this idea to develop a dialog system, adjustable to the user, limited but intelligent enough to satisfy several types of requests redirected toward the modules of a platform allowing a civil servant for accessing on line services in the public administration domain. The platform is currently developed by the European TERREGOV project (see section 6 for more information).
This dialog system must let civil servants enter various questions, using natural language, and has to elaborate an appropriate answer. The system may query the user to obtain complements or precisions in order to prepare the answer. In this sense, we can say that a dialog occurs between the system and the user.
Our dialog system has a Question/Answer behavior, used for searching information in different modules considered as information sources. A successful human-machine interface involves two steps: understanding what is being said and controlling the dialog. Understanding the user intention is the subject of many research efforts in the field of Natural Language Processing (NLP), for example

Mark Last et al. (Eds.): Advances in Web Intelligence and Data Mining (SCI) **23**, 151-160 (2006)
www.springerlink.com

with techniques of analyzing natural language for modeling and handling dialogs. Such techniques include either network structures [9], [7] or grammars [1].
However in the context of the TERREGOV project, the human machine interface does not require a full understanding of what is being said. It is only necessary to extract enough information for determining the information source that can give the answer and for building the request. Our dialog system is organized as a multi-agent system composed of a dialog agent in charge of the analyze of users' entries and some agents encapsulating the platform modules. The dialog analyze relies on an automaton whose states try to match user entries against pattern strings, and an ontology ensuring a common vocabulary to the modules of the global system.
The aim of this paper is not to present the dialog analyze but to describe the role of the action agents in direction of the modules of the platform. The paper is organized as follows: Section 2 presents the architecture of the system. Section 3.1 gives an overview of the platform ontology. Sections 3, 4 and 5 give the details of agent roles.

2 Architecture Overview

Several data structures of the platform can play the role of information sources and can be used to answer users' questions. They are controlled by four modules: the semantic registry of web services, the eProcedure module, the Knowledge Base manager or the Ontology manager (see Figure 1).
The semantic registry manages the OWL-S [11] semantic descriptions of the services. They contain information like service categories, input and output data-types. The eProcedure Core Module supports the dynamic discovery of semantically enriched web services and their dynamic composition in workflows. The selection of web services is currently based on matching of inputs/output types, which are formally specified in the web services descriptions and defined in the global ontology of the platform. The Knowledge base manager indexes documents on the ontology objects (concepts, relations, individuals) and allows their discovery. The ontology contains the definitions of all of its objects and the ontology manager supports queries against the ontology elements.
The presence of these modules explains why the answers can comprise service or procedure description elements, URLs of documents indexed in the knowledge base or descriptions of ontology elements. The conversational process starts when receiving a question. The dialog agent [3] analyzes the question and extracts the key words or expressions that it contains. According to this analysis the destination of the question is defined. It is a specific agent (Ontology agent, KB agent or eProcedure agent accessing either the semantic registry or the eProcedure core module), that must build a qualified request comprehensible by the information sources. See [2] for a good introduction to multi-agent systems. The behavior of agents depends on the module answers. Sometimes, it is needed to send different queries when answers are empty.
Sometimes, the system needs to clarify a question and then a dialog between the

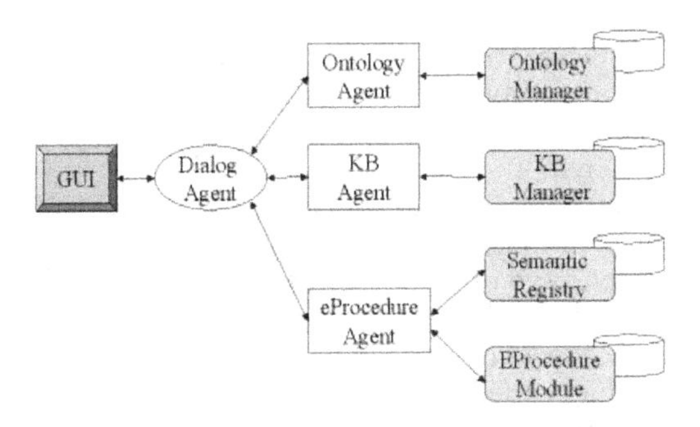

Fig. 1. Architecture of the environment.

user and the system is launched. A dialog ends either with a success i.e. it gives the answer of the user's question or a failure.

3 Ontology Agent

3.1 Ontology Overview

The platform ontology is built to be the glue between main platform modules. We developed an ontology language, simpler and less verbose than OWL from which a parser can generate the OWL format used for interoperability between tools. Other files are also generated: (i) an HTML file given an easy reading of the ontology; (ii) a graph description file allowing a graphical presentation of the ontology. The set of concepts is divided in chapters and sections for a better classification. Only direct relations are designed; the parser generating all inverse relations (for object properties only). This provides a better way of designing relations between concepts and having all inverse relations is useful for reasoning on knowledge bases.
The parser creates the OWL identifiers from the elements labels insuring a normalization of the ids and keeping them in a legible way. It is interesting when building service description and when using editors where concept tree windows are based on ids. The parser also generates in the OWL file an index based on labels in each language of the project (French, English, Italian and Polish). It is used by the ontology agent.

3.2 The Role of the Ontology Agent

To explain the role of the ontology agent in our dialog system, let's consider the following scenario: an unemployed person, citizen of France, asks a social care service for an allowance (e.g. RMA or RMI). For example, to be able to receive the RMA allowance, candidates must fulfill three conditions: be a resident

in France, be older than 25, and have income not higher than 417.88 euro per month.
In its OWL [10] representation the ontology has a concept identified by Z-Program and two sub-concepts Z-SocialProgram and Z-ResearchProgram, representing two types of programs. Z-Program concept also has a data-type properties called "hasAcronym".
When the civil servant writes a question like "what is RMA?", the dialog agent extracts words or expressions ("RMA"), determines the question type and the agent concerned by the type of question. In this case, the question is redirected toward the ontology agent which receives the word "RMA". This agent has to look for the ontology objects concerned by this word. It finds that "RMA" is an acronym of an individual called "z-revenuMinimalDActivité" as shown in Figure 2 designing a specific French social program. The ontology agent sends queries to

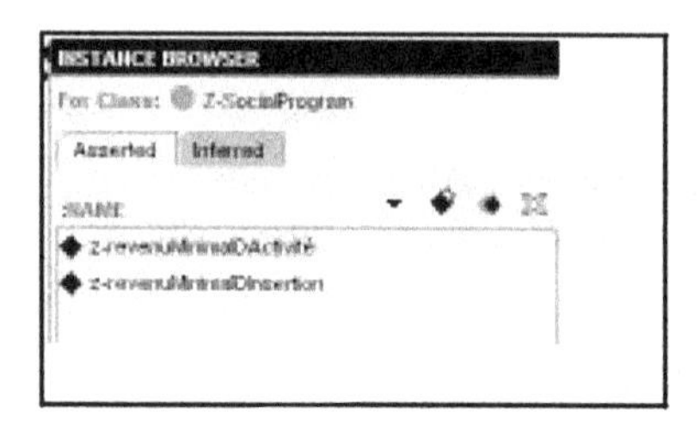

Fig. 2. Individuals of social program concept.

the ontology manager. Currently RDQL queries [13] are used, but as SPARQL [14] seems to become a recommended standard, our system can evolve without any problem.
First, it uses the index ontology features and searches for elements, concepts or else, having the user entry as an index entry. Indeed, in the multi-lingual project ontology, each object label is considered as an individual of the "Index" concept and is associated to the element that it denotes through relations like "isEnIndexOf" or "isFrIndexOf" that depend on a specific language. The first query is:

```
SELECT ?x
WHERE  (tg:Rma tg:isEnIndexOf ?x)
USING tg FOR
   <http://www.terregov .../terregov#>
```

A search failure, means that no ontology object is directly associated with the word "RMA". In this case, the agent searches for elements that might contain the user entry as an acronym:

```
SELECT ?x
WHERE  (?x tg:hasAcronym ?y)
AND ?y =~ /(?i)rma/
USING tg FOR
  <http://www.terregov .../terregov#>
```

Note that the Jena[1] inference engine used in our system runs RDQL queries which may contain Perl like regular expressions. The previous queries are not directly built by the agent, because they have to integrate user fragments like "RMA". The agent adapts parameterized queries and thus its behavior is more declarative. We defined a particular variable "(?user)" representing the user entry. It is substituted by the actual content before querying the ontology.

```
SELECT ?x
WHERE  (tg:(?user) tg:isEnIndexOf ?x)
USING tg FOR
 <http://www.terregov  /terregov#>

SELECT ?x
WHERE  (?x tg:hasAcronym ?y)
AND ?y =~ /(?i)(?user)/
USING tg FOR
 <http://www.terregov .../terregov#>
```

4 KB agent

The KB agent is in charge of retrieving documents corresponding to users' queries. For example, let's consider the following question: "What are the texts of law concerning allowances". The dialog agent extracts two parts: "texts of law" and "allowances" and determine that the query is interesting the KB agent. It procures these elements to the KB agent that has to retrieve corresponding documents and to send them back to the dialog agent. The task of the KB agent is to establish a relation between expressions extracted from the user query, concepts of the ontology and indexed documents.
First, we remind what is semantic indexing of documents. In a very technical sense, a document can be seen merely as a collection of words[2]. Document indexing aims to provide access to a document starting from one or more words (i.e. a query), and then retrieving all documents in which these words occur. Accordingly, it generates the so called "inverted indices" in which each word points to a collection of documents - such index is "inverted" in the sense that instead of presenting a document as a collection of words, it rather presents each word as a collection of associated document links.
Unlike standard keyword indexing designed to generate inverted indices containing every word that occurs in a document, "semantic" indexing associates documents only with items from a predefined vocabulary - the so called "controlled vocabulary". Actually, the term "semantic" is justified only if such controlled vocabulary is provided by ontology, i.e. as labels associated with formal concepts, which are defined by a hierarchy of concepts, and the relations that hold between them. In such case, a concept label is merely a verbal representation of a semantic unit formally defined in the underlying ontology, and thus a technical means to associate (i.e. index) a document with that semantic unit. Accordingly, semantic indexing allows identification of a document by

[1] Hewlett-Packard Jena framework: http://jena.sourceforge.net/index.html
[2] Statistical approaches to information retrieval often see a document as a "bag of words".

its meaning (hence the term "semantic"), rather than merely by its vocabulary[3].
Starting with concept labels extracted from ontology, semantic indexing must then attempt to match these labels with a document text. An extensive body of literature exists presenting the enormous difficulties that occur when attempting to merely "match" words to a document text formulated in a natural language. The common source of these difficulties is the enormous diversity of the natural languages that enable humans to formulate similar ideas in a potentially uncountable number of ways.
The simplest of these difficulties are probably the so called "morphological inflections" (e.g., plural noun forms that differ from singular). This type of difficulties is typically overcome by "normalizing" the original document text. Words are "normalized" to their basic grammatical form (the so called "base form") in order to match them.
On the other end of the complexity range in the encountered difficulties is probably the lexical ambiguity. Overcoming this wide range of difficulties evidently assumes a great deal of human knowledge that computers simply do not have, and need to have it somehow provided. The formal methodology for encapsulating human knowledge in a machine readable, and (to a certain extent) machine understandable format is ontology.
Therefore, it appears that the most appropriate approach to the semantic indexing is to utilize ontology - not only as the source of the controlled vocabulary, but also as the encapsulated lexical, semantic, grammatical, etc. knowledge required to resolve the complexity encountered in the natural languages. This is indeed the approach undertaken by TERREGOV.
Its ontology provides the semantic indexing with the required controlled vocabulary. To that purpose a mechanism was developed allowing specification of concepts in ontology that are to be used for indexing. Evidently, this is not sufficient in the light of the difficulties encountered when attempting to match the labels of these concepts to the document text formulated in a natural language. Accordingly, an additional, auxiliary ontology was developed in TERREGOV that encapsulates the lexical, semantic, grammatical, etc. knowledge necessary to correctly detect item from that controlled vocabulary in the document texts.
The central concept in this auxiliary ontology is "token". Unlike the concepts in the TERREGOV ontology that can specify labels consisting of multiple words, the auxiliary ontology must enable text processing that operates on single units. This segmentation of a text formulated in a natural language into single units is typically referred to as "tokenization", and hence the concept "token".
The concept "token" has five basic properties: (1) "reference", (2) "target", (3) "position", (4) "action", and (5) "operator". We shall briefly present each of them.
"Reference" holds the identifier of a concept in the TERREGOV ontology. It is merely an association between a token, and a corresponding concept in the TERREGOV ontology. It thus allows association of documents in which tokens were correctly detected with the adequate concepts in ontology, and consequently transforms the ontology into an "inverted index" in which every relevant concept points to a collection of associated documents.
"Target" property is significantly more complex. It can have five other concepts as its value: (1) "literal", (2) "base", (3) part of speech ("POS"), (4) "role", and (5) "regular expression". The problem handled with this property is that in text processing a single token can have multiple aspects: it can be presented as it occurred in the original

[3] Alternative terminology for "semantic indexing" is "semantic annotation" that emphasizes the fact that through association with ontology a document text is enriched (annotated) with the semantic description of its content.

document text (i.e. simply quoted in its "literal" form), or it can be presented in its normalized form (i.e. as "base" form), or it can be presented by its grammatical definition (e.g., "common noun in first person singular nominative form" - this specification is typically referred to as the "part of speech"), etc. When matching a token from the controlled vocabulary to the tokens encountered in a document text, it is necessary to specify which of these aspects should be targeted by matching. The auxiliary ontology developed in TERREGOV aims at enabling the maximal flexibility in specifying the intended target of the matching. Accordingly, in addition to the literal, base form and part of speech, it also provides roles that tokens can have in a document text (e.g., "acronym", "date", "punctuation", etc.), and regular expressions (e.g., the so called "Kleene's Star" usually symbolized by "*").

This flexibility in the target specification is necessary in particular when multiple word phrases need to be detected. For example, the natural language phrases (in English) "reporting of the municipality to the regional authority" and "reporting by the municipality to the regional authority" have evidently the same intended meaning, but they are phrased differently (using "of" and "by" alternatively). Accordingly, in order to detect the concept label (in English) "reporting of the municipality" correctly in both phrases, it is necessary to "rephrase" it in a more flexible way: "reporting" + any word (i.e. "Kleene's Star") + "the" + "municipality". An even better "rephrasing" would be to target "reporting" in its base form (i.e. "report"), to replace "Kleene's Star" with the POS specification (i.e. "simple prepositional adposition"), to replace "the" with the regular expression for a potential single word (possibly specifying "determiner" as its POS), and to target "municipality" in its base form. In this case, even a phrase such as "reports of municipalities" would be correctly identified with the desired concept label (i.e. "reporting of the municipality").

The example above was provided in order to demonstrate how critical is the flexibility in the target specification when matching controlled vocabulary to natural languages. The auxiliary ontology developed in TERREGOV was thus designed specifically to enable the required flexibility, and thus to maximize the precision of such matching. The problem of matching multiple word phrases from the example above is also related to the next property of the concept "token": "position". Although text processing operates on individual tokens, in order to correctly identify multiple word phrases, it is necessary to specify the position of a token within such phrase. This property works in conjunction with another property of the same concept: "action". "Action" provides instructions to the text processor regarding what needs to be done when a particular match is encountered at a particular position. For example, as long as the end of the phrase "reporting of the municipality" is not reached (i.e. as long as the position is less than 4), the text processor should simply keep track of the matched tokens, and proceed. When the last token of the phrase is correctly matched, the text processor should record the association of the currently processed document with the concept referred to by the matched phrase.

Finally, the property "operator" of the concept "token" enables specification of the basic logical operators (the so called "Boolean operators": "and", "or", "not", etc.). For example, this property provides means to combine several alternative tokens in the same position: "municipality" or "city" or "town", etc.

However, this auxiliary ontology developed in TERREGOV specifically for semantic indexing provides also effective means for querying in natural languages. As mentioned above, semantic indexing transforms the TERREGOV ontology into an inverted index. Nevertheless, querying such index is not trivial. Queries can contain inflected or ambiguous keywords that may retrieve undesired results. The approach followed by the

KB agent is to treat a query just as any other natural language input (i.e. as a very short document). The same methodological approach that is applied to the semantic indexing of documents is thus also applied to the queries: the same concept labels that were successfully detected in the indexed documents will be consequently successfully detected in the queries too. This approach ensures a complete analogy between the query processing results, and the indexing of the documents that are targeted by this query.

5 EProcedure Agent

The eProcedure agent may encapsulate the functionalities of TERREGOV eProcedure Core Module, which supports the dynamic discovery of semantically enriched web services and their dynamic composition in workflows. Notice that we call eProcedure a process of a public administration that can be initiated by a civil servant. In a digital environment an eProcedure is represented by a formal description of a workflow, whose steps involve the execution of web services.
In the following sections we describe two interesting scenarios (interaction between a civil servant and the dialog agent) that involve the eProcedure agent, and show how it can support a civil servant with a higher degree of flexibility.

5.1 First Scenario

In the first scenario we assume that a civil servant types the following question: "what are the services concerning allowance?". The dialog agent identifies the presence of the keywords "services" and "allowance" in the question. The keyword "services" identifies the question as a task that can be delegated to the eProcedure agent. The keyword "allowance" is used to perform a query returning a concept URI defined in the ontology (Z-Allowance).
The system asks the eProcedure agent to find all eProcedures related to the "allowance" concept (specified as its URI). In this scenario the eProcedure agent assumes that "allowance" is the requested output of an eProcedure, and its goal is to find an eProcedure that produces an output whose type matches with the "allowance" concept. The eProcedure agent can fulfill this task using the eProcedure Core Module to dynamically discover all services whose output type matches with "allowance".
The eProcedure Core Module will answer such a request querying a semantic registry. The queries performed by the eProcedure Core Module will ultimately run on an RDF [12] store, which contains the OWL-S descriptions of the known web services [5]. The dynamic discovery performed by the eProcedure Core Module may take advantage of a reasoner that exploits both the semantics of OWL-S/OWL/RDF, and (potentially) additional semantics expressed by custom rules. For example, the eProcedure Core Module exploits subsumption among concepts used as inputs/output in the OWL-S descriptions (see for more details [4]).

5.2 Second Scenario

Another interesting scenario is the following: a civil servant asks the question "is there any service to find out exemptions for a person?". This question is more complex then the one described before. The keyword "service" again identifies the question as a task

that can be delegated to the eProcedure agent; the system identifies that "exemption" is the requested output, and "person" is the available input (as before both "exemption" and "person" are traced back to URIs pointing to OWL concepts in the ontology). In this scenario the eProcedure agent may ask to the eProcedure Core Module to find out a service (or a sequence of services) that accept "person" as an input and return "exemption" as an output.
The interesting aspect of this scenario is that the eProcedure Core Module may dynamically compose known web services to build a new eProcedure. This may happen if the semantic registry has only a reference to a web service (WS_n) which returns "exemption" as output, but which requires one (or more) input that do not match with "person". In this case, the eProcedure Core Module may compose a sequence of web services $\{WS_1, WS_2, \ldots WS_n\}$ where WS_1 requires only "person" as input, and WS_n returns "exemption" as output.

6 Acknowledgments

The TERREGOV project is an integrated project cofunded by the European Commission[4] under the IST (Information Society Technologies) Program, eGovernment unit, under the reference IST-2002-507749.

7 Conclusion

We have presented a way to add a dialog system, structured as a multi-agent system, to a platform dedicated to eGovernment services. The objective was not to describe the dialog itself but to show the role of three main agents: (i) the ontology agent in charge of searching concepts and their definitions in the ontology; (ii) the KB agent in charge of searching documents in a document base using semantic indexing techniques; (iii) the eProcedure agent in charge of searching services from their semantic descriptions. We have shown that searching documents from expressions extracted from users' entries relies on the same concepts that semantic indexing of texts.
This work is a first step toward the conception of personal assistant for civil servants. As we can imagine, with the introduction of new technologies and integration of administrative processes in eGovernment systems, the role of civil servants will evolve. Their tasks will become more complex and will require assistant agents for helping them.

References

1. Chomsky, N., "Three Model for the Description of Language". IRI Transactions on Information Theory, 2(3), pp. 113-124, 1956.
2. Jennings, N.R., Sycara K. and Wooldridge, M., A Roadmap of Agent Research and Development, Journal of Autonomous Agents and Multi-Agent Systems, Kluwer Academic Publishers, vol. 1, no1, p. 7-18, 1998.

[4] The content of this paper is the sole responsibility of the authors and in no way represents the views of the European Commission or its services.

3. Johnson W., Shaw E., Marshall A. and Labore C. Evolution of User Interaction : the Case of Agent Adele, Proceedings of Intelligent User Interfaces (IUI-03), ACM Press, pp. 93-100, 2003.
4. Moulin, C. and Sbodio, M., Using Ontological Concepts for Web Service Composition. in IEEE/WIC/ACM international conference on Web Intelligence 2005, (Compiègne, France, 2005), pp. 487-490.
5. Paolucci, M., Kawmura, T., Payne, T., Sycara, K.: Semantic matching of web services capabilities. In: Proc. of the First International Semantic Web Conference, Sardinia, Italy (2002).
6. Paraiso E. C. and Barthès J. P., "An Intelligent Speech Interface for Personal Assistants Applied to Knowledge Management", WEB Intelligence and Agent Systems, 2005.
7. Sack W., "Conversation Map: An Interface for Very-Large-Scale Conversations", Journal of Management Information Systems, 17(3), pp. 73-92, 2000.
8. Shriver S., Toth A., Zhu X., et al., "A Unified Design for Human-Machine Voice Interaction". The CHI 2001 Conference on Human Factors in Computing Systems, Washington, 2001.
9. Winograd T. and Flores F. Understanding Computers and Cognition. Norwood N.J.: Abblex. 1986.
10. Web Ontology Language (OWL). http://www.w3.org/2004/OWL/
11. OWL-S, OWL-based Web Service Ontology. http://www.daml.org/services/owl-s/
12. Resource Description Framework (RDF). http://www.w3.org/RDF/
13. RDQL, W3C Recommendation, 9 January 2004, http://www.w3.org/Submission/2004/SUBM-RDQL-20040109/
14. SPARQL, W3C Recommendation, 23 November 2005, http://www.w3.org/TR/2005/WD-rdf-sparql-query-20051123/

Using Consensus Methodology in Processing Inconsistency of Knowledge

Ngoc Thanh Nguyen

Institute of Information Science and Engineering, Wroclaw University of Technology, Poland thanh@pwr.wroc.pl

1 Introduction

Inconsistent knowledge may be considered on two levels: *syntax level* and *semantic level* [14]. For the syntax level an approach of knowledge inconsistency representation and solving is presented in [7], [8], [9], where knowledge is a set of logic formulae which has no model. For example, consider the knowledge as a set of formulae $\{\alpha \vee \neg\beta, \neg\alpha, \beta\}$ with the inference engine of standard logic, for which one can deduce simultaneously β and $\neg\beta$.

There are known main four approaches for inconsistency leaving: The first relies on knowledge base revision; the second is based on paraconsistent logics; the third is related to Boolean reasoning and the fourth concerns measuring up the inconsistency level. The first approach, most often used in deductive databases, removes data from the base to produce a new consistent database [4], [10], [11], [12]. Apart from the above mentioned approaches, there are a number of attempts to resolving inconsistent data in databases by labeling [1], [6]. The disadvantage of this approach is that it is hard to localize the inconsistency and perform an optimal selection, that is with minimal loss of useful information. In the second approach paraconsistent logics have been defined. These logics give sensible inferences from inconsistent information. Hunter [7] has defined the several logics, such as weakly-negative logics, four-valued logics, quasi-logical logic, and argumentative logics.

On the semantic level formulae set should be considered referring to a concrete real world. Inconsistency arises if in this world a fact and its negation may be inferred. In this paper we present some conceptions for representing inconsistency of knowledge on syntactic and semantic level. The main subject is the semantic level. We present two approaches for representing inconsistent knowledge on this level. The first is based on using relational structures and the second relies on using logical (clause-based) structures. We also show how to use consensus methods for solving the inconsistency.

Consensus methods were known in ancient Greece and were applied mainly in determining results of voting [3]. Along with the development of software

Mark Last et al. (Eds.): Advances in Web Intelligence and Data Mining (SCI) **23**, 161-170 (2006)
www.springerlink.com

methods consensus has found many fields of applications, especially in solving conflicts and reconciling inconsistent data.

In short the scheme of using consensus methods in a process of solving conflict or data inconsistency may be presented as follows [13]:

1. Defining the set of potential versions of data
2. Defining the distance function between these versions
3. Selecting a consensus choice function
4. Working out an algorithm for consensus choice

Some comments should be done to these steps. Referring to first point notice that up to now the following structures of data versions have been investigated: rankings, partitions, ordered partitions and coverings, sets, n-trees, semilattices. The logical structure presented in this paper has been adopted from work [14]. The definition of distance functions is dependent on the specific structure. It seems to be impossible to define a universal distance function for all structures. Referring to selecting a consensus choice function there are known 2 consensus choice functions. The first of them is the median function defined by Kemeny [2], which minimizes the sum of distances between the consensus and given inconsistent versions of data. The second function minimizes the sum of these distances squared. As the analysis has shown, the first function in the best way represents the conflict versions while the second should be a good compromise of them. We denote the consensuses chosen by these functions by O_1-consensus and O_2-consensus, respectively. The choice of a consensus function should be dependent on the conflict situation. With assumption that the final version which is to be determined represents an unknown solution of some problem then there two cases:

- In the first case the solution is independent on the opinions of conflict participants. Thus the consensus should at best represent the conflict versions of data. For this case the criterion for minimizing the sum of distances between the consensus and the conflict versions should be used, thus O_1-consensus should be determined.
- In the second case the solution is dependent on the opinions of conflict participants. Then the consensus should be a compromise acceptable by the conflict participants. For this case an O_2-consensus should be determined.

However, it is worth to notice that determining an O_2-consensus is often a complex problem and requires working out heuristic algorithms.

2 Inconsistency on Syntactic Level

We assume that there is given a finite set A of agents which work in a distributed environment. The term "agent" is used here in very general sense: as an agent we may understand an expert or an intelligent and autonomous computer program. An example of such environment is a multi-agent system [5]. We assume that these agents have their own knowledge bases. In general, by a state of agent knowledge we understand a state of the agent knowledge base. Such a state may be treated as a view or opinion of the agent on some subject or matter.

We assume that a knowledge state of an agent is represented by a logic clause, the predicate symbols for building clauses belong to a finite set S. We assume that these agents work on a finite set of subjects (or matters) of their interest. Inconsistency appears if for some subject agents generate different knowledge states. Below we give an example:

- Set of agents: $A = \{a_1, a_2, a_3\}$,
- Set of predicate symbols: $S = \{u, v, x, y, z\}$,
- For a common subject the agents generate the following set of knowledge states:

Agent	*Knowledgestate*
a_1	$x \vee \neg y \vee v$
a_2	$\neg x \vee y \vee u$
a_3	$x \vee y \vee \neg v$

Thus inconsistency exists because of the difference between knowledge states of the agents. The inconsistency is on syntactic level because it will be solved only on the basis of the difference between clauses, which is measured wit using the set S of symbols, without using interpretation of the clauses.

The aim of solving this kind of inconsistency is to generate a possible knowledge state (that is a clause) so that it at best represents the given clauses. In other words, it is needed to determine the consensus of given agent knowledge states.

First, we define the distance function d between clauses. Notice that a clause c may be represented by a pair of 2 sets: the first consists of these symbols in the clause, which are not negated, and the second consists of remain symbols. For example, the representation of $c = x \vee \neg y \vee v$ is pair (c^+, c^-) where $c^+ = \{x, v\}$ and $c^- = \{y\}$. Next we define the distance of two clauses c_1 and c_2 as follows:

$$d(c_1, c_2) = \frac{w_1 \cdot d_1(c_1^+, c_2^+) + w_2 \cdot d_2(c_1^-, c_2^-)}{w_1 + w_2}$$

where

- $d_1(c_1^+, c_2^+)$: distance function between sets of non-negated symbols in clauses c_1 and c_2, and is equal $\frac{card(c_1^+ \div c_2^+)}{card(c_1^+ \cup c_2^+)}$,
- $d_1(c_1^-, c_2^-)$: distance function between sets of non-negated symbols in clauses c_1 and c_2, and is equal $\frac{card(c_1^- \div c_2^-)}{card(c_1^- \cup c_2^-)}$,
- w_1 and w_2 are the weights distances d_1 and d_2 in distance d, respectively, $w_1 + w_2 = 1$ and $0 \leq w_1, w_2 \leq 1$.

It is known that a clause represents a rule, where its negated symbols represent the pre-condition, while non-symbols represent the post-condition. Notice that using values w_1 and w_2 we can distinguish the weights of the distances between the pre-conditions and post-conditions. The above defined distance function d is of course a metric.

For consensus choice we will use the criterion O_1, the consensus problem is defined as follows: Let $X = \{c_1, c_2, \ldots, c_n\}$ be a given set of clauses with symbols from a finite set S. By C_S we denote the set of all possible clauses with symbols from set S. A clause $c^* \in C_S$ is called a consensus of set X if it satisfies the following condition:

$$\sum_{i=1}^{n} d(c^*, c_i) = \min_{c \in C_S} \sum_{i=1}^{n} d(c, c_i)$$

The choice of clause c^* reduces to the choice of sets of symbols c^{*+}, c^{*-} which minimize sums $\sum_{i=1}^{n} d_1(c^{*+}, c_i^+)$ and $\sum_{i=1}^{n} d_2(c^{*-}, c_i^-)$, respectively. These choices can be done in an independent way. These problems are known to be NP-complete, so we propose some heuristic algorithms. The heuristic algorithm for determining c^{*+} is presented as follows:

Algorithm 1. Computing c^{*+}.
`Given:` Finite set $X^+ = \{c_1^+, c_2^+, \ldots, c_n^+\}$.
`Result:` Set $c^{*+} \subseteq S$ minimizing sum $\sum_{i=1}^{n} d_1(c^{*+}, c_i^+)$.
`BEGIN`

1. For each symbol $s \in S$ create a number $q(s)$ being the number of appearances of s in sets $c_1^+, c_2^+, \ldots, c_n^+$;
2. Set $c^{*+} = \emptyset$;
3. For each symbol $s \in S$ if $q(s) \geq n/2$ then $c^{*+} := c^{*+} \cup \{s\}$

`END.`

The computational complexity of this algorithm is $O(n \cdot m)$ where n is the number of elements of X^+ and m is the cardinality of S. The algorithm for determining set c^{*+} is identical.

3 Inconsistency on Semantic Level

3.1 Relational Structure

Relational structures are very useful for representing knowledge consistency. We present here a specific kind of multi-valued relations. Formally, we assume that a real world is described by means of a finite set A of attributes and a set V of attributes *elementary values*, where $V = \bigcup_{a \in A} V_a$ (V_a is the domain of attribute a). In short, pair (A, V) is call a real world. Let $\prod(V_a)$ denote the set of subsets of set V_a and be called the *super-domain* of attribute a. Let $\prod(V_B) = \bigcup_{b \in B} \prod(V_b)$. We accept the following assumption: For each attribute a its value is a set of elementary values from V_a, thus it is an element of set $\prod(V_a)$. By an elementary value we mean a value which is not divisible in the system. Thus it is a relative notion, for example, one can assume the following values to be elementary: time units, set of numbers, partitions of a set etc.

We define the following notions [13]: Let $T \subseteq A$,

- A tuple of type T is a function $r : T \to \prod(V_T)$. Instead of $r(t)$ we will write r_t and a tuple of type T will be written as r_T. The set of all tuples of type T is denoted by $TYPE(T)$. A subset of $TYPE(T)$ is called a relation of type T. A tuple is elementary if all attribute values are empty sets or 1-element sets. The set of elementary tuples of type T is denoted by $E\_TYPE(T)$. Empty tuple, whose all values are empty sets, is denoted by symbol ϕ. Partly empty tuple, whose at least one value is empty, is denoted by symbol θ.
- A non-empty and finite set R of tuples of type T is called a relation of type T, thus $R \subseteq TYPE(T)$.
- A sum of two tuples r and r' of type T is a tuple $r"$ of type $T(r" = r \cup r')$ such that $r"_t = r_t \cup r'_t$ for each $t \in T$.
- A product of two tuples r and r' of type T is also a tuple $r"$ of type $T(r" = r \cap r')$ such that $r"_t = r_t \cap r'_t$ for each $t \in T$.

3.2 Logical Structure

Conjunctive Structure

The relational structure presented above can be replaced by a logical structure based on the standard logic. For this aim we use logical variables to represent logical values of expressions $(a = \{v\})$ where $a \in A \setminus \{Agent\}$ and $v \in V_a$. A complete event z can be represented by means of the following formula:

$$z = (x_1 \wedge x_2 \wedge \ldots \wedge x_n) \wedge (y_1 \wedge y_2 \wedge \ldots \wedge y_m)$$

where logical variables $x_1, x_2, \ldots, x_n$ are related to attributes from set T_{PS} and their values, and variables $y_1, y_2, \ldots, y_m$ are related to attributes from set T_{PC} and their values. Incomplete events included in this event are represented by formula

$$(x_1 \wedge x_2 \wedge \ldots x_n) \wedge (s_1 \wedge s_2 \wedge \ldots s_k),$$

where $\{s_1, s_2, \ldots, s_k\} \subseteq \{y_1, y_2, \ldots, y_m\}$.

Notice that a formula $(x_1 \wedge x_2 \wedge \ldots \wedge x_n) \wedge (y_1 \wedge y_2 \wedge \ldots \wedge y_m)$ represents an opinion of an agent, in which sub-formula $(x_1 \wedge x_2 \wedge \ldots \wedge x_n)$ refers to the *subject* and sub-formula $(y_1 \wedge y_2 \wedge \ldots \wedge y_m)$ represents the *content* of the opinion. A conflict takes place if for the same subject several agents have different contents of opinions.

A complete event occurs if and only if the formula $(x_1 \wedge x_2 \wedge \ldots \wedge x_n) \wedge (y_1 \wedge y_2 \wedge \ldots \wedge y_m)$ is true. From this truth it implies that formula $(x_1 \wedge x_2 \wedge \ldots \wedge x_n) \wedge (s_1 \wedge s_2 \wedge \ldots \wedge s_k)$ is also true, thus an incomplete event included in an occurring event also occurs. A complete event does not occur if and only if formula $(x_1 \wedge x_2 \wedge \ldots \wedge x_n) \wedge (\neg y_1 \wedge \neg y_2 \wedge \ldots \wedge \neg y_m)$ is true. It means that any incomplete events included in this event should not occur.

The above mentioned consistency of an agent means that if in its opinion 2 formulae $(x_1 \wedge x_2 \wedge \ldots \wedge x_n) \wedge (y_1 \wedge y_2 \wedge \ldots \wedge y_m)$ and $(x_1 \wedge x_2 \wedge \ldots \wedge x_n) \wedge (\neg y'_1 \wedge \neg y'_2 \wedge \ldots \neg \wedge y'_m)$ are true (i.e. an event should occur and another should not occur) then there should be $\{y_1, y_2, \ldots y_m\} \cap \{y'_1, y'_2, \ldots y'_l\} \neq \emptyset$ (that is these events should not include any common event). Below we consider another logical structure.

Alternative Structure

In this structure an agent opinion has the following form:

$$(x_1 \wedge x_2 \wedge \ldots \wedge x_n) \wedge (y_1 \vee y_2 \vee \ldots \vee y_m)$$

where formula $(x_1 \wedge x_2 \wedge \ldots \wedge x_n)$ represents the subject and formula $(y_1 \vee y_2 \vee \ldots \vee y_m)$ represents the content. We assume that a logical variable from set $\{y_1, y_2, \ldots y_m\}$ represents the logical value of expression $(a = v)$ *or* $(a \neq v)$ where $a \in A \setminus \{Agent\}$ and $v \in \prod(V_a)$. Notice that owing to this structure an agent may express other type of its opinion than in conjunctive structure, viz an agent can now give its opinion in form of an alternative of a number of events.

It is known that in a conflict all opinions should refer to the same subject. Thus we can leave the formula $(x_1 \wedge x_2 \wedge \ldots \wedge x_n)$ in an agent opinion assuming that the conflict subject is known. Notice that formula $(y_1 \vee y_2 \vee \ldots \vee y_m)$ can be treated as a clause where each variable from set $\{y_1, y_2, \ldots y_m\}$ is treated as a literal. Literal $(a = v)$ is a positive one and literal $(a \neq v)$ is negative because it is equivalent to $\neg(a = v)$. We can write

$$(y_1 \vee y_2 \vee \ldots \vee y_m) \equiv (t_1 \vee t_2 \vee \ldots \vee t_k) \vee (\neg u_1 \vee \neg u_2 \vee \ldots \vee \neg u_l)$$

where $t_1, t_2, \ldots t_k$ are positive literals and $\neg u_1, \neg u_2, \ldots \neg u_l$ are negative literals among $y_1, y_2, \ldots y_m$. Further we have

$$\begin{aligned}(y_1 \vee y_2 \vee \ldots \vee y_m) &\equiv (t_1 \vee t_2 \vee \ldots \vee t_k) \vee \neg(u_1 \wedge u_2 \wedge \ldots \wedge u_l) \\ &\equiv (u_1 \wedge u_2 \wedge \ldots \wedge u_l) \rightarrow (t_1 \vee t_2 \vee \ldots \vee t_k) \\ &\equiv u_1, u_2, \ldots, u_l \rightarrow t_1, t_2, \ldots, t_k\end{aligned}$$

It is the well-known form of clauses. The above transformation shows that the alternative structure should be very useful in practice for agents to express their opinions.

We now define the semantic of bodies and heads of clauses on the basis of some real world. A clause c is based on a real world (A, V) (or (A, V)-based for short) if each its literal has the form: (a, v) where $a \in A$ and $v \in \prod(V_a)$. Then a body of a (A, V)-based clause as a conjunction should have the following form: $(a_1, v_1) \wedge (a_2, v_2) \wedge \ldots \wedge (a_k, v_k)$ for some $k \in N$. By Body we denote the set of all bodies of (A, V)-based clauses. The semantic of clause bodies is defined by the following function:

$$S_B : Body \rightarrow 2^{\bigcup_{T \subseteq A} E\_TYPE(T)},$$

such that $S_B(b) = \{r \in \bigcup_{B \subseteq T \subseteq A} E\_TYPE(T) : r_B \prec b'$ and $r_a = \emptyset$ iff $b'_a = \emptyset$ for all $a \in B\}$ where $b = (a_1, v_1) \wedge (a_2, v_2) \wedge \ldots \wedge (a_k, v_k)$; $b' =< (a_1, v_1), (a_2, v_2) \wedge \ldots \wedge (a_k, v_k) >$ and $B = \{a_1, a_2, \ldots, a_k\}$.

The body expresses the condition (in the opinion of an agent) which, if fulfilled, will cause occurrence of some other event (represented in the head of the clause). It follows that an event satisfies the condition b of the agent if it belongs to set $S_B(b)$.

Notice that the body

$$b^* = (a_1, V_1) \wedge (a_2, V_2) \wedge \ldots \wedge (a_n, V_n)$$

where $a_1, a_2, \ldots, a_n \subseteq A$ should represent the statement "*everything is possible*" because $S_B(b^*) = \bigcup_{T \subseteq A} E\_TYPE(T)$. Note that because $S_B(B) = S_B(B')$ we have the following:

Theorem 1. *Body* $B = (a_1, v_1) \wedge (a_2, v_2) \wedge \ldots \wedge (a_k, v_k)$ *is equivalent to body* $B' = (a_1, v_1) \wedge (a_2, v_2) \wedge \ldots \wedge (a_k, v_k) \wedge (a, V_b)$ *where attribute a does not occur in B.*

A notice should be made referring to empty values in bodies. If in a body an attribute has empty value then it is understood that the attribute should not appear in the condition. This situation may take place when some attribute expresses a contradictory feature referring to another. A head of a clause is in the form of an alternative $(a_1, v_1) \wedge (a_2, v_2) \wedge \ldots \wedge (a_l, v_l)$. By *Head* we denote the set of all heads of (A, V)-based clauses. The semantic of clause heads is defined by function:

$$S_H : Head \rightarrow 2^{\bigcup_{T \subseteq A} E\_TYPE(T)},$$

such that $S_H(h) = \{r \in \bigcup_{H \subseteq T \subseteq A} E\_TYPE(T) : r_H \prec h'$ and $r_{H' \setminus H} = \phi$ where H' is the type of $r\}$ for $h = (a_1, v_1) \vee (a_2, v_2) \vee \ldots \wedge (a_l, v_l)$; $h' =< (a_1, v_1), (a_2, v_2) \wedge \ldots \wedge (a_l, v_l) >$ and $H = \{a_1, a_2, \ldots, a_l\}$.

A head expresses the result (in the opinion of an agent) which should occur if the condition in the body is fulfilled.

Theorem 2. *According to the semantic defined above a head* $H = (a_1, v_1) \wedge (a_2, v_2) \wedge \ldots \wedge (a_k, v_k)$ *is equivalent to head* $H' = (a_1, v_1) \wedge (a_2, v_2) \wedge \ldots \wedge (a_k, v_k) \wedge (a, \emptyset)$ *where attribute* a *does not occur in* B.

4 Consensus versus Inconsistency

4.1 Consensus Determining for Conjunctive Structure

For the conjunctive structure (or its equivalent relational structure) a consensus method has been proposed in works [13]. In these works a methodology for consensus choice and its applications in solving conflicts in distributed systems is presented. In the methodology consensus problem has been considered on 2 levels. On the first level general consensus methods which may effectively serve to solving multi-valued conflicts are worked out. For this aim a consensus system, which enables describing multi-valued and multi-attribute conflicts is defined and analyzed. Next the structures of tuples representing the contents of conflicts are defined as distance functions between these tuples. Two distance functions (ρ and σ) have been defined. Finally the consensus and the postulates for its choice are defined and analyzed. For defined structures algorithms for consensus determination are worked out. Besides the problems connected with the susceptibility to consensus and the possibility of consensus modification, are also investigated. The second level concerns varied applications of consensus methods in solving of different kinds of conflicts, which often take place in distributed environments.

4.2 Consensus Determining for Alternative Structure

For opinions of alternative structure first we propose a method for measuring up the distance between 2 clauses

$$c_1 = u_1^{(1)}, u_2^{(1)}, \ldots, u_m^{(1)} \rightarrow t_1^{(1)}, t_2^{(1)}, \ldots, t_m^{(1)}$$

and

$$c_2 = u_1^{(2)}, u_2^{(2)}, \ldots, u_m^{(2)} \rightarrow t_1^{(2)}, t_2^{(2)}, \ldots, t_m^{(2)}.$$

Generally, it is needed to measure the distance between the bodies and the distance between the heads of these clauses. Firstly, we deal with the distance between clause bodies. Referring to the semantics of clause bodies defined above, for calculating the distance d_B between 2 bodies B and B' the following procedure should be used:

1. Complete these bodies by adding to them these attributes which appear in only one of them, as their values take their domains;
2. For each attribute calculate the distance between its values in the bodies using function (ρ or σ).

3. The distance between the bodies will be equal to the average of the distances calculated in step 2.

For determining a consensus for a clauses' set it is needed to complete the clauses by adding attributes to their bodies and heads so that the same attributes occur in each clause. The rule of attribute completing is give above in distance determining way.

Assuming that the attributes from the real world (A, V) are independent, for determining a O_1-consensus we can adopt the algorithm given in [13], in which value of each attribute in the consensus can be calculated in an independent way. The adopted algorithm is presented as follows:

Algorithm 2. Computing O_1-consensus ingredients c_b where b is an attribute occurring in the clauses.
`Given:` Finite set C (with repetitions) of (A, V)-based clauses and distance function $\partial \in \{\rho, \delta\}$ between attribute values.
`Result:` O_1-consensus ingredient c_b.
`BEGIN`

1. For attribute b create set: $profile(b) = \{v_b : (b, v_b)$ appears in a clause of set $C\}$;
2. Let $X := \emptyset; S_b := \sum_{y \in profile(b)} \partial(X, x)$;
3. Select from $V_b \backslash X$ an element x such that the sum $\sum_{y \in profile(b)} \partial(X \cup \{x\}, y)$ is minimal;
4. If $S_b < \sum_{y \in profile(b)} \partial(X \cup \{x\}, y)$ then Begin $S_b := \sum_{y \in profile(e)} \partial(X \cup \{x\}, y); X := X \cup \{x\}$; End;
5. If $V_b \setminus (X \cup \{x\}) \neq \emptyset$ then GOTO 3;
6. Let $c_b = X$;

`END.`

For determining an O_2-consensus for set C it is impossible to treat the attributes in an independent way. For many structures of attribute values determining a consensus on the basis of this criterion is a NP-complete problem. In this case a heuristic algorithm should be worked out. Besides, genetic algorithms should also be useful.

5 Conclusions

This paper presents two structures for representing inconsistency of knowledge on semantic level. The first structure is based on relational model, which is multi-valued and multi-attribute. This structure has been analyzed in the earlier literature. For the second structure (called *logical structure*) in the form of clauses this work presents its semantics, the way for distance calculating and the algorithm for consensus computing. The future work should concern working out algorithm for determining O_2-consensus for inconsistent knowledge in logical structure.

References

1. Balzer, R.: Tolerating Inconsistency. In: Proceedings of the 13th International Conference on Software Engineering, IEEE Press (1991) 158-165
2. Barthelemy, J.P., Janowitz, M.F.: A Formal Theory of Consensus. SIAM J. Discrete Math. **4** (1991) 305-322
3. Day, W.H.E.: Consensus Methods as Tools for Data Analysis. In: Bock, H.H. (ed.): Classification and Related Methods for Data Analysis. North-Holland (1987) 317-324
4. Doyle: A Truth Maintenance System. Artificial Intelligence J. **12** (1979) 231-272
5. Ferber, J.: Multi-Agent Systems, Addison Wesley, New York (1999)
6. Fehrer, D.: A Unifying Framework for Reason Maintenance. In M. Clark et. al (eds) Sym-bolic and Qualitative Approaches to Reasoning and Uncertainty. LNCS **747** (1993) 113-120. Springer-Verlag
7. Hunter, A.: Paraconsistent Logics. In: D. Gabbay, P. Smets (eds), Handbook of Defeasible Reasoning and Uncertain Information. Kluwer Academic Publishers (1998) 13-43
8. Hunter, A.: Evaluating the Significance of Inconsistencies. In: Proceedings of the Interna-tional Joint Conference on AI (IJCAI'03). Morgan Kaufmann (2003) 468-473
9. Kifer, M., Lozinskii, E.L.: A Logic for Reasoning with Inconsistency. Journal of Automatic Reasoning **9** (1992) 179-215
10. De Kleer, J.: An Assumption-based TMS. Artificial Intelligence **28** (1986) 127-162
11. Knight, K.: Measuring Inconsistency. Journal of Philosophical Logic **31** (2002) 77-98
12. Naqvi, S., Rossi, F.: Reasoning in Inconsistent Databases. In: Logic Programming: Pro-ceedings of the North American Conference. MIT Press (1990)
13. Nguyen, N.T.: Consensus System for Solving Conflicts in Distributed Systems. Journal of Information Sciences **147** (2002) 91-122
14. Nguyen N.T.: Processing Inconsistency of Knowledge on Semantic Level. Journal of Uni-versal Computer Science **11**(2) (2005) 285-302

Creating Synthetic Temporal Document Collections for Web Archive Benchmarking

Kjetil Nørvåg* and Albert Overskeid Nybø
Norwegian University of Science and Technology
7491 Trondheim, Norway

Abstract. In research in web archives, large temporal document collections are necessary in order to be able to compare and evaluate new strategies and algorithms. Large temporal document collections are not easily available, and an alternative is to create synthetic document collections. In this paper we will describe how to generate synthetic temporal document collections, how this is realized in the *TDocGen* temporal document generator, and we will also present a study of the quality of the document collections created by TDocGen.

1 Introduction

In this paper we will describe how to make document collections to be used in development and benchmarking of web archives, and how this is realized in the *TDocGen* temporal document generator.

Aspects of temporal document databases are now desired in a number of application areas, for example web databases and more general document repositories:

- The amount of information made available on the web is increasing very fast, and an increasing amount of this information is made available *only* on the web. While this makes the information readily available to the community, it also results in a low persistence of the information, compared to when it is stored in traditional paper-based media. This is clearly a serious problem, and during the last years many projects have been initiated with the purpose of archiving this information for the future. This essentially means crawling the web and storing snapshots of the pages, or making it possible for users to "deposit" their pages. In contrasts to most search engines that only store the most recent version of the retrieved pages, in these archiving projects all (or at least many) versions are kept, so that it should also be possible to retrieve the contents of certain pages as they were at a certain time in the past. The most famous project is this category is probably the Internet Archive Wayback Machine [1], but in many countries similar projects also at the national level, typically initiated by national libraries or similar organizations.
- An increasing amount of documents in companies and other organizations is now only available electronically.

* Email of contact author: Kjetil.Norvag@idi.ntnu.no
[1] `http://archive.org/`

Mark Last et al. (Eds.): Advances in Web Intelligence and Data Mining (SCI) **23**, 171-180 (2006)
www.springerlink.com

Support for temporal document management is not yet widespread. Important reasons for that are issues related to 1) space usage of document version storage, 2) performance of storage andretrieval, and 3) efficiency of temporal text indexing. More research is needed in order to resolve these issues, and for this purpose test data is needed in order to make it easier to compare existing techniques and study possible improvements of new techniques. In the case of document databases test data means document collections. In our previous work [8], we have employed versions of web pages to build a temporal document collection. However, by using only one collection we only study the performance of one document creation/update pattern. In order to have more confidence in results, as well as study characteristics of techniques under different conditions, we need test collections with different characteristics.

Acquiring large document collections with different characteristics is a problem in itself, and acquiring *temporal* document collections close to impossible. In order to provide us with a variety of temporal document collections, we have developed the TDocGen temporal document generator. TDocGen creates a temporal document collection whose characteristics are decided by a number of parameters. For example, probability of update, average number of new documents in each generation, etc., can be configured. A synthetic data generator is in general useful even when test data from real world applications exists, because it is very useful to be able to control the characteristics of the test data in order to do measurements with data sets with different statistical properties.

Creating synthetic data collections is not a trivial task, even in the case of "simple" data like relational data. Because one of our application areas of the created document collections is study of text-indexing techniques, the occurrence of words, size of words, etc., have to be according to what is expected in the real world. This is a non-trivial issue that will be explained in more detail later in the paper. In order to make temporal collections, the TDocGen document generator essentially simulates the document operation by users during a specific period, i.e., creations, updates, and deletes of documents. The generator can also be used to create non-temporal document collections when collections with particular characteristics are needed.

The organization of the rest of this paper is as follows. In Section 2 we give an overview of related work. In Section 3 we define the data and time models we base our work on. In Section 4 we give requirements for a good temporal document generator. In Section 5 we describe how to create a temporal document collection. In Section 6 we describe TDocGen in practice. In Section 7 we evaluate how TDocGen fulfill the requirements. Finally, in Section 8, we conclude the paper.

2 Related work

For measuring various aspects of performance in text-related contexts, a number of document collections exist. The most well-know example is probably the TREC collections [2], which includes text from newspapers as well as web pages. Other examples are the INEX collection [6] which contains 12,000 articles from IEEE transaction and

[2] `http://trec.nist.gov/`

magazines in XML format, and documents in Project Gutenberg [3], which is a collection of approximately 10,000 books.

A number of other collections are also publicly available, some of them can be retrieved from the UCI Knowledge Discovery in Databases Archive [4] and the Glasgow IR Resources pages [5]. We are not aware any temporal document collections suitable for our purpose.

Several synthetics document generators have been developed in order to provide data to be used by XML benchmarks, however, these do not create document versions, only independent documents. Examples are ToXgene [1], which creates XML documents based on a template specification language, and the data generators used for the Michigan benchmark [9] and XMark [10]. Another example of generator is the change simulator used to study the performance of the XML Diff algorithm proposed in [3], which takes an XML document as input, do random modification on the document, and outputs a new version. Since the purpose of that generator was to test the XML Diff algorithm it does no take into account word distribution and related aspects, thus making it less suitable for our purpose.

In the context of web warehouses, studies of evolution of web pages like those presented in [2,4] can give us guidelines on useful parameters to use for creating collections reflecting that area.

3 Document and time models

In our work we use the same data and time model as is used in the V2 document database system [8].

A document version V is in our context seen as a list of words, i.e., $V = [w_0, w_1, ..., w_k]$. A word w_i is an element in the vocabulary set W, i.e., $w_i \in W$. There can be more than one occurrence of a particular word in a document version, i.e., it is possible that $w_i = w_j$. The total number of words n_w in the collection is $n_w = \sum_{i=0}^{n} |V_i|$.

In our data model we distinguish between documents and document versions. A temporal document collection is a set of document versions $V_0...V_n$, where each document version V_i is one particular version of a document D_j. Each document version was created at a particular time T, and we denote the time of creation of document version V_i as T_i. Version identifiers are assigned linearly, and more than one version of different documents can have been created at T, thus $T_i \geq T_{i-1}$. A particular document version is identified by the combination of document name N_j and T_i. Simply using document name without time denotes the most recent document version.

A document version exists (is valid) from the time it is created (either by creation of a new document or update of the previous version of the document) and until it is updated (a new version of the document is created) or the document is deleted (the delete is logical, so that the document is still contained in the temporal document database). The collection of all document versions is denoted C, and the collection of all document versions valid at time T (a snapshot collection) is denoted C_T. A temporal

[3] http://www.gutenberg.net

[4] http://kdd.ics.uci.edu/

[5] http://www.dcs.gla.ac.uk/idom/ir_resources/test_collections

document collection is a document collection that also includes historical (non-current versions, i.e., deleted documents and versions that were later updated) documents. The time model is a linear (non-branching) time model.

4 Requirements for a temporal document generator

A good temporal document generator should produce documents with characteristics similar to real documents. The generated documents have to satisfy a number of properties:

- Document contents: 1) number of unique words (size of vocabulary) should be the same as for real documents, both inside a document and at the document collection level, 2) size and distribution of word size should be the same as for real documents, and 3) average document size as well as distribution of sizes should be similar to real documents.
- Update pattern: 1) a certain number of document in the start, i.e., when the database is first loaded, 2) a certain number of documents created and deleted at each time instant, 3) a certain number of documents updated at each time instant, 4) different documents have different probabilities of being updated, i.e., dynamic versus relatively static documents, and 5) the amount of updates to a document, including inserting and deleting words.

Many parameters of documents depend on application areas. The document generator should be used to simulate different application areas, and has to be easily reconfigurable. We will now in detail describe some of the important parameters and characteristics.

4.1 Contents of Individual Documents and a Document Collection

Documents containing text will in general satisfy some statistical properties based on empirical laws, for example size of vocabulary will typically follow Heaps' law [5], distribution of words are according to Zipf's law [11], and have a particular average length of words.
Size of Vocabulary: According to Heaps' law, the number of *unique* words $n_u = |W|$ (number of elements in vocabulary) in a document collection is typically a function of the total number of words n_w in the collection: $|W| = Kn_w^{\beta}$, where K and β are determined empirically. In English texts typical values are $10 < K < 100$ and $0.4 < \beta < 0.6$ (cf. `http://en.wikipedia.org/wiki/Heaps'_law`). Note that Heaps' law is valid for a *snapshot collection*, and not necessarily valid for a complete temporal collection. The reason is that a temporal collection in general will contain many versions of the same documents, contributing to the total amount of words, but not many new words to the vocabulary.
Distribution of Words: The distribution of the words in natural languages and typical texts is Zipfian, i.e., the frequency of use of the nth-most-frequently-used word is inversely proportional to n: $P_n = \frac{P_1}{n^a}$, where P_n is the frequency of occurrence of the nth

ranked item, a is close to 1, and $P_1 \approx 0.1$.[6]
Word Length: Average word length can be different for different languages, and we have also two different measures: 1) average length of words in vocabulary, and 2) average length of words occurring in documents. Because the most frequent words are short words, the latter measure will have a lower value. The average word length for the words in the documents we used from the Project Gutenberg collection was 4.3.

4.2 Temporal Characteristics

The characteristics of a snapshot collection as described above is well studied during the years, and a typical document collection will obey these empirical laws. Temporal characteristics, on the other hand, are likely to be more diverse, and very dependent of application area. For example, in a document database containing newspaper articles the articles themselves are seldom updated after publication. On the other hand, a database storing web pages will be very dynamic.

It will also usually be the case that some documents are very dynamic and frequently updated, while some documents are relatively static and seldom or never updated after they have been created. Because these characteristics are very application area dependent, a document generator should be able to create documents based on specified parameters, i.e., update ratio, amount of change in each document, etc., as listed earlier in this section.

5 Creating a temporal document collection

In this section we describe how to create a temporal document collection. We describe first the basis of creating non-temporal documents, before we describe how to use this for creating a temporal document collection.

Each snapshot collection C_T, or "generation", should satisfy properties as described in the previous section. The basis for creating the first generation as well as new texts to be inserted into updated documents is the same as if creating a non-temporal collection.

5.1 Creating Synthetic Non-Temporal Documents

Several methods exists for creating synthetic documents, we will here describe the methods we considered in our research, which we call the *naive*, *random-text*, *Zipf-distributed/random words*, and the *Zipf-distributed/real words* methods.
Naive: The easiest method is probably to simply create a document from a random number of randomly created words. Although this could be sufficient for benchmarking when only data amount is considered, it would for example not be appropriate for benchmarking text indexing. Two problems are that occurrence distribution of words and vocabulary size is not easily controllable with this method. Although the method can be improved so that these problems are reduced, there is also the problem that because words in real life are not created by random, and frequent words are not necessarily uniformly distributed in the vocabulary, some of them can be close to each other.

[6] For our test collections we have measured $P_1 = 0.05$.

One example is some frequently occurring words starting with common prefixes, or different forms of the same word (for example "program" and "programs"), especially the case when stemming (where only the root form of a words is stored in the index) is not employed.[7]
Random-text: If a randomly generated sequence of symbols taken from an alphabet S where one of the symbols are blank (*white space*), and the symbols between two blank spaces are considered as a word, the frequency of words can be approximated by a Zipf distribution [7]. The average word size will be determined by the number of symbols in S. Such sequences can be used to create synthetic documents. However, the problem is that if the average length of words should be comparable to natural languages like English, the number of symbols in S have to be low. Another problem is that the distribution is only an approximation to Zipf: it is stepwise distribution, all words with same length has same probability of occurrence. Both problems can be fixed by introducing bias among different symbols. By giving a sufficient high probability for blanks the average length of words even with a larger number of symbols (for example, 26 in the case of the English language) can be reduced to average length of English words, and by giving different probabilities for the other symbols a smoother distribution is achieved. It is also possible to introduce cut-off for long words. The advantage with this methods is that an unlimited vocabulary can be created, but the problem with lexicographically closer words as described above remain.
Zipf-distributed/random-words: A method that will create a document collection that follow Heaps' law and has a Zipfian distribution, is to first create $n = n_u$ random words with an average word length L. The number of n can be determined based on Heaps' law with appropriate parameters. Then, each word is assigned an occurrence probability bases on Zipfian distribution. This can be done as follows: As described in Section 4.1, the Zipfian distribution can be approximated to $P_n = \frac{P_1}{n}$. The sum of probabilities should be 1, so that:
$\sum_{i=1}^{n_u} P_i = 1 \Rightarrow \sum_{i=1}^{n_u} \frac{P_1}{i} = 1 \Rightarrow P_1 \sum_{i=1}^{n_u} \frac{1}{i} = 1 \Rightarrow nP_n \sum_{i=1}^{n_u} \frac{1}{i} = 1$
$\Rightarrow P_n = \frac{1}{n \sum_{i=1}^{n_u} \frac{1}{i}}$
In order to select a new word to include in a document, the result r from a random generator producing values $0 \leq r \leq 1$ are used to select the word ranked k that satisfies $\sum_{j=1}^{k-1} P_j \leq r < \sum_{j=1}^{k} P_j$
Using this method will create a collection with nice statistical properties, but still have the problem of not including the aspect of lexicographically close words as described above.
Zipf-distributed/real words: This actually the approach we use in TDocGen, and is an extension of the Zipf-distributed/random-words approach. Here a *real-world vocabulary* is used instead of randomly created words. In order to make any improvement, these words need to have the same properties as in real documents, including occurrence probability and ranking. This is achieved by first making a histogram of word frequency (i.e., frequency/word tuples) based on real texts and rank words according to this. The result will be documents that include the aspect of lexicographically close words as well as following Heaps' law and having a Zipfian distribution of words.

[7] This is typically the case for web search engines/web warehouses.

5.2 Creating Temporal Documents

The first event in a system containing the document collection, for example a document database, is to load the initial documents. The number of documents can be zero, but it can also be a larger number if an existing collection is stored in the system. The initial collection can be made from individual documents created as described above.

During later events, a random number of documents are deleted and a random number of new documents are inserted. The next step is to simulate operations to the document collection: inserting, deleting, and updating documents.

Inserting documents. New documents to be inserted into the collection are created in the same way as the initial documents.

Deleting documents. Documents to be deleted are selected from the documents existing at a particular time instant.

Updating documents. The first task is to decide *which* documents to be updated. In general, the probability of updates to files will also in general follow a Zipfian distribution. A commonly used approximation is to classify files into dynamic and static files, where the most updates will be to dynamic files, and the number of dynamic size is smaller than the number of static files. A general rule of thumb in databases is that 20% of the data is dynamic, but 80% of the updates are applied to this data. This can be assumed to be the case in the context of document databases as well, and in TDocGen documents are characterized as being static or dynamic, and which category a document belongs to is decided when it is created. When updates are to be performed, it is first decided whether the update should be to a dynamic or static file, and which document in the category that is actually updated, is chosen at random (i.e., uniform distribution).

After it is decided what documents to update, the task is to perform the actual update. Since we do not care about structure of text in the documents, we simply delete a random number of lines, and insert a random number of new lines. The text in the new lines are created in the same way as the text to be included in new documents.

One of the goals of TDocGen is that it should be able to create temporal document collections that can have characteristics for chosen application areas. This is achieved by having a number of parameters that can be changed in order to generate collections with different properties. The table on the next page summarizes the most important parameters. Some of them are given a fixed value, while other parameters are given as average value and standard deviation. The table also contains the values for two parameter sets in our experiments which are reported in Section 7.

6 Implementation and practical use of TDocGen

TDocGen has been implemented according to the previous description, and consists of two programs: one to create histograms from an existing document collection, and a second program to create the actual document collection. Creating histograms is a relatively time-consuming task, but by separating this into a separate task this only have to be performed once. Histograms are stored in separate histogram files that can also be distributed, so that it is it is not actually necessary for every user to retrieve a large collection. This is a big saving, because a histogram file are much smaller than the document collection it is made from, for example, the compressed size of the document collection we use is 1.8 GB, while the compressed histogram file is only 10 MB.

Parameters	Pattern I		Pattern II	
	Avg. or Fixed	Std. dev.	Avg. or Fixed	Std. dev.
Number of files that exist the first day	1000	-	10	-
Percentage of documents being dynamic	20	-	20	-
Percent of updates applied to dynamic documents	80	-	80	-
Number of new documents created/day	200	5	2	1
Number of deleted documents/day	100	2	1	1
Number of updated documents/day	500	20	5	2
Number of words in each line in document	10	-	10	-
Number of lines in new document	150	10	150	10
Number of new lines resulted from update	25	5	25	5
Number of deleted lines resulted from update	20	5	20	5

The result of running TDocGen is a number of compressed archive files. There is one file for each day/generation, and the file contains all document versions that existed during that particular time instant. The words in the documents will follow Heaps' and Zipf's laws, but because the vocabulary/histogram has a fixed size, Heaps' law will only be obeyed as long as the size of a the documents in a particular generation is smaller than the data set which the vocabulary was created from.

7 Evaluation of TDocGen

The purpose of the output of a document generator is to be used to evaluate other algorithms or system, and it is therefore important that the created documents have the quality in terms of statistical properties as expected. It is also important that the document generator has sufficient performance, so that the process of creating test document does not in itself become a bottleneck in the development process. In this section, we will study the performance of TDocGen, and the quality of the created document collection.

TDocGen creates documents based on a histogram created from a base collection. In our measurements we have used several base collections. The performance and quality of results when using these is mostly the same, so we will here limit our discussion to the largest collection we used. This collection is based on a document collection that is available from Project Gutenberg [8]. The collection contains approximately 10,000 books, and our collection consists of most of the texts there, except some documents that contain contents we do not expect to be typical for document databases., e.g., files contains list of words (for crosswords), etc.

Cost: In order to be feasible in use, a dataset generator has to provide results within "reasonable time". The total run time for creating a collection using the actual parameters in this paper is close to linear wih respect to number of days.

[8] `http://www.gutenberg.net`

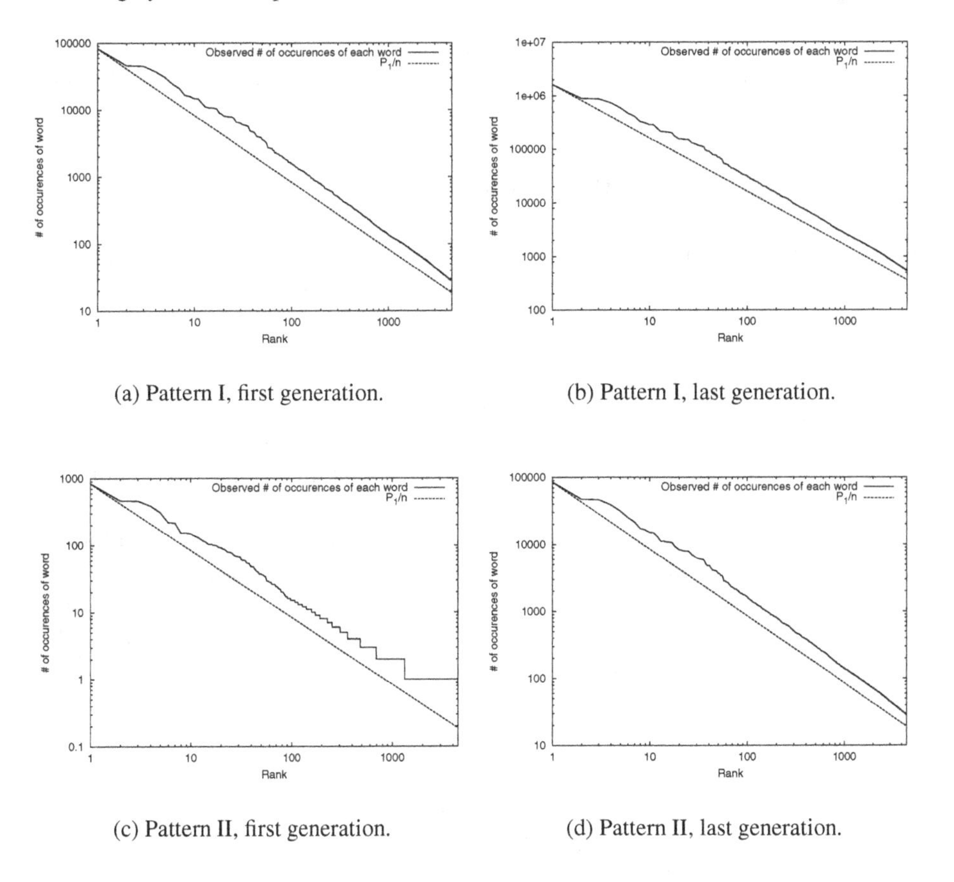

(a) Pattern I, first generation.

(b) Pattern I, last generation.

(c) Pattern II, first generation.

(d) Pattern II, last generation.

Fig. 1. Comparison of ideal Zipf distribution and the words in the actual created document collections.

The total collection created in this experiment is quite large, a total of 1.6 million files are created, containing 13.3 GB of text. The size of the last generation is 159MB of text in 18,000 files. The elapsed time is less than half an hour, which should be low for most uses of such a generator.

Quality of Generated Document Collections: We will study the quality of the generated document collections with respect to word distribution and number of unique words.

The first study is word distribution in the created collections, and we perform this study on the first and last snapshot collections created during the tests using the two patterns in the previous table. As Figure 1 show, words distribution is Zipfian in the created collections. It should also be mentioned that an inspection of the highest ranked words shows that the most frequently occurring words are "the", "of", "and", and "to". This is as expected in a document collection that is based on documents that are mostly in English.

We also studied the value K for the created collections. We saw that K is between 60 and 70 which is well within reasonable bounds, and hence confirmed that the document collections are according to Heaps' law.

8 Conclusions and further work

In research in web archiving, different algorithms and approaches are emerging, and in order to be able to compare these good test collections are important. In this paper we have described how to make temporal document collections, how this is realized in the TDocGen temporal document generator, and we have provided a study of the quality of the document collections that are created by TDocGen.

TDocGen have been shown to meet the requirements for a good temporal document collection generato. Also available are ready-made histograms, including the one used for the experiments in this paper, based on 4 GB of text documents from Project Gutenberg.

If users want to generate temporal document collections especially suited for their own domain, it is possible to use own existing documents as basis for building the histograms used to generate the temporal document versions. It should also be noted that the generator can also be used to create non-temporal document collections when collections with particular characteristics are needed.

References

1. D. Barbosa et al. ToXgene: a template-based data generator for XML. In *Proceedings of the 2002 ACM SIGMOD International Conference on Management of Data*, 2002.
2. B. E. Brewington and G. Cybenko. How dynamic is the Web? *Computer Networks*, 33(1-6):257–276, 2000.
3. G. Cobena, S. Abiteboul, and A. Marian. Detecting changes in XML documents. In *Proceedings of the 18th International Conference on Data Engineering*, 2002.
4. D. Fetterly, M. Manasse, M. Najork, and J. L. Wiener. A large-scale study of the evolution of Web pages. *Software - Practice and Experience*, 34(2):213–237, 1996.
5. H. S. Heaps. *Information Retrieval: Computational and Theoretical Aspects*. Academic Press, Inc., 1978.
6. G. Kazai et al. The INEX evaluation initiative. In *Intelligent Search on XML Data, Applications, Languages, Models, Implementations, and Benchmarks*, 2003.
7. W. Li. Random texts exhibit Zipf's-law-like word frequency distribution. *IEEE Transactions on Information Theory*, 38(6), 1992.
8. K. Nørvåg. The design, implementation, and performance of the V2 temporal document database system. *Journal of Information and Software Technology*, 46(9):557–574, 2004.
9. K. Runapongsa et al. The Michigan Benchmark: A microbenchmark for XML query processing systems. In *Efficiency and Effectiveness of XML Tools and Techniques and Data Integration over the Web*, 2002.
10. A. Schmidt et al. XMark: a benchmark for XML data management. In *Proceedings of VLDB'2002*, 2002.
11. G. K. Zipf. *Human Behaviour and the Principle of Least Effort: an Introduction to Human Ecology*. Addison-Wesley, 1949.

Predicting Stock Trends with Time Series Data Mining and Web Content Mining

Gil Rachlin[1] and Mark Last[1]

[1] Ben Gurion University of the Negev,
Department of Information Systems Engineering,
Beer- Sheva 84105, Israel,
{ gilrach, mlast } @ bgu.ac.il

Summary. This paper presents a new methodology for predicting stock trends and making trading decisions based on the combination of Data Mining and Web Content Mining techniques. While research in both areas is quite extensive, inference from time series stock data and time-stamped news stories collected from the World Wide Web require further exploration. Our prediction models are based on the content of time-stamped web documents in addition to traditional Numerical Time Series Data. The stock trading system based on the proposed methodology (ADMIRAL) will be simulated and evaluated on real-world series of news stories and stocks data using several known classification algorithms. The main performance measures will be the prediction accuracy of the induced models and, more importantly, the profitability of the investments made by using system recommendations based on these predictions.

Key words: Financial Intelligence, Data Mining, Web Content Mining, Text Mining, Classification, time-series analysis, decision trees, Efficient Market Hypothesis.

1 Introduction

The Efficient Market Hypothesis (EMH), as stated by Fama ([4, 5]), asserts that 'Stock Prices fully reflect all their relevant information at any given point in time'. As the basis for growth and development of a modern economy this means that no information or analysis can be expected to result in out performance of the market and that stock prices follow 'Random Walks' ([1]), where a change in stock price over time is purely random and statistically independent of the stock price in the past. However, to this day no one can explain the anomalies in the market, which can be used to assure some short term predictive power ([1, 7, 9]).

In making their own forecasts most financial specialists try to exploit the time gap of the market's adjustment to new information. They reduce their risk by combining both technical (base future price predictions on past prices) and fundamental (base predictions on real economy factors , such as inflation,

Mark Last et al. (Eds.): Advances in Web Intelligence and Data Mining (SCI) **23**, 181-190 (2006)
www.springerlink.com

trading volume, organizational changes in the company etc.) analysis strategies, which were mentioned by Gidófalvi ([6]) and are fully explained in the literature by [1]. In order to obtain the data required for both strategies one can refer to various publicly available resources like the stock market itself, the companies, news papers or others.

A rather new source for information in the late 20th and the 21st centuries is, of course, the Internet. In order to exploit this relatively new media as an additional helping tool in our forecasting task, we need to combine techniques from both time series data mining and web content mining.

Most studies ([8, 11, 12]) agree that the process of Knowledge Discovery in Databases (KDD) involves iterating over four general steps, each using independent tools: 1) Data cleaning and preprocessing (create a common data representation from different sources and different data types, e.g: relational, transactional and spatial databases to large repositories of unstructured data such as the World Wide Web), 2) Test and discover relationships in the data using Artificial Intelligence and Statistical Analysis tools, 3) Postprocessing of discovered patterns, 4) Use the model to perform actions on real world data.

When adding the aspect of time to the Data Mining process, it is understood ([12]) that database records are time stamped and meaningful only as part of a time segment or time series.

In [12], Last et al. use a signal processing technique to pre-process the raw time series data. Then they construct an information theoretic connectionist network (IFN) to induce time series prediction rules, which are later reduced using fuzzification and aggregation. Finally, the rules are presented in natural language and used to predict future behavior of the time series. Their purpose is to predict the timing of Change Points. A Change Point is the point where a specific trend of values in the data is changed, for example, the time point from which a stock price starts to steadily increase after a steady decrease. Last et al. mention two common methods for finding change points: one is by recursive binary partitioning of the time segment, where likelihood criteria are used as the underlying model for each segment. A second method to finding the optimal number of linear segments in time series starts with large number of equal size segments (with 3 time points each), and proceed by merging two adjacent segments, which minimize the balance of error, which is calculated as a standard deviation of errors in all the segments.

Applying Data Mining techniques on the World Wide Web is usually referred to as Web Mining. It introduces new challenges to the Data Mining process on how to clean, categorize and utilize the information in order to create useful models, which can help in the future decision making process. Studies ([2, 10]) divide Web Mining into discovering intra-document structure (Web Content Mining) and to discovering inter-document structure (Web Usage Mining and Web Structure Mining). Web Content Mining is more relevant to our prediction task. It refers to finding relevant information on the Web, mainly, by using Information Extraction (IE) and Information Retrieval (IR) techniques for semi structured data. IE is viewed as trying to find the structure or representation of a document, for example, putting it in the form of a database. On the other hand, IR is the automatic retrieval of all relevant documents, while discarding as many non relevant ones as possible. Mining Multimedia data can be important in many areas such as Medical Informatics or in the case of analyzing stock charts but we decided to leave it out of our scope.

Most of the studies done in order to combine inference from time stamped news stories and time series stock data are different in their concepts and methods. Each group uses a different time series, text classifier, features, target attributes, time window length, weighting method etc. However, they do go through the following common stages: 1) Define stock trends from the raw Time Series stock data (using the above mentioned methods, which were used by [12]). 2) Define a Window of Influence (WOI), which is a time frame taken before and after the publication time, *t*, of a web article ([Window Start Point], [Window End Point]): In [6] it is [-20, 20] minutes. In ([13]) it is [-300, 0] minutes and in [15] a [-60, 0] minute window is used. 3) Align time stamped news articles to stock trends according to the WOI and score them: [6] Scores news articles relative to the price changes in the stock and the index it belongs to (Δstock-price, Δindex-price) and labels them in reference to a threshold. Both [15] and [17] compare an expert's predefined list of key words, to their occurrence in the text. They normalize the words weights (values between 0 to 1). Weights are calculated using one of three methods, which were already mentioned in the literature ([2, 10]): Boolean (Occurrence/non-occurrence of words), TF*IDF (Term Frequency and Inverse Document Frequency), TF*CDF (Term Frequency and Category Discriminating Factor). 4) Induce a model, which learns how to classify an article with a predefined trend and use the model to detect future trend occurrences: [6] uses the Rainbow Naïve Bayesian text classifier package to get three probabilistic indicators for each document, *d*: P(Up| d), P(Down| d) and P(Exp| d). [13] also uses the Bayes theorem to find trend relevance probability, but they compare arriving documents against a representative 'language model' of five possible trends. Both [15] and [17] follow a rule base approach to create probabilistic classification data-log rules, which are applied to one or more backward time periods (one hour to one day). 5) Evaluate their model prediction ability: [6] showed low predictive power, which he explained to be due to duplicate stories in the dataset. After running a 40 day simulation on the real market, [13] showed better results than random actions. Predictions made by [15] outperformed conventional time series analysis, two different neural nets and random guessing. [17] tried to predict the indices of five global markets and the importance of their results is in showing that the best accuracy (sometimes over 60%) was achieved in the US market indices.

In this paper, we present a new method for detecting stock trends based on the combination of Data Mining and Web Content Mining techniques, which attempts to overcome some of the problems of previous studies. Thus, the main contributions of our research are: 1) The creation of a "melting pot" of numeric and textual data before running an induction algorithm, 2) The automatic extraction of key words/phrases instead of using a prior Expert list of phrases, 3) The elimination of the need for word independence assumption by using Decision Trees instead of Naïve Bayes, 4) Extension of the WOI of news articles in the prediction task (from minutes to days). A prototype system called ADMIRAL is currently under development in order to achieve the objectives of this study.

The rest of this paper is organized as follows: In Section 2 we describe the stages needed for ADMIRAL to be operationally useful. Section 3 describes the planned simulation and evaluation of the system. Section 4 offers some conclusions for this article.

2 ADMIRAL - Performing the Prediction Steps

The ADMIRAL system aims to become a full cycle prediction system for stock trends according to past numeric values of the stocks as well as its related textual web articles. ADMIRAL goes through six steps, as shown in Fig. 1, which are: 1) Data Collection, 2) Feature Extraction, 3) Term Weighting, 4) Data-Set Structuring, 5) Decision Tree Induction, 6) Trading Simulation.

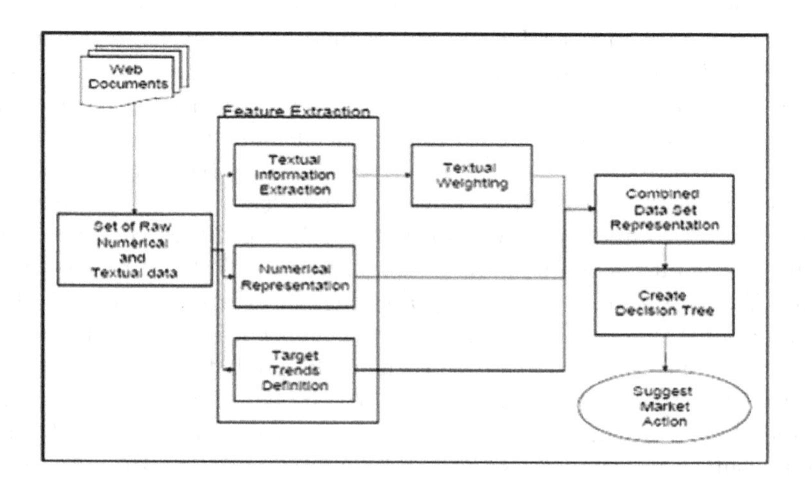

Fig. 1. Prediction Scheme of a Mixed Numerical and Textual Data

Most steps offer various tools to perform the task at hand. The preferred configuration of tools for ADMIRAL's task can either be set manually or automatically. We aim to find the best configuration for the stock prediction task. Eventually a set of recommendations is created, which will later be compared against real time market results in order to evaluate the predictive power of each set of prior definitions. Following is a semi detailed description for each step in the prediction scheme.

2.1 Step 1: Data Collection

For the scope of our research, our system needs to collect data from Financial web sites, which are considered to have good real-time data (Numerical and Textual). Good choices of financial sites, which are also being used by financial professionals and can be used in our system are: www.forbes.com, finance.yahoo.com, www.bloomberg.com, today.reuters.com and others. In order to collect the relevant data from each one of the selected sites, we decided to create a configuration XML file, which includes its structural characteristics. Thus our code becomes generic and whenever we want to add a new financial site to be monitored and modeled the only addition will be the definition of an appropriate XML file for it.

The granularity of the data collection for the textual training data is once a day after the end of trade. For each textual article we keep its publication time stamp, its header and substance. Numerical data is collected three times each day: after the trade opens, in the middle of the day and after the trade ends. We note that we are looking to create long term prediction models; hence we skip the need to collect numerical data in 10-15 minutes intervals as was done in previous studies.

2.2 Step 2: Feature Extraction

The feature Extraction is done for both Textual and Numerical data. Feature Extraction for Textual Data includes two activities: first is the automatic extraction of key words and key phrases from a predefined Window of Influence. Second is the creation of a word dictionary, which includes the most influential words in all our past web articles. Feature Extraction for the Numerical data also includes two activities: first is the calculation of additional commonly used financial values. Second is the long term stock price trend discovery.

The Automatic Word Extraction is done by the Extractor software package ([3]). The Extractor is a text summarization engine, which uses a patented genetic extraction algorithm, GenEx. GenEx analyzes the recurrence of words and phrases, their proximity to one another, and the uniqueness of the words to a particular document. It removes all stop words from the document, applies a stemming procedure and selects a finite number of the most influential words in the document. In ADMIRAL we construct each document, *d*, to be analyzed by the Extractor, as a set of the web articles related to a specific stock inside a backward WOI. Here, the WOI is a time frame from the current time point of measurement, *t*, until a predefined time in the past, *t-i*. So the interval [t-i, t] represents the first WOI behind point *t*. Textual feature extraction table is created, where each row consists of: the measurement time stamp, the related stock, some general information about the articles in the WOI (e.g: the number of articles in it or the number of words in it) and of course the influential words in the WOI with their initial given Extractor weights as well as other commonly used term weights (e.g: TF or Boolean). The terms extracted in each Time Window are added to the term dictionary, where they undergo a separate generic weight calculation procedure.

The Term Dictionary has a predefined number of words, which we define as N. It is rebuilt each time a new training set is evaluated for creating a new classification model. Each word in the term dictionary receives a score, which determines its degree of membership inside the dictionary and only the K highest ranking words will eventually be used from the final dictionary (K $\leq$ N). As seen in Equation 1, the score, S, gives eqaul importane to TF against the time frame each term appeared in.

$$\mathbf{S} = \frac{1}{2} \times \left(\frac{TF}{N}\right) + \frac{1}{2} \times \left(\frac{P}{L} \times \frac{B}{L}\right) \tag{1}$$

Where:

L: Time frame, in days, for the word dictionary.
B: Time window between the first and last appearance of a word.
P: Number of days to the last occurrence of a word.
TF: Number of days to the last occurrence of a word.
N: Number of words in the dictionary.

As an example, let's define a final number of terms in the term dictionary as $N = 20,000$ and the total number of days for creating the term dictionary as $L = 30$. The Term Frequency of the phrase "High Volume" in all the previously collected documents is $TF = 8$. The phrase last occurred 3 days ago, $P = 3$. Its first occurrence was 10 days ago, $B = 7$. Thus Equation 1 generates the Grade of Membership S, for the key phrase "High Volume" as 0.0118.

Entries in the original numerical database representation need additional commonly used financial ratios calculations like the β value, the Sharp Index value or Capital Ratio value ([1]). The reason for adding the Capital Ratio, for example, is because it is a known fact in the capital market ([1]) that one of the anomalies in the EMH states that if the Capital Ratio for a given stock is low, than the return on investing in it will be higher. If the EMH is correct we will not find any evidence of that fact in building our models. However, if the Efficient Market Hypothesis is affected by its anomalies, than our model will be influenced by it.

The Target Trends Definition is obtained by following the method introduced by Last et al. in [12]. We are interested in the value (stock rate) for each stock at each point of measurement, t. Every such point is part of a trend of values (mostly increasing, mostly decreasing, mostly remain the same) which has a starting point and an end point (the length of the trend is determined in day units), a slope degree and a fluctuation of the values, which constitutes the trend.

2.3 Step 3: Term Weighting

In order to later use the extracted textual features inside each Window of Influence, [t-i, t] we need to apply a normalized value between zero and one to each key phrase. The normalization is done by dividing the weights, which were either automatically given by the Extractor or manually counted according to TF and Boolean methods ([2, 10]), by the overall grade of membership in the term dictionary, which was also calculated in section 2.2 above.

2.4 Step 4: Data Set Structuring

After having prepared both numerical and textual data and assigned a trend to each one of the stock's prices, we need to combine the data, which we want to include in our prediction task.

We would like to predict the forthcoming trend, which will last more than a predefined time length. Hence, our target attribute is the trend, which can take the following five values: Up, Slight-Up, Expected, Slight-Down and Down. We assume that the target trend is influenced by the two time periods of textual data prior to its occurrence: [t-2i, t-i] and [t-i, t] and the extracted numerical trend information. We create a dataset by concatenating all that extracted data. Table 1 shows an example of two entries in the final data set representation.

2.5 Step 5: Decision Tree Induction

We use two Decision Tree Induction algorithms, which do not assume word independence and compare their results against Naïve Bayes Classification, which assumes word independence and is widely used in the field.

The two algorithms are: IFN (Info Fuzzy Network) developed by Last et al. ([12, 14]) and C4.5 developed by Quinlan in [16]. In [12], IFN showed better performance than C4.5 in the stock prediction task based only on numerical data. Each algorithm will yield a set of trends and lengths predicting rules on which we can rely in order to perform our next step of recommendation. In order to show the effect of the combination between Numerical and Textual data, we will also try running the algorithms on each separated type of data.

2.6 Step 6: Trading Simulation

After running the induction algorithms on the training set, the system can run on a test set in order to simulate real time activities. New data collected from the web sites are put through the above mentioned steps 1-4. After predicting a trend and its length from the induced model, a good and objectively

Table 1. The final Data-Set combines the extracted trend info and textual info for two consecutive WOI ([t-i, t] and [t-2i, t-i]) as well as the required target trend type.

Type of Data	Table Column	TF Method Example	...	Bool Method Example
General Entry Info.	Date/Time Stamp	01/01/2006 09:45	...	01/10/2006 17:45
	Stock Symbol	MSFT	...	GOOG
Current Trend Info.	Trend Slope	0.03	...	0.095
	Trend Fluctuation	0.008	...	0.037
General Text Info.	Articles in time window	11	...	9
	:			
	Average size of article	87	...	56
Text in [t-2i, t-i].	Effiecient	0.64	...	0.89
	:			
	Volume	0.22	...	0.15
Text in [t-i, t].	Effiecient	0.46	...	0.83
	:			
	Volume	0.78	...	0.376
Target Trend Data	Trend Length	1.75	...	0.8
	Trend	Expected	...	Slight-Up

measurable trading strategy could be followed. We currently formulated three different, non random, trading strategies to be compared: 1) Buy and Hold – according to the recommendation of the model, at the beginning of a predefined time period, we buy the recommended stocks and hold on to them until the end of the predefined time period. 2) Automatic – We follow the exact recommendations of the system's rules. 3) Semi Automatic – We combine activities based on the recommendations of the system with our own assessment, which is based on our prior knowledge and guesses. However, this method can not be objectively evaluated. Hence it will only be used informally.

As seen in table 2, when an automatic strategy is used the action to be used is set according to the trend. If the trend is Up we will buy the stock expecting to earn from the increase. If the trend is Down we sell short, hoping to make money on the ability it gives us to sell at a higher price than the expected future price. If the trend is Expected we do nothing.

Table 2. Real Time Market action on a stock based on the model trends

Trend in the Model	Action
Up	Buy
Down	Sell Short
Expected	No Action

3 System Evaluation

In order to perform our evaluation for the proposed method we will introduce our prediction model from stocks in the US NASDAQ index. Starting from January 1st 2006, we collect data on 20 stocks and later choose the 12 stocks for which we have the largest amount of textual data. We chose to collect our raw data (both Numerical and Textual) from two leading Financial web sites: today.reuters.com and www.forbes.com. In addition to the reason mentioned in Section 2.1 two more reasons made us choose them: first is that at the time of our experiments the articles on each web site are mutually exclusive, which means that a comparison between the information gathered from them will not overlap. Second, they allow data filtering according to requested stocks, which will reduce our IR efforts.

In order to simulate the steps described in Sections 2.1 to 2.5 we define each WOI as 24 hours. This means that if the raw data is checked on January 4th 2006 at 9:45 than its two consecutive Windows of Influence according to the above mentioned definition: [t-2i, t-i] and [t-i, t] are [02/01/2006 9:45, 03/01/2006 9:45] and [03/01/2006 9:45, 04/01/2006 9:45] respectively.

We will measure several parameters, which will give us a good estimation on the quality of our method against previously used methods: 1) The predictive accuracy of the model calculated as: the percent of days in which the classification was correct. 2) The CPU time to build the model. 3) The number of attributes used by the classification model. 4) The expected profit over time by applying the model predictions on real time stock data (according to [1] 12% is the yearly return on investing in the American stock market).

The first three measurement parameters can be calculated from our previously collected information. However, in order to estimate the expected profit we intend to apply our investment strategies, which were mentioned in Section 2.6. Avoiding overnight risk during hours when the market is closed can be achieved if we trade according to a Single Day Trading Strategy (SDTS) as was mentioned in the literature ([6, 13]). This means that at the end of each day all our short term stocks holdings are sold. Our model predicts the forthcoming trend, with its predicted duration; therefore, we will only use entries in which the predicted duration for a specific trend is more than one or two days. We will compare our results against a random activity scheme, where an arbitrary action will be performed/not performed on each stock each day.

4 Conclusions

The ADMIRAL system is currently in the stages of collecting data from the Financial web sites. The data is stored on an SQL server. The code, which performs the other tasks, is currently under development.

This study hopes to show an improvement in the stock trend profitability by finding the best configuration of prediction techniques and trading strategies. The methods to improve the profitability include:

- Combination of both Numeric and Textual Data.
- The use of an Automatic Text Extraction.
- The use of Decision Tree Prediction Models.
- The use of smart trading strategies (initial results from the ADMIRAL system will be reported at the AWIC Conference).

References

1. Bodie Z, Kane A, Marcus AJ (2001) Investments, 4th Edition. McGraw Hill.
2. Cooley R, Mobasher B, Srivastava J (1997) Web Mining: Information and Pattern Discovery on the World Wide Web. Proceedings of the IEEE International Conference on Tools with Artificial Intelligence (ICTAI'97), Newport Beach, CA.
3. Extractor DBI technologies (2003) [http://www.dbi-tech.com].
4. Fama EF (1970) Efficient Capital Markets: A Review of Theory and Empirical Work, Journal of Finance 25: 383-417.
5. Fama EF (1991) Efficient Capital Markets: II, Journal of Finance 46: 1575-1617.
6. Gidófalvi G (2001) Using News Articles to Predict Stock Price Movement, Online at: [http://citeseer.nj.nec.com/517027.html] or [http://www.cs.aau.dk/~gyg/docs/financial-prediction.pdf].
7. Huagen RA (1995) The New Finance: The Case Against Efficient Markets. Prentice-Hall.
8. Jain AK, Murty MN, Flynn PJ (1999) Data Clustering: A Review, ACM Computing Surveys, Vol 31 No. 3.
9. Kaboudan MA (2000) Genetic Programming Prediction of Stock Prices, Computational Economics 16: 207-236.
10. Kosala R, Blockeel H, Web Mining Research: A Survey, SIGKDD Explorations, Vol 2, Issue 1.
11. Landry R Jr., Debreceny R, Grey GL (2004) Grab Your Picks and Shovels! There's Gold in Your Data, Strategic Finance, January (85, 7).
12. Last M, Klein Y, Kandel A (2001) Knowledge Discovery in Time Series Databases, IEEE Transactions on Systems, Man and Cybernetics – Part B: Cybernetics, Vol 31 No. 1.
13. Lavrenko V, Schmill M, Lawrie D, Ogilvie P, Jensen D, Allan J (2000) Language Models for Financial News Recommendation, CIKM 2000, McLean, VA USA, ACM 2000.
14. Maimon O, Kandel A, Last M (2000) Knowledge Discovery and Data Mining, The Info-Fuzzy Network (IFN) Methodology, Norwell, MA: Kluwer.
15. Peramunetilleke D, Wong RK (2002) Currency Exchange Rate Forecasting from News Headlines, Thirteenth Australasian Database Conference (ADC2002), Melbourne, Australia, Conferences in Information Technology, Vol 5.
16. Quinlan JR (1993) C4.5: Programs for Machine Learning, Morgan Kaufman Publishers Inc., San Francisco, CA.
17. Wuthrich B, Cho V, Leung S, Permunetilleke D, Sankaran K, Zhang J, Lam W (1998) Daily Stock Market Forecast from Textual Web Data, IEEE International Conference on Systems, Man. and Cybernetics, Vol 3: 2720 -2725.

CatS: A Classification-Powered Meta-Search Engine

Miloš Radovanović and Mirjana Ivanović

University of Novi Sad
Faculty of Science, Department of Mathematics and Informatics
Trg D. Obradovića 4, 21000 Novi Sad
Serbia and Montenegro
{radacha,mira}@im.ns.ac.yu

Summary. CatS is a meta-search engine that utilizes text classification techniques to improve the presentation of search results. After posting a query, the user is offered an opportunity to refine the results by browsing through a category tree derived from the dmoz Open Directory topic hierarchy. This paper describes some key aspects of the system (including HTML parsing, classification and displaying of results), outlines the text categorization experiments performed in order to choose the right parameters for classification, and puts the system into the context of related work on (meta-)search engines. The approach of using a separate category tree represents an extension of the standard relevance list, and provides a way to refine the search *on need*, offering the user a non-imposing, but potentially powerful tool for locating needed information quickly and efficiently. The current implementation of CatS may be considered a baseline, on top of which many enhancements are possible.

1 Introduction

Modern general-purpose Web search engines like Google and Yahoo, although invaluable, could benefit from improvements in many areas, including coverage (only a portion of the Web is indexed by any one search engine), relevance ranking of Web pages with regards to a query, and presentation of results. For this reason many meta-search engines have been set up, which address coverage and ranking issues by aggregating results returned by several search engines for a specific query (Dogpile, Mamma), and may also attempt (more rarely, however) to give an alternative presentation of results using clustering techniques (Vivissimo, KartOO). This paper will describe CatS (http://stribog.im.ns.ac.yu/cats/) – a meta-search engine that focuses on presentation of search results by automatically sorting them into a hierarchy of topics employing *text categorization* [23], instead of the widely used clustering methods.

To illustrate the benefits of rendering search results in a form different from the now standard relevance list model, whatever technique is used, consider two scenarios at the opposite extremes of user intention:

Mark Last et al. (Eds.): Advances in Web Intelligence and Data Mining (SCI) **23**, 191-200 (2006)
www.springerlink.com

(i) a user looking for a very specific piece of information ("I need some information about this particular model of guitar."), and
(ii) a user attempting to learn about, or simply browse pages related to some broader topic ("I want to read something about animals living in England.").

In both cases a search engine may fail the user – for (i) a good example is given in [6], where the query 'martin d93 guitar,' which refers to a legendary model of guitar manufactured by C. F. Martin, returned dozens of unrelated pages ranked higher than any truly relevant answer[1]. If the relevant search results were sorted, for example, into topic Arts → Music, they would have been easier to trace.

For case (ii) the benefits of sorting are even more obvious – for instance, pages about Animals, the band, would have been separated from pages about English wildlife.

The common denominator of both described cases is *query ambiguity*. In the first case the ambiguity was introduced by the content and structure of the Web, coupled with the ranking algorithms of search engines. Relying on *inverse document frequencies* made search engines rank any page which contained the rare term 'd93' unrealistically high, disregarding the term's connection to 'martin' or 'guitar.' Furthermore, "fancy hits," such as pages containing query terms as part of e-mail addresses, introduced additional mixup. In the second case the ambiguity simply resulted from different possible meanings of the word 'animals.' Resolution of ambiguity, like the one described above, is given additional importance by the finding that the majority of queries posted to search engines consist of only one word [2].

Practically all meta-search engines that provide an alternative presentation of search results rely on clustering techniques, which may generate excellent topic hierarchies for some queries, but prove not so satisfactory for others (see Sections 4 and 5 for a more comprehensive discussion). Although classification was often mentioned as a possible method for enhancing the presentation of search results [11, 23], it has seldom been attempted in practice. Some general-purpose search engines do indicate topics of their Web search results, but only for pages which are actually included in the employed Web-page directories – no *automatic* classification is performed. To our knowledge, CatS is the first freely available meta-search engine to rely solely on classification in the result sorting task.

The rest of the paper is organized as follows. The next Section describes the CatS meta-search system: its usage and implementation, including HTML parsing, classification, and presentation of results. Section 3 outlines the text categorization experiments performed in order to choose the right combination of document representation, feature selection method and classification algorithm. Section 4 discusses the system in the context of related work on (meta-)search engines, and the last Section concludes with an analysis of the pros and cons of the chosen approach, and points out relevant directions for future work.

---

[1] In the meantime, a couple of D93 guitars have appeared on online shopping sites, but the vast majority of results still refer to models of electronic devices and obscure e-mail addresses.

2 The CatS Meta-search Engine

CatS operates by forwarding the user query to a major Web search engine, and displaying the returned results together with a tree of topics which can be browsed to sort and refine the results. Figure 1 shows the top 100 results for query 'animals england,' returned from Yahoo, displayed according to their original order of relevance in root category *All*, and Fig. 2 depicts the portion of the results classified into category *Arts* $\rightarrow$ *Music*. Therefore, this model of presentation, from a user interface point of view, can be considered an extension of the standard relevance list.

2.1 CatS Structure

The system is implemented as a Java servlet running on an Apache Tomcat server. The basic structure and flow of information is illustrated in Fig. 3. The user posts a query (1) and chooses the search engine to be used via a simple HTML form. The CatS servlet reads the query, forwards it verbatim to the search engine, and retrieves the results in HTML format (2). It then parses the HTML, extracting the title, hyperlink and excerpt of each result. The results are then classified based on their titles and excerpts, and the category tree (3), together with the results, is sent to the user for viewing.

HTML Parsing. An open-source HTML parsing library simply called HTMLParser (http://htmlparser.sourceforge.net/) was used to extract the titles, links and excerpts of search results, using its convenient filter mechanism. The filter for Yahoo search results, which builds a `NodeList` data structure of `A` tags containing titles and links, and `DIV` tags with the excerpts, is defined as follows

```
new OrFilter(
    new AndFilter(
        new TagNameFilter("A"),
        new HasAttributeFilter("class", "yschttl")),
    new HasAttributeFilter("class", "yschabstr"));
```

The list is built from `A` tags which have the attribute 'class' set to the value 'yschttl,' and any tags (in this case `DIV`s) with 'class' set to 'yschabstr.' Concrete text is then trivially extracted from the formed list of alternating `A`s and `DIV`s.

The filter approach is robust to changes in HTML content of search results pages that are unrelated to the properties described above. Even if a change breaks the parser, the administrator is notified and the filter can quickly be adjusted using the visual tool for filter construction provided with HTMLParser. Currently, the system's parsers support Yahoo, Teoma and AllTheWeb search engines.

Classification. The categories for sorting search results are based on topics from the dmoz Open Directory (as of July 2, 2005), and are organized in a hierarchical manner. Two levels were considered enough for the sorting task – eleven top-level categories were chosen, namely *Arts*, *Business*, *Computers*, *Games*, *Health*, *Home*, *Recreation*, *Science*, *Shopping*, *Society* and *Sports*; and an additional category *Other*

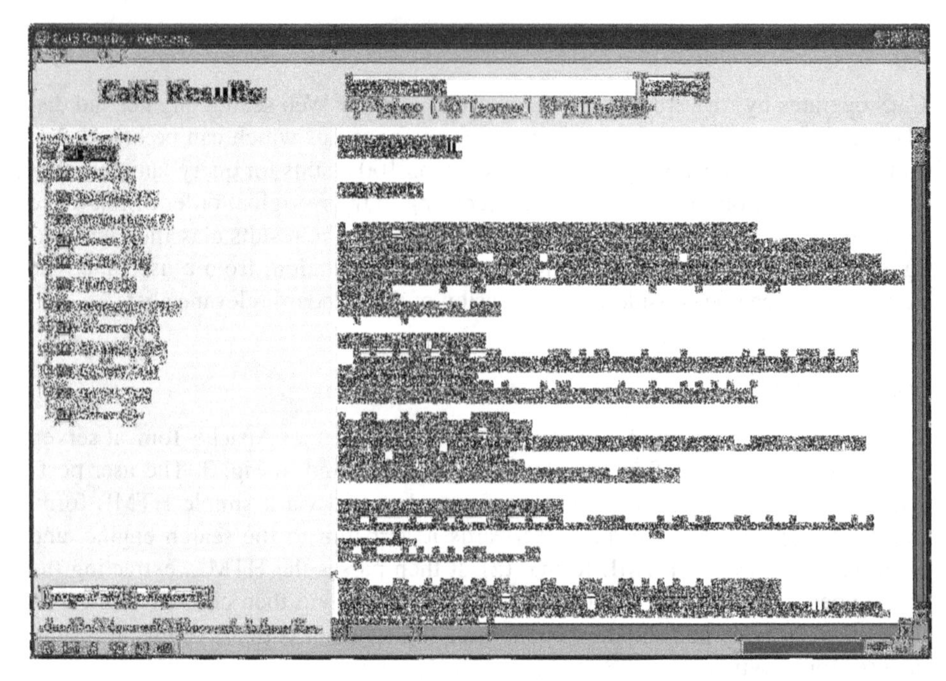

Fig. 1. All results displayed by CatS for query ‘animals england’

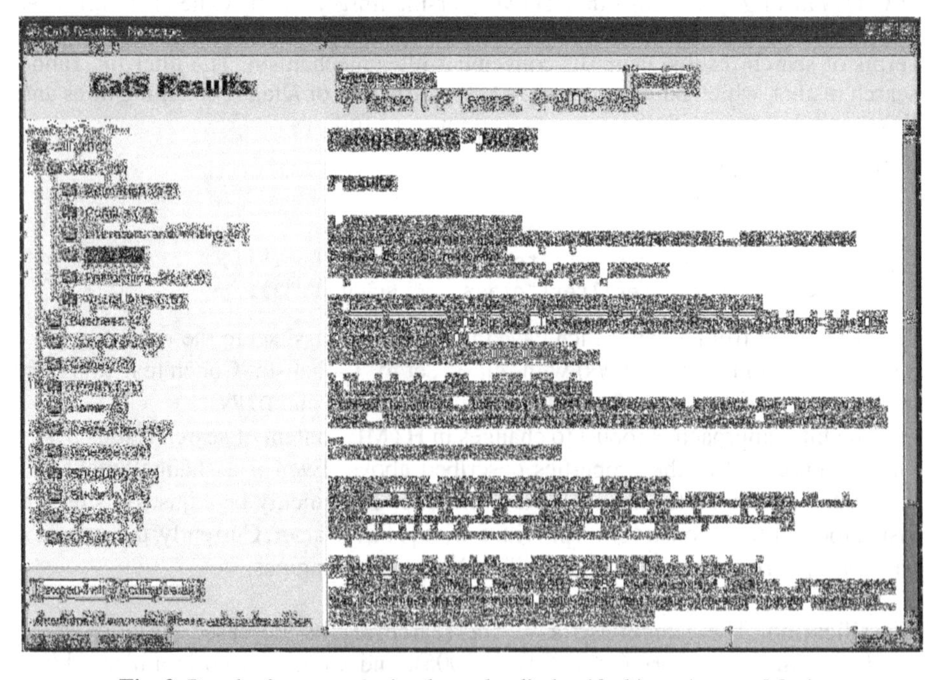

Fig. 2. Results for query ‘animals england’ classified into *Arts* → *Music*

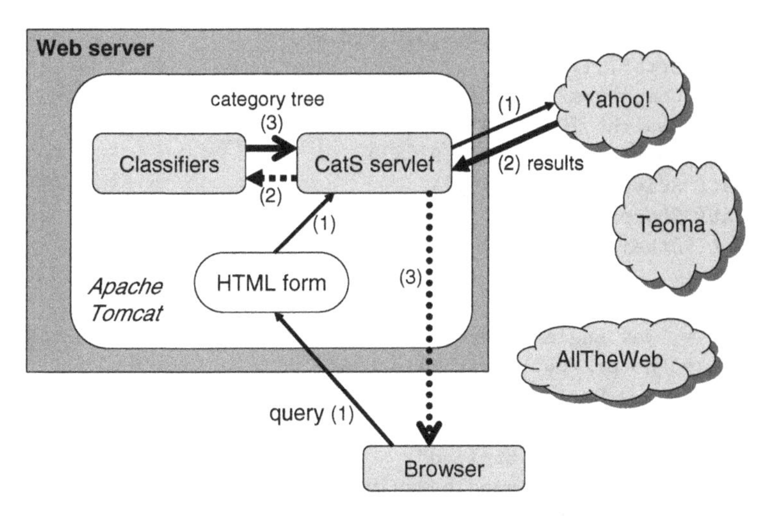

Fig. 3. CatS structure and information flow

was added for unclassified documents. The second level of dmoz contains simply too many categories to be employed in unmodified form, therefore many categories needed to be merged or discarded. The criteria that were followed when performing these operations were the apparent relatedness of certain topics (like *Artificial Intelligence* and *Robotics*), their specificity (*Volleyball, Handball* etc. were all merged into *Ball Sports*), and small number of examples (such categories were either discarded or merged with a related topic). Since English is a highly predominant language on the Web, at this time the system focuses on English language documents only.

For each category, a binary classifier was trained using Web-page titles and descriptions from dmoz data. The employed classification algorithm is the ComplementNaiveBayes classifier from the WEKA machine learning environment [24]. The document representation, feature selection method and classification algorithm were not chosen ad hoc, but were based on a controlled set of experiments (see Section 3).

During classification, the first step is to decide which results belong in each top-level category. Then, if a category is populated by more than 10 results, classification is done at the second level. Since the classifiers are binary, each search result may belong to multiple categories, or none, in which case it is sorted into *Other*.

The Category Tree. Upon classification, the servlet generates a JavaScript tree data structure containing the names of all categories, and array indexes of results classified into each. Figure 4 shows the category tree of 10 results returned by Yahoo for query 'animals england,' and a sample entry for arrays of tiles, links and excerpts. The tree data, together with the arrays, is embedded into the header of the HTML page presented back to the user, and the category tree is displayed in the left frame of the page using Treeview (http://www.treeview.net/) – a JavaScript tree menu library.

Using this approach, all browsing is done "offline" on the user's computer, maximizing the responsiveness of the system, and freeing the Web server from session management and additional query processing.

```
tree = new Cat("All",[0,1,2,3,4,5,6,7,8,9],
[  new Cat("Arts",[1,2],[]),
   new Cat("Computers",[2],[]),
   new Cat("Health",[0,1,3,4,7,9],[]),
   new Cat("Recreation",[1,3,5,9],[]),
   new Cat("Science",[0,5,7],[]),
   new Cat("Shopping",[2,9],[]),
   new Cat("Other",[6,8],[])
]);
  . . .
title   [0] = "New England Physical Therapy for Animals > About NEPTA";
link    [0] = "http://www.pt4animals.com/";
excerpt[0] = "New England Physical Therapy for Animals provides...";

title   [1] = "Animals in England";
link    [1] = "http://www.woodlands-junior.kent.sch.uk/customs/...";
excerpt[1] = "... Animals of England and the rest of Britain. ...";

title   [2] = "Animals CD's and Videos";
link    [2] = "http://www.chordsandtab.com/animals_cd's_and_videos.htm";
excerpt[2] = "Animals CD's and Videos brought to you by Chords And...";
```

Fig. 4. JavaScript data structures generated for classified search results

3 Text Categorization Experiments

Prior to implementing the meta-search engine, two rounds of text categorization experiments were performed, in order to determine the best combination of classifiers, document representations and feature selection methods for the task of sorting search results. The experiments involved five classifiers from WEKA: ComplementNaiveBayes – a variant of the classic Naïve Bayes algorithm [21, 8], SMO – an implementation of Platt's Sequential Minimal Optimization method for training Support Vector Machines [15, 7], VotedPerceptron – introduced by Freund and Schapire [5], IBk – a form of k-Nearest Neighbor [1], and J48 – an implementation of revision 8 of the C4.5 decision tree learner [17].

The first round of experiments dealt with document representations – variations of the bag-of-words model – and included all possible combinations of stemming (using the Porter Stemmer [16]), normalization, term frequencies and their logarithms, and inverse document frequency transformations. Stopword elimination (using the standard stopword list from [22]) was done as a preprocessing step for all texts.

The second round concentrated on determining the best feature selection method and reduction rate, and included chi-square, information gain, gain ratio, ReliefF and symmetrical uncertainty criteria [23, 11, 9, 24]. All experiments were performed in 5 runs of 4-fold cross validation on datasets extracted from dmoz, and the values of standard evaluation measures – accuracy, precision, recall, F_1 and F_2 [23] – were compared using the the corrected resampled t-test [12] supplied with WEKA.

Table 1. Performance of classification (in %) using the best document representations on the *Home* dataset without feature selection, together with improvements over the worst representations (statistically significant ones are in **boldface**)

	CNB	SMO	VP	IBk	J48
Accuracy	82.56 (**5.26**)	83.19 (1.67)	78.38 (**5.12**)	74.93 (**21.96**)	71.77 (**3.64**)
Precision	81.24 (**8.66**)	85.67 (**3.86**)	80.45 (**7.85**)	71.32 (**14.32**)	90.24 (1.60)
Recall	83.91 (1.81)	78.93 (3.80)	74.06 (0.96)	81.66 (**45.20**)	47.59 (**10.59**)
F_1	82.48 (**3.64**)	82.07 (2.17)	77.02 (**4.23**)	76.07 (**33.90**)	62.12 (**9.09**)
F_2	83.31 (2.19)	80.14 (3.30)	75.20 (2.16)	79.31 (**39.72**)	52.48 (**10.41**)

The first round of experiments demonstrated that there are statistically significant differences between document representations for every classifier with at least one measure. Table 1 illustrates this finding for one extracted dataset in which positive examples are taken from dmoz's *Home* category, and negative examples uniformly (in a stratified fashion) from all other first-level topics.

The overall result of the experiments, concerning the problem of sorting search results, was that the combination of the normalized term frequency representation, with stemming, but without any feature selection method, and the ComplementNaiveBayes classifier was best suited for the task. For reasons similar to those in [11], the F_2 measure was valued most, which gives precedence to recall over precision, i.e. it prefers false positives to false negatives. What this means for categorization of search results is that it is more desirable to overpopulate categories to a certain extent, than leave results unclassified in the *Other* category. Furthermore, ComplementNaiveBayes exhibited an order of magnitude faster training times than all other classifiers, and was among the quickest at classification as well.

Detailed accounts of the experiments are presented in [20, 18].

4 Related Work

As for general-purpose search engines, information available on meta-search engines is rather sparse, but research has been gaining momentum in recent years. SnakeT (http://snaket.di.unipi.it/) is the most recent implementation of a meta-search engine which sorts results by clustering Web-page snippets[2], and also provides Web interfaces for books, news and blog domains [4]. Carrot2 (http://carrot2.sourceforge.net/) is an open-source "research framework for experimenting with automated querying of various data sources (such as search engines), processing search results and their visualization," which also relies on clustering [13]. CiiRarchies (http://www.cs.loyola.edu/~lawrie/hierarchies/) is a hierarchical clustering engine for Web search results described in [10], while Highlight (http://highlight.njit.edu/) provides the option to sort results at the outermost level using classification, before resorting to clustering for deeper levels of the topic hierarchy [25]. All these systems provide Web interfaces *and* published results, unlike leading commercial clustering

[2] A Web-page snippet consists of the page's title, link and excerpt.

engines like Vivissimo.com and KartOO.com, which keep the details about the algorithms they employ in hiding. According to [4], no meta-search engine, research or commercial, has outperformed Vivissimo with regards to the quality of generated clusters, and the majority of engines with known internal workings are rather slow, which limits their usefulness in practice.

A pure classification approach to sorting search results was described in [3], where a closed-environment study involving Internet users of different profiles showed that their category style of presentation was generally preferred over the list model. The authors chose to break up the list of results between categories right at the initial displaying of results, showing only several examples from each category, based on which the user could choose to "expand" a particular topic. This makes their approach a *replacement*, rather than an *extension* of the standard relevance list. This differentiates it from the approach taken in CatS, which is more in line with the majority of clustering meta-search engines mentioned above.

Another way to utilize classification in sorting search results is by means of focused, or *vertical* search: the user is first asked to navigate and fix a particular category of interest, and then post a query [23]. The results would not only be restricted to the chosen (and related) categories, but also ranked in accordance with that particular line of search, instead of using only global link graph analysis [14]. But today, when "googling" is becoming a synonym for "searching the Web," changing the user's habits may prove a difficult, if not an impossible task. Thus, asking the user to *first* fix a topic (by choosing a topic explicitly, or choosing the search engine itself) and *then* post a query, as required by the majority of vertical search engines (like LookSmart.com), can be viewed with reluctance by many users.

5 Conclusion

Using classification for sorting search results has both its advantages and disadvantages. Categories are chosen by humans, rather than constructed by some clustering algorithm, so it is clear what their meanings are, and the user will most probably find browsing comfortable and intuitive. But, this clarity makes classification errors obvious, possibly reducing the user's faith in the system. This effect was emphasized for CatS by our choice to prefer overpopulating categories over leaving results unclassified. Furthermore, a fixed set of topics may not suit every query: for example, CatS sorts almost all results for 'java servlets' under *Computers* → *Programming*, which is unlikely to be particularly useful if computer programming is precisely what the user is interested in. Clustering approaches to sorting search results tackle this issue by dynamically generating topics from the list of results for every individual query. But, while for some queries the identified topics look natural and helpful, for others they may seem rather ambiguous – unclear as to what makes them "topics."

The categories currently used by the presented meta-search system were chosen and engineered by the authors in a rather ad hoc fashion. Further work on the system may include a more systematic evaluation of the practical utility of every individual category, based on user feedback and usage statistics. Fine-tuning of classifiers and adding some simple heuristics to improve classification performance is also on the agenda. Aggregating results from several search engines, instead of (or in addition to the option of) using only one, presents another possible path for development. Introducing additional levels of categories could improve their suitability to queries, with the danger of compromising classification accuracy at deeper levels of the hierarchy.

Perhaps the most interesting and novel direction of research would be to investigate a mixture of classification and clustering techniques for sorting results, presenting a marriage of the clarity and familiarity of human-engineered topics, with the dynamic nature and adaptability of automatically generated clusters. To our knowledge, in the research community this has only been attempted with Highlight, but its choice to use classification only at the first category level seems to be a rigid way to accomplish this. Delving into the Semantic Web arena, the marriage might be achieved by utilizing ontologies both for Web pages (like dmoz) *and* human users (with well-identified simple, but universal needs, intentions, profiles etc.). The ontologies could then impose constraints on the execution of a clustering algorithm, or subsequent mapping of independently generated clusters to ontologies could be performed using classification.

In the long run, on the scale from a feasibility demonstration to a full-fledged search engine, CatS leans more to the former. The ideal scenario would be the integration of classification technology into existing, or newly crawled Web-page collections like the ones maintained by dominant general-purpose search engines. Then, classification performance would be enhanced not only by the availability of full Web-page texts, but also by consulting the link data used to calculate page rankings. Furthermore, in the spirit of vertical search, results could be re-ranked depending on the category being viewed, but without requiring the user to pre-determine the topic of the search.

As it is, CatS presents a unification of the style of search result displaying which is employed by many clustering meta-search engines (like Vivissimo), with text categorization techniques. Using the category tree as an extension of the relevance list provides a way to refine the search *on need*, by browsing topics which are clear and familiar, that way offering the user a non-imposing, but potentially powerful tool for locating needed information quickly and efficiently.

References

1. D. Aha, D. Kibler, and M. K. Albert. Instance-based learning algorithms. *Machine Learning*, 6(1):37–66, 1991.
2. D. Butler. Souped-up search engines. *Nature*, 405:112–115, May 2000.
3. H. Chen and S. T. Dumais. Bringing order to the Web: Automatically categorizing search results. In *Proceedings of CHI00, Human Factors in Computing Systems*, pages 145–152, 2000.
4. P. Ferragina and A. Gulli. A personalized search engine based on Web-snippet hierarchical clustering. In *Proceedings of WWW05, 14th International World Wide Web Conference*, pages 801–810, Chiba, Japan, 2005.

5. Y. Freund and R. E. Schapire. Large margin classification using the perceptron algorithm. *Machine Learning*, 37(3):277–296, 1999.
6. P. Jackson and I. Moulinier. *Natural Language Processing for Online Applications: Text Retrieval, Extraction and Categorization*. John Benjamins, 2002.
7. S. S. Keerthi, S. K. Shevade, C. Bhattacharyya, and K. R. K. Murthy. Improvements to Platt's SMO algorithm for SVM classifier design. *Neural Computation*, 13(3):637–649, 2001.
8. A. M. Kibriya, E. Frank, B. Pfahringer, and G. Holmes. Multinomial naive bayes for text categorization revisited. In *Proceedings of AI2004, 17th Australian Joint Conference on Artificial Intelligence*, LNAI 3339, pages 488–499, Cairns, Australia, 2004.
9. I. Kononenko. Estimating attributes: Analysis and extensions of RELIEF. In *Proceedings of ECML97, 7th European Conference on Machine Learning*, pages 412–420, 1997.
10. D. Lawrie and W. B. Croft. Generating hierarchical summaries for Web searches. In *Proceedings of SIGIR03, 26th ACM International Conference on Research and Development in Information Retrieval*, Toronto, Canada, 2003.
11. D. Mladenić. *Machine Learning on non-homogenous, distributed text data*. PhD thesis, University of Ljubljana, Slovenia, 1998.
12. C. Nadeau and Y. Bengio. Inference for the generalization error. *Machine Learning*, 52(3), 2003.
13. S. Osiński and D. Weiss. A concept-driven algorithm for clustering search results. *IEEE Intelligent Systems*, 20(3):48–54, 2005.
14. L. Page, S. Brin, R. Motwani, and T. Winograd. The PageRank citation ranking: Bringing order to the Web. Unpublished manuscript, 1998.
15. J. Platt. Fast training of Support Vector Machines using Sequential Minimal Optimization. In B. Scholkopf, C. Burges, and A. Smola, editors, *Advances in Kernel Methods – Support Vector Learning*. MIT Press, 1999.
16. M. F. Porter. An algorithm for suffix stripping. *Program*, 14(3):130–137, 1980.
17. R. Quinlan. *C4.5: Programs for Machine Learning*. Morgan Kaufmann Publishers, 1993.
18. M. Radovanović. Machine learning in Web mining. Master's thesis, Department of Mathematics and Informatics, University of Novi Sad, Serbia and Montenegro, 2006. To appear.
19. M. Radovanović and M. Ivanović. Search based on ontologies. In *Proceedings of PRIM2004, 16th Conference on Applied Mathematics*, Budva, Serbia and Montenegro, 2004.
20. M. Radovanović and M. Ivanović. Document representations for classification of short Web-page descriptions. To appear, 2006.
21. J. D. M. Rennie, L. Shih, J. Teevan, and D. R. Karger. Tackling the poor assumptions of naive Bayes text classifiers. In *Proceedings of ICML03, 20th International Conference on Machine Learning*, 2003.
22. G. Salton, editor. *The SMART Retrieval System: Experiments in Automatic Document Processing*. Prentice-Hall, 1971.
23. F. Sebastiani. Machine learning in automated text categorization. *ACM Computing Surveys*, 34(1):1–47, 2002.
24. I. H. Witten and E. Frank. *Data Mining: Practical Machine Learning Tools and Techniques*. Morgan Kaufmann Publishers, 2nd edition, 2005.
25. Y.-F. Wu and X. Chen. Extracting features from Web search returned hits for hierarchical classification. In *Proceedings of IKE03, International Conference on Information and Knowledge Engineering*, Las Vegas, Nevada, USA, 2003.

A Decision Tree Framework for Semi-Automatic Extraction of Product Attributes from the Web

Lior Rokach[1] and Roni Romano[2], Barak Chizi[2], Oded Maimon[2]

[1] Department of Information Systems Engineering, Ben-Gurion University of the Negev, Beer-Sheva, Israel liorrk@bgu.ac.il
[2] Department of Industrial Engineering, Tel Aviv University, Israel ronir,barakc,maimon@eng.tau.ac.il

Abstract. Semi-Automatic extraction of product attributes from URLs is an important issue for comparison-shopping agents. In this paper we examine a novel decision tree framework for extracting product attributes. The core induction algorithmic framework consists of three main stages. In the first stage, a large set of regular expression-based patterns are induced by employing a longest common subsequence algorithm. In the second stage we filter the initial set and leave only the most useful patterns. In the last stage we represent the extraction problem (in which the domain values are not known in advance) as a classification problem and employ an ensemble of decision trees. An empirical study performed on a real-world extraction tasks illustrates the capability of the proposed framework.

1 Introduction

Web content mining refers to mining, extraction and integration of useful data, information and knowledge from web page contents. Reliable extraction of product attributes from the web is a well-known issue which is crucial for comparison-shopping agents [8].

Due to the variety of sources and the dynamic nature of the internet, shopping comparison services can not rely on manual development of extraction agents and an automatic framework for developing these agents is required. This is especially true because most of the information on the web is stored in a semi-structured manner and sometimes the information is even provided as free-text. Different sites have different structures and might refer to different attributes of the same product.

A general engine for extracting product attributes should be able to extract information from structured pages (in which the entities' attributes values are

Mark Last et al. (Eds.): Advances in Web Intelligence and Data Mining (SCI) **23**, 201-210 (2006)
www.springerlink.com

composed in a certain structure) and from non-structured pages (for example if the entities' attributes values are included in some free-text sentences). Roughly speaking there are two approaches for *structured data extraction*: *wrapper induction* and *automatic data extraction*.

In wrapper induction a human user manually label the target items in a few training Web pages. This set of labeled pages is given to an induction algorithm that learns extraction rules or patterns. The rules are applied to extract target items from other pages. There are several systems that use wrapper induction methodology [1].

In automatic data extraction a set of positive examples is given. The idea is to generate from these pages, the extraction patterns. Thus no manual labeling is required, but instead a set of positive pages of the same template is required [2].

KnowItAll is a Web-based, unsupervised domain-independent information extraction system [3]. Given a set of relations of interest, it instantiates relation-specific generic extraction patterns into extraction rules which find candidate facts. Then it outputs a probability associated with each fact by using a Naive Bayes classifier. Integrating information from multiple web sources is complicated because the same data objects can exist in inconsistent text formats across sites or because a search for a particular object can return multiple results.

When developing a general engine for product data extraction no assumption should be made regarding the attributes because each product class has different attributes. Moreover the same html page usually contains additional information other than the interested product attributes (such as information regarding the store itself or information regarding similar or complementary products) and it is possible that some pages contain errors. Previous works [5] that have extracted product attributes by solving classification problems, have assumed that the attributes value domains are known in advance.

The goal of this research is to develop a simple learning framework for extracting product attributes. Our assumption is that arranging a very large number of simple extraction patterns in a hierarchical manner by using a decision tree induction algorithm can provide a high accuracy.

2 Proposed System

We are given a training set which consists of instances that describe different product instances taken from a single product class (for instance laptop computers). Each product is characterized by the same set of attributes (i.e. the class properties). In addition to that a non empty set of URLs is attached to each product. Each product instance may have a different number of URLs from different sites. The URL contains an html document that provides information about the product. Note that it is not necessarily true that all product attributes can be extracted from a single document.

2.1 Preprocessing

The preprocessing step consists of the following three phases:

Normalization The main goal of normalization is to convert all writing styles of the attribute values into a single style. For instance a number and a capital letter located in the same tag and possibly separated by one or more white space or special sign (such as comma), is converted into a number and text having one space only (i.e. 512MB will be represented as a 512 MB).

Zoning The learning algorithm presented below requires that the html document will be broken into extraction zones. A good strategy for breaking the HTML into its ingredients is to search for the smallest string that contains the extracted value and much as possible of relevant support tokens (such as the attribute name) and less as possible of non-relevant tokens.

Labelling Because the training set does not provide a specific labelling of the product attributes in each URL but only a general list of attributes. For this reason, we are required to associate the zones with the relevant label. For this purpose we go over all products. For each product we go over all its URL documents. Each zone in the document is labelled according to the product attribute to which it refers. Note that some zones might have no match for a certain attribute (we refer to this class as "undefined") while other may refer to more than a single property.

2.2 Learning of Regular Expression Patterns

We suggest the following regular expression learning procedure. This procedure consists of three phases: Creating a set of extraction patterns over the labeled text; Patterns selection; and Creating a classifier that employs the patterns in a hierarchical manner.

Creation of Extraction Patterns

Each extraction pattern (p) consists of three items:

- A regular expression denoted as $p.t$. Each zone will be matched to this regular expression. It may contain one or more groups of tagged expressions to be used by the extraction expressions.
- Associated product attribute (denoted as $p.r$).
- An extraction expression to be used for extraction the value assuming that pattern has been matched (denoted as $p.c$).

We identify several types of extraction patterns. The simplest set of patterns will be based on the attribute name followed by a value. This kind of patterns can be used to extract the value for other products of the same class and from the same URL prefix. This pattern is relevant only to tabular structure.

Another simple set of patterns will be based on the actual values in each attribute. This can be useful for attributes that have small set of values. For instance the fact that the token Intel or AMD appears in the text indicates that this is the manufacture. However in some cases the same value can be used in different attributes (for example the value 333 MHz can be used for the attribute "RAM Speed" or for the attribute "Bus Speed"). Moreover the value may appears in the text but it refers to unhandled attribute, for instance the value "1024MB" in the zone "User upgradeable up to 1024MB" refer to the potential RAM size but not the actual RAM size.

Additional set of patterns will be based on the generalized attribute name. The basis for discovering a regular expression is a method that compares two strings with the same meaning (concept). By employing the LCS (*longest common subsequence*) algorithm (Myers, 1986) for each part in the zone (i.e. the substrings that precede the value and the substrings that follow the value), we can obtain a regular expression that fits these two strings is created. For instance if we obtain the following three zones of the same attribute (Installed RAM):

```
<TR><TD>Installed RAM:</TD><TD>512 MB</TD></TR>
<TR><TD>Memory Size</TD><TD>1024 MB</TD></TR>
<TR><TD>Installed Memory</TD><TD>1024 MB</TD></TR>
```

Each pair of zones is combined using the LCS algorithm that was revised to compare tokens as opposed to comparing characters in its classic implementation. Moreover whenever there was only insertion (or only deletion) we added a wild card string of minimum length of 0 and maximum length of the inserted string (including the leading and trailing spaces). On the other hand whenever there was simultaneously insertion and deletion, we added a wild card string with the minimum length of the shortest string and maximum length of the largest string (without leading and trailing spaces because they are part of the common substring). In this case the following patterns are created:

```
<TR><TD>Installed .{4,6}</TD><TD>\([0-9 MGB]{6,7}\)</TD></TR>
<TR><TD>.{0,10}Memory.{0,5}</TD><TD>\([0-9 MGB]{6}\)</TD></TR>
<TR><TD>.{11,14}</TD><TD>\([0-9 MGB]{6,7}\)</TD></TR>
```

For instance by applying this algorithm on zone 1 and zone 3 we obtain the first pattern. The substring "\([0-9 MGB]{6,7}\)" indicates a group expression that refers to the extracted value (in this case "512 MB" or "1024 MB"). The set of possible characters ([0-9 MGB]) in this group is defined according all values of this attribute in the training set. The extraction command will be in this case \1. This indicates that the first matched group is the output. The third pattern presented is too general and obviously will not contribute to the extraction's performance. This kind of patterns will be filtered out as part of the pattern selection procedure that is presented in the following section.

Note that the same approach can be used to extract information from zones that contain unstructured data. For instance given the following two zones:

```
This superb laptop computer has 128 MB of fast SDRAM memory.
```

```
Acer has packed on an incredible 512 MB of onboard DDR
memory along with a huge 100GB hard drive.
```

If the anchor attribute is still the memory size, then in this case we execute the Longest Common Subsequence by comparing the substrings that precede the value and the substrings that follow the value and get the following regular expression:

```
.{32,33}\([0-9 MGB]{6}\) of .{10,13} memory.{1,36}
```

Another set of patterns will be based on the generalized attribute value. For instance given the above example we obtain the following patterns:

```
\([0-9][0-9][0-9][0-9]?\) MB
\([0-9][0-9][0-9]\) MB
```

The first pattern is generated using zone 1 and zone 2. Because in zone 1 the memory size has three digits and in zone 2 it has four digits then the combined group have three mandatory digits and additional one optional digit. The rest of the pattern is based on the common string which represents the units in which the value is measured (in this case MB). Additional set of pattern is created by combining the patterns of type 3 and 4, for instance:

```
<TR><TD>Installed .{4,6}</TD><TD>\([0-9][0-9][0-9][0-9]?\)
MB</TD></TR>
```

Patterns Selection

Obviously there are many patterns that can be created (each pair of zones with the same attribute). Moreover each pattern will have a different generalization capability. Thus we need a criterion to select the pattern that best differentiate one attribute from the other.

We are using the following two heuristic procedures sequentially in order to obtain the best discriminating patterns:

(a) Generalizing from similar patterns– Many of the generated patterns differ only in the distance of important keywords from the seed attribute. Grouping such patterns taking only maximum distances eliminates many patterns;

(b) Using existing feature selection methods. Feature selection is the process of identifying relevant features in the dataset and discarding everything else as irrelevant and redundant. In this research we are using a non-ranker filter feature selection algorithm. Filtering means that the selection is performed independently of any learning algorithm. Non-ranker means that the algorithm does not score each pattern but provides only which pattern is relevant and which is not.

Two measures are used sequentially for evaluating the patterns as part of the feature selection process. The first one is the information gain and its main goal is to find the patterns that can identify the most relevant zones for a certain attribute.

Additionally to the information gain, we evaluate the likelihood l_t of a certain pattern to extract the correct value. The aim of this measure is to find the pattern that can extract the correct values for a certain attribute. Given a certain pattern t for attribute r and a corresponded extraction command c, this likelihood can be simply estimated by dividing the number of products in which the pattern $p.t$ was matched and the corresponded command $p.c$ has extracted the correct value of attribute $p.r$ divided by the number of products in which pattern $p.t$ was matched.

Creating the Classifier Extractor Tree

The filtered patterns for the last section are then fed into a decision tree inducer which creates a classification decision tree. Note that we are creating a different decision tree for each product attribute.

For this purpose we need to represent our data as regular training set, such that existing decision trees learning methods can be used. Note however that our problem is not a classification one nor a regression one, but an extraction one. This is because we do not know in advance all possible nominal values that can be extracted (for instance the name of the companies that produced the product).

Each pattern represents a different column in the dataset. Note that for each product attribute we are creating a different dataset because each attribute has a different set of patterns (recall that each pattern is associated with only one product attribute).

Each zone is represented by one or more rows according the following procedure. Lets say that the we are looking for the attribute r and we are given a filtered set of patterns $p_1,\ldots,p_n$ to be checked . Given a zone s_j we check all available patterns. For each pattern p_i that is matched we extract the value according to its extraction expression. For each distinct extracted value v_k we create a different row in the training set. In this row each of the column representing the patterns will have either "1" (if the pattern is matched and it extracted the value v_k), "0" (if the pattern is matched but it extracted a value that is different from v_k) or "-1" (if the pattern is not matched). The value of the class column will have either "1" (if the v_k is equal to the real attribute value) or "0" (otherwise).

The following example illustrates this procedure. We are given the zone:

```
<TD>Installed RAM:</TD><TD>256 MB (512 MB optional)</TD>
```

We are interested to extract the "installed RAM" attribute and we are given the following three patterns:

```
\([0-9][0-9][0-9]\) MB
<TD>Installed .{4,6}</TD><TD>\([0-9][0-9][0-9][0-9]?\) MB
<TD>.{11,14}</TD><TD>\([0-9][0-9][0-9]\) MB</TD></TR>
```

The extraction expression is "\1 MB" for all patterns. By using these patterns on the given zone we extract the following two values v_1="256 MB" from p_1 and p_2 and the value v_2="512 MB" from p_2. Note that p_3 has not been matched in this zone. In this case the zone will be represented in the training set as two rows (see Table 1). The first row refers to the value "256 MB" (which is the correct one, thus $y = 1$) and the second row refers to the value "512 MB".

Table 1. Illustration the mapping procedure

p1	p2	p3	y
1	1	-1	1
0	1	-1	0

Using a decision tree as a classifier in this case has several advantages: (1). the zone is not classified based on a single pattern, but on set of patterns, i.e. this classifier can be used to indicate that a zone is classified to a certain label only if this value was extracted by two patterns and was not extracted by a third pattern. This is more expressive than the classical approach in which the classification is based on a single pattern; (2). As opposed to other classifiers (such as neural networks) the meaning of the classifier can be explained.

2.3 Using Patterns for Extracting Information of New Products

Given a new product with a set of URLs, we first perform the abovementioned pre-processing procedure. Then for each attribute needed to be extracted we employ its corresponded decision tree in the following way. Each zone s_j (from all URLs) is matched against the list of patterns. Then a set of rows is created according to the procedure presented above, i.e. for each distinct value v_k that was extracted, a different row is created. We then execute each row against the decision tree and obtain the probability $\Pr_{j,k}$for obtaining the corresponded value v_k by zone s_j.

Following [3] we are using Naïve Bayes approach to combine all probabilities from all zones. Then we normalize the result to get the posteriori probability. The a-priori probability is calculated based on the frequency of the value v_k in the training set. However using the frequency as-is will typically over-estimate the probability. In order to avoid this phenomenon it is useful to perform the Laplace correction. According to Laplace's law of succession, the probability of having v_k which has been observed n times of m instances is $(n+1)/(m+d)$ where d denotes the distinct count of the extracted values. Thus, value that has never been seen in the training set before still has a small but positive a-priori probability.

3 Experimental Study

The potential of the proposed methods for use in real word applications was studied. In this study we have examined the proposed system in two class of products: Strollers and PC Laptops. Strollers are characterized by attributes such as "Overall Weight", "Number Of Harness Points" etc. PC Laptops are characterized by attributes such as Processor Type, Processor Speed etc.

For each product class, there were 600 available instances. Each instance may have one or more URLs from which the attributes can be extracted. On average there were 7.7 URLs for each instance. The attributes in every product instance have been extracted manually. Almost 20% of the values are missing. It is also estimated based on previous statistics that data is noisy and only 90% of the values are correct. We performed several experiments in order to determine the classifier sensitivity to the following parameters: (a) Different training set sizes; (b) the effect of using feature selection (c) the effect of using ensemble of decision trees – it is well known that ensemble can improve accuracy. We examined here in what extent it can improve the results.

3.1 Measures Examined

The first measure used is the well-known accuracy rate, indicating the portion of instances' attributes that has been extracted with the correct value. Another measure that was examined is the so-called "Non-Extracted Ratio" portion of instances' attributes that has not been extracted at all. Note that extracting an incorrect value is considered to be more severe problem then not extracting a value at all.

For examining the above measures the product instances have been split into two sets: (1) Training set (2/3); (2) Test set (1/3). All results presented below refer to the test set. Additionally missing values in the test set has been excluded from the evaluation. Only half of the instances in the training set have been used for the creation of the extraction patterns. While for creating the classifier extractor tree we have used all instances in the training set. The motivation behind this procedure is two-folds. First, we need that our framework will be capable to extract from domain values that have not been defined in advance. Thus the creation of the classifier extractor should include instances with domain values that have not been seen in the phase of pattern creation. By that it simulates the real extraction task we are facing. The second reason relies on the fact that if many instances are used in the pattern creation then more extraction patterns are created (as potentially each pair of instances can create several patterns for each product's attribute). This can lead to the problem known as "curse-of-dimensionality". By using a smaller training set we can reduce this effect.

3.2 Algorithm Used

We have used the well-known C4.5 algorithm as the main decision tree inducer [9]. For creating the ensemble of decision trees we used the AdaBoost algorithm [4].

In this paper we are using the Correlation-based Feature Subset Selection (CFS) as a subset evaluator [6]. CFS Evaluates the worth of a subset of attributes by considering the individual predictive ability of each feature along with the degree of redundancy between them. Subsets of features that are highly correlated with the class while having low inter-correlation are preferred. The search method used is the Forward Selection Search with Gain Ratio as the scoring criterion.

3.3 Results

Table 2 presents the mean Accuracy and Non-Extracted Ratio obtained by the Regular expressions classifier in each one of the products. The results indicate that the regular expressions classifier usually obtain high accuracy. Moreover the portion of the incorrect extracted values is relatively small. In the product Stroller about 6.8% of the values (100%-87.2%-6%) have been extracted incorrectly. While in product Laptop about 5.4% of the values have been extracted incorrectly.

Table 2. Overall Results

Method	Non-Extracted Ratio	Accuracy
Stroller	6%	87.2%
Laptop	4%	90.61%

We compared the mean accuracy obtained by the proposed method before and after CSF features selection as described above. The filter eliminates on average 92% of the features while achieving substantial accuracy gain. Moreover the filtering improves the accuracy by 6% on average. This can imply that the "curse-of-dimensionality" problem does exist in this problem.

We also examined how the ensemble size affects the accuracy. We have checked seven ensemble sizes. It seems that the accuracy of the proposed method is improved from 84.2% (single model) to 88.9 % (ensemble size of 6). Note that the optimal value is reached in the ensemble of size "6" from that value there is a minor deterioration.

4 Conclusion and Future Research

A new pattern based learning framework for extraction of product attributes from the web is presented. It has been shown that the new approach has relatively high performance. The pattern based approach manages to learn and utilize structural features and distances in addition to keywords. Further work is required to tune and refine the method. Moreover one might use other algorithms in this framework (for instance using a different decision tree algorithm or a different feature selection algorithm).

Moreover in the current implementation the extraction of each attribute has been performed independently to the other features. However in some cases we might benefit from this dependency and improve the extraction results. For instance there is a strong relationship between the attribute "Processor Type" and the attribute "Processor Speed".

References

1. Cohen W., Jensen L., and Hurst M. (2002), A flexible learning system for wrapping tables and listsin HTML documents. In Proceedings of the Eleventh International World Wide Web Conference, Honolulu, Hawai, pp. 232-241
2. Crescenzi V., Mecca G., and Merialdo P. (2001), RoadRunner: Towards Automatic Data Extraction from Large Web Sites, Proceedings of 27th International Conference on Very Large Data Bases, September 11-14, 2001, Roma, Italy, pp. 109-118.
3. Etzioni, O., Cafarella, M., Downey, D., Kok, S. Popescu, A. Shaked, T., Soderland, S., Weld, D. and Yates, A. (2005). Unsupervised named-entity extraction from the web: An experimental study. Artificial Intelligence, 165(1):91–134.
4. Freund Y. and Schapire R. E., Experiments with a new boosting algorithm. In Machine Learning: Proceedings of the Thirteenth International Conference, pages 325-332, 1996.
5. Ghani R. and Fano A. E. (2002), Using Text Mining to Infer Semantic Attributes for Retail Data Mining, IEEE International Conference on Data Mining, December 9-12, 2002. Maebashi, Japan.
6. Hall, M. Correlation- based Feature Selection for Machine Learning. Ph.D. Thesis, University of Waikato, 1999.
7. Myers E. (1986), An O(ND) difference algorithm and its variations, Algorithmica 1(2):251-266..
8. Perkowitz , M. Doorenbos, R., Etzioni, O. and Weld S. (1997). Learning to Understand Information on the Internet: An Example-Based Approach, Journal of Intelligent Information Systems 8(2), pp 133-153
9. Quinlan, J. R. C4.5: Programs for Machine Learning. Morgan Kaufmann, 1993.

A Dynamic Generation Algorithm for Meta Process in Intelligent Platform of Virtual Travel Agency

Qi Sui[1] and Hai-yang Wang[1]

School of Computer Science and Technology, Shandong University
sq@dareway.com.cn, why@sdu.edu.cn

Abstract. Intelligent Platform of Virtual Travel Agency (IPVita) is a platform which can intelligently and automatically compose all kinds of travel web services into a satisfactory travel for tourists. This paper introduced concepts of Meta Services and Meta Process, which can largely simplify the complexity of web services composition. And corresponding to these concepts, a Dynamic Generation Algorithm for Meta Process is presented instead of predefining a rigorous workflow model. And differently with some AI planning method, this approach deals with the various requirements of tourists more effectively and flexibly. Additional this approach also can resolve these similar problems in other domain.
Key words: Services Composition, Meta Service, Meta Process

1 Introduction

Web services have been are announced as the next wave of Internet-based business applications that will dramatically change the use of Internet[1]. Now tourists may get a satisfactory travel by invoking travel services one by one, however when it come to the question how to compose these services into a satisfactory travel, they become puzzled. Today a group including authors, is developing IPVita, which is short for the Intelligent Platform of Virtual Travel Agency [13], and the aim of IPVita project is to dynamically and intelligently organize the travel services as a travel process, then offer a satisfactory travel to tourists.

Nowadays, many trade or open organizes are dedicated to helping the travel industry take full advantage of the near universal access to the Internet, such as The Open Travel Alliance (OTA)[1], Travel Technology Initiative(TTI)[2], the UN/CEFACT[3]. At the same time, some research projects such as METEOR-S[4], iBOM[5] which have similar goal with IPVita are in develop-

Mark Last et al. (Eds.): Advances in Web Intelligence and Data Mining (SCI) **23**, 211-219 (2006)
www.springerlink.com

ing and gain wonderful successes. Differently with these efforts, IPVita tries to provide some novel features.

(a) In IPVita a travel process is automatically generated instead of being pre-defined by designer or user[4], which is more adept to the complex and uncertain environment in Internet.

(b) Instead of traditional planning algorithm[10][11] to compose web services straight forward, IPVita used a Dynamic Generation Algorithm for Meta Process (MPGA) to generate a meta process, and then transform it to a travel workflow model after matching for meta services with an appropriate travel service.

To provide a satisfactory travel for tourists, first problem is how to deal with so flexible users' demands and so many travel services with different forms. Paper[7] classified services based on their operating properties and presents the concept of service community, which simplified the complexity of service composition. Nevertheless, paper[7] hasn't a further discuss on how to use service community to deal with the user demands and how to create the integrated process automatically. Paper[8][9][10][11] introduced some automatically services composition methods using AI planning algorithm, which are most on the basis of the operating properties of atomic service using intelligent planning technology. But the fact that operating properties of service cannot express its functions completely, and the immense number of web services in Internet made these algorithms doubtable. In author's another paper[12], concepts of meta service and meta process were presented to simplify the complexity of travel services composition in IPVita and had a good effect. But in paper[12] a meta process model is represented by a simple n-matrix and author used first logic rules as generation conditions, which had some limits in the ability of describing a travel process and the performance of generation algorithm. In this paper, author advanced the definition of meta process and presented a new meta process generation algorithm having got more effective results in stochastic and dynamic nature environment.

The remainder of this paper is organized as follows: In Sec.2 the new definition of meta process is introduced and a short introduction about how to use meta process in IPVita is generalized. In Sec.3, we introduced the Meta Process Generation Problem at first, then an improved dynamic Generation Algorithm for Meta Process (MPGA) is introduced. In Sec. 4, we compared the result of two algorithm and analyze the time and space complexity of this algorithm. And in last section, there are the conclusion and future work.

2 Meta Service and Meta Process in IPVita

In practice, travel services can be classified into some categories according to their functions. Services in a same category have so similar function that it is very sound to look them as a virtual service. In IPVita we present a concept

of meta service[12], presents a set of similar services in operation properties.
Definition 1: Meta Service

A meta service $[S_1, S_2, S_3, .. S_n]$, is a set of services with similar operation properties.

A meta service can be seen as a delegate of some similar services. For example in travel industry, meta service of Ticket Booking delegates all kinds of bus, railway and airline ticket booking services.

The ideal of meta service illumines from Services Community[7], which classes all web services into some groups to simplify the services composition. The author of this paper advances Services Community to meta service, which is important to services registration, services discovery, tourists' requirements and process generation automatically. Using meta services, IPVita's complexity is largely decreased and reliability is largely increased.
Definition 2: Meta Process

A meta process is a tri-tuple ¡C, M, Arc¿,where

C = $[c_1, c_2, c_3, ..., c_n]$, is a set of resources container;

M = $[m_1, m_2, m_3, ..., m_n]$, is a set of meta services;

Arc = (C × M) $\bigsqcup$ (M × C), is a transition from C to M or from M to C, but noticed that those transition from C to C or from M to M is prohibit;

And there are at least one Arc from c_1 to M, there are not arc from M to c_n.

A meta process is an abstract process model composed by meta services. A meta process describes which kinds of services this travel need and which control flow of this travel process is. In fact, a meta process is a simplified Petri net whose places are empty and transitions have not been bounded with specific services. After services matching, a meta process becomes a travel process to be executed in workflow engine.

There are four kinds of basic routers in meta process: sequence, parallel, choice and loop showed in Fig.1.

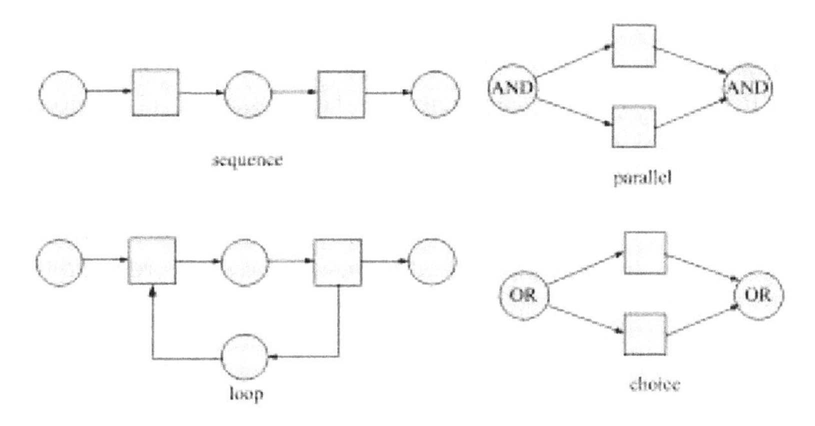

Fig. 1. 4 basic routers in meta process

Fig.2 showed the running of IPVita more detailed in paper[12]. The requirement template which lists all requirement with null value is created by requirement template generator automatically based on meta service. Then Meta Process Generator creates the meta process according to the requirement of tourists.

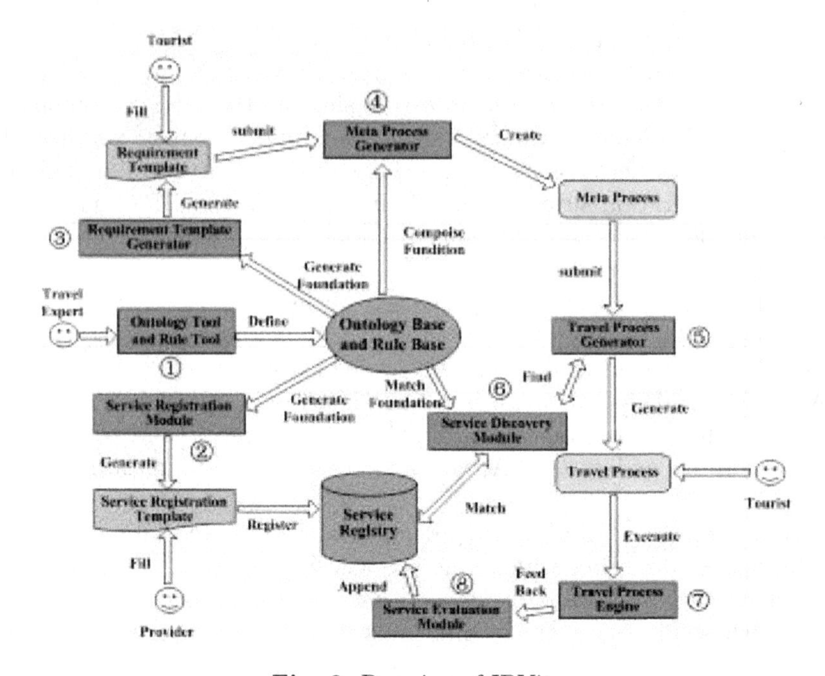

Fig. 2. Running of IPVita

In IPVita, tourists' requirement for travel is transformed to a tree name as Requirement Constraint Tree showed in Fig.3.

Requirement Constraint Tree is composed by Description Class which is consists of two parts, Key Requirements and Extended Requirements.

Key Requirements is requirements for key properties of service, corresponding to the properties in the meta service ontology.

Extended Requirements is accessional demands of tourists, which is very interesting for tourists but is not necessary to be fulfilled.

Mentioned in above paragraphs also, the most novel feature in IPVita is automatically and intelligently generating meta process according to tourists' requirements, which is named as Meta Process Generation Problem.
Definition 3: Meta Process Generation Problem.

Input: a Requirement Constraint Tree as the tourists' requirements, and a initial meta process MP_0 which all meta services are parallel;

Output: a target meta process MP_t using an dynamic generation algorithm for meta process.

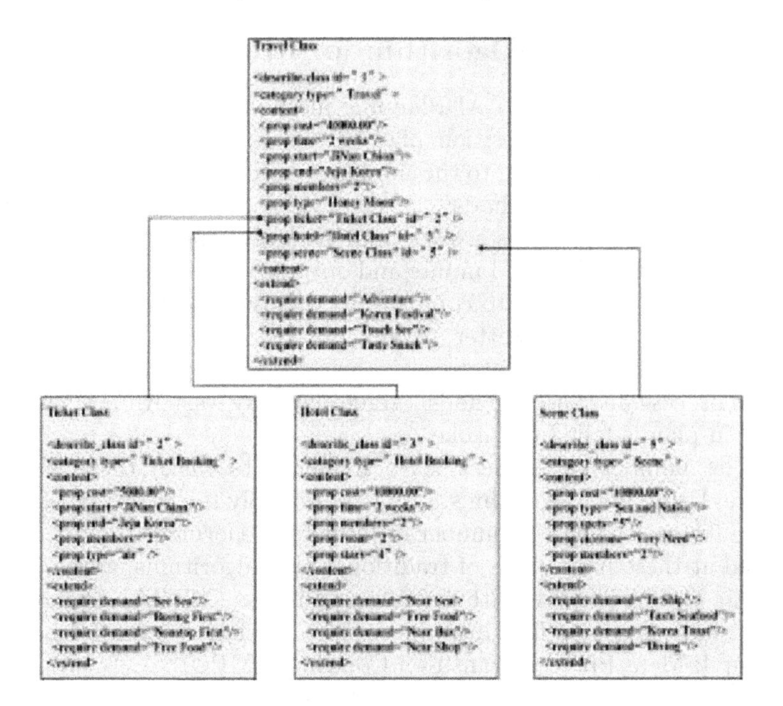

Fig. 3. Example of Requirement Constraint Tree

Fig.4 showed the meta process Generation Problem, a meta process generator generates a target meta process whose inputs are a requirement constraint tree and a initial meta process. Next paragraph the author presents a Dynamic Generation Algorithm for Meta Process (MPGA) to resolve meta process Generation Problem.

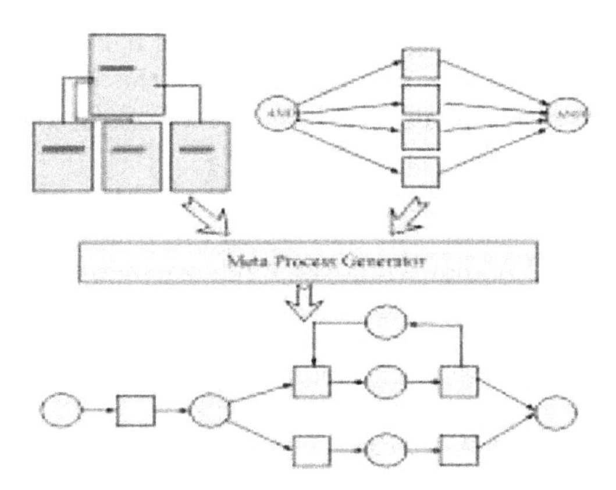

Fig. 4. Meta Process Generation Problem

3 Auto Generation Algorithm for Meta Process

Today most of methods used AI Planning methods to dynamically compose web services using a composition plan whose inputs and outputs satisfied users' requirements according to the similarity of services operation. But there are some limits in these methods:

(a) Operations of services always can't represent the function of services. When a programmer designed inputs and outputs of a web service, he usually considered more about how to complete its functions better, but not how to describe service's function better.

(b) It is not flexible enough to make services binding and plan generation as a same process. In these methods, target plan may waste time when services bounded in plan is failed to invoke.

(c) The number of services is so immense to make these algorithms doubtable. Using plan algorithms, there are possibly too many satisfied plans to choose for user when the number of services is enormous.

Aimed at these limitations of traditional plan algorithms, author presents a Dynamic Generation Algorithm for Meta Process (MPGA). Before this algorithm is advanced, we first introduce some correlative definitions.

Definition 4: Meta Process Transition Condition (MPC)

MPC is a tree whose every leaf of MPC is a tri-tuple ¡D, C, V¿, where

D = $[d_1, d_2, d_3 d_n]$, is a sequences of attributes of meta service, d_i is a simple attribute or a complex attribute such as another meta service;

C = $[c_1, c_2, c_3 c_n]$, is a sequences of compare operators, such as "=", ">", "<", and so on, if c_i is a complex attribute, ci is only "=";

V = $[v_1, v_2, v_3 v_n]$, is a sequences of value compared whit D, and if d_i is a complex attribute, v_i is a sub-MPC.

MPC is a condition of meta process transition is invoked, if MPC_i is satisfied meta process transmit from MP_i to $MP_i + 1$. In fact, MPC is a condition tree whose every leaf is a simple condition. Nowadays, most papers see conditions as first order logic expressions, but it is well-known that logic rules have many limits in practical application. In this paper, author describes the condition using a tree related with meta service's attributes more simple and specific than logic rules.

Definition5: Meta Process Generation Action (MPA)

A MPA is a action which transmits a meta process to another one. There are 3 kinds of basic MPA showed in Fig.5: remove, add sequence, add loop. If MPC_i is satisfied, MPA_i is invoked and meta process transmit from MP_i to $MP_i + 1$.

4 Analyze of MPGA

In MPGA every MPC are validated, and a MPC validated Algorithm travel a MPC then DFS or MFS a CT with hash code, so this is a O(n*n*n*logn) = O(n^3logn).

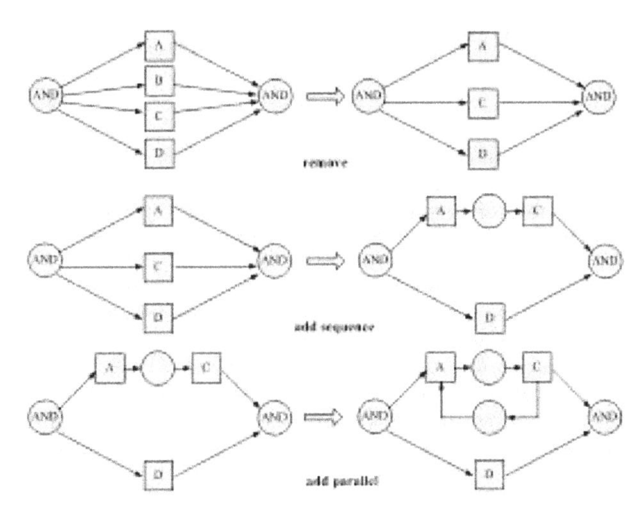

Fig. 5. three basic MPA

MPGA

Input: a Requirement Constraint Tree (*CT*), Initial Meta Process(MP_0) and a set of Meta Process Condition (MPC), a set of Meta Process Action (MPA)
Output: Target Meta Process(MP_t)
Step1 For every MPC do:
    Invoke MPC Validate Algorithm, calidate MPC;
    If MPC is satisfied, invoke Meta Process Action(MPA) corresponding, transition Meta Process from Mp_i to Mp_{i+1}
Step4 return MP_t.

MPC Validate Algorithm

Input: a Requirement Constraint Tree (*CT*) and a Meta Process Condition(MPC)
Output: Succeed or Failed
Step1 For every attributes of root of MPC
    If attribute is a simple condition, validate whether it can be satisfied in CT.
    If not satisfied, return Failed.
    If attribute is a complex condition, validate whether it can be satisfied using MPC Validate Algorithm(CT, attribute).
Step2 return Succeed.

Tab.1 show different results between MPGA and traditional planning algorithm. There are 6 meta services and 67 real travel services, and the server's configuration is P4 2.8G, 512M memory and Windows 2003 Server. Because the Service Discovery Module of IPVita hasn't been completed, and meta process can not be transformed to real travel process now, the following compare is not very convinced. But we also can see the number of results is greatly decreased using MPGA and the cost of time is only 1/10 of traditional plan algorithm.

Table 1. Contrast Between MPGA and traditional plan algorithm

	Number of Result Plans	Cost of Time (s)	Number of Satisfied Plans
MPGA	6	15	2
traditional plan	107	163	53

Compared with the old MPGA in paper[12], the new MPGA cost more little time get the same result showed in Fig.7 and the new definition of meta process is more convenience to be transformed a travel process.

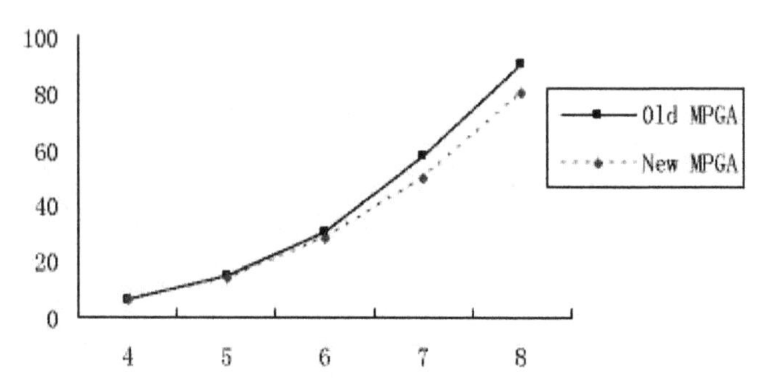

Fig. 6. Compared of old MPGA and new MPGA

5 Conclusions and Future Work

This paper introduced concepts of meta services and meta process, which can largely simplify the complexity of web services composition. Corresponding to these concepts, a dynamic generation algorithm for meta process is presented instead of predefining a rigorous workflow model. And differently with some AI planning method, this approach deals with the various requirements of tourists more effectively and flexibly in Internet.

In conclusion, this method has some advantages than other dynamic web services composition methods. Classifying web services into meat services largely simplify the complexity of services composition. In fact management for user requirements and services registration is more simple using meta services.

In future work, definition of meat process may be adjusted with the transformation from abstract process to workflow model. And composition constructs presented in BPEL4WS will also be further investigated in future work.

References

1. http://www.opentravel.org/ 2005
2. http://www.tti.org/ 2005
3. http://www.unece.org/ 2005
4. http://lsdis.cs.uga.edu/projects/meteor-s/ 2005
5. EMalu Castellanos, Fabio Casati, Ming-Chien Shan, Umeshwar Dayal: iBOM: A Platform for Intelligent Business Operation Management. ICDE 2005: 1084-1095
6. Naoki Fukuta, Tetsuya Osawa, Tadashi Iijima, Takahira Yamaguchi: On Implementing Ontology-Based Semantic Application Integration Framework for Web-Based Portals. SAINT Workshops 2005: 320-323
7. Boualem Benatallah, Quan Z. Sheng, Anne H. H. Ngu, Marlon Dumas: Declarative Composition and Peer-to-Peer Provisioning of Dynamic Web Services. ICDE 2002: 297-308
8. Sheila McIlraith, Tran Cao Son, and Honglei Zeng. Semantic web services. IEEE Intelligent Systems, 16(2):46-53, March/April 2001
9. Brahim Medjahed, Athman Bouguettaya, and Ahmed K. Elmagarmid. Composing web services on the semantic web. The VLDB Journal, 12(4), November 2003
10. Jinghai Rao, Peep Kungas, and Mihhail Matskin. Logic-based web services composition:from service description to process model. In The Third International Conference on Web Services, San Diego, USA, July 2004. IEEE
11. Evren Sirin, James Hendler, and Bijan Parsia. Semi-automatic composition of web services using semantic descriptions. In Web Services: Modeling, Architecture and Infrastructure" workshop in conjunction with ICEIS2003, 2002
12. Qi Sui, Hai-yang Wang, Meta Service in Intelligent Platform of Virtual Travel Agency, SKG2005, 2005
13. Qi Sui, Hai-yang Wang, IPVita: An Intelligent Platform of Virtual Travel Agency, APWeb2006 2006

Linguistic Summaries of Standardized Documents

Piotr S. Szczepaniak[1,2] and Joanna Ochelska[1]

[1] Institute of Computer Science, Technical University of Lodz, Sterlinga 16/18, 90-217 Lodz, Poland
[2] Systems Research Institute, Polish Academy of Sciences Newelska 6, 01-447 Warsaw, Poland

Summary. Automatic summarization of databases has become indispensable in a number of tasks involving information exchange or strategic decision making. It is also important when huge bases of documents must be clustered. The present paper deals with summarization of standardized databases containing both numerical and textual records. The method and its variations are described and explained on illustrative examples.

Key words: Summarization, linguistic summaries, textual documents, fuzzy similarity, text comparison.

1 Introduction

According to the definition given in [12], *summarization* is the process of distilling the most important information from a source (or sources) to produce an abridged version for a particular user (or users) and task (or tasks).

Information contained in databases includes a variety of forms, namely numerical data, text, images, sound, and the like. The number and size of documents may be, and in fact frequently are, too large to be handled and interpreted effectively by a human. In many situations, compression is of greatest value here than the accuracy of formulation (although wrong statements are obviously unacceptable). Information is manageable for people only if it is condensed and given in a natural language. The explosion of information sources, their size and variability, make the use of computer for automatic summarization necessary [5]—[9].

Computers prove to be highly efficient while dealing with numbers. However, when databases involve many textual records to be summarized, the situation becomes more complicated. It is due to the fact that, in such a case, classification and comparison of natural language expressions are needed. These

Mark Last et al. (Eds.): Advances in Web Intelligence and Data Mining (SCI) **23**, 221-232 (2006)
www.springerlink.com

in turn are not straightforward tasks for machines designed to process numbers. In recent years, the fuzzy sets theory by Zadeh [19, 20] has been proposed as the instrument in effective dealing with this problem. Nevertheless, summarization of full text documents has not been solved satisfactorily yet. However, when the documents are domain-restricted and standardized with respect to their structure and form of information entities (records), a recognizable progress is observed.

In the first part of this paper, a brief description of a method for linguistic summarization of databases containing standardized textual records is given, while in the second part — some possible variations and refinements of the method are presented.

2 Yager's linguistic summaries

The Yager's concept of the linguistic summary of a database [17] is as follows:

Definition 1. *The linguistic summary of a database is of the form*

$$Q\ P\ are\ (have)\ S\ [T] \tag{1}$$

or, in an extended version

Q of objects being P are S (have property S)
[and correctness of this is of degree T].

The symbols are interpreted as follows:
Q — amount determination,
P — subject of the summary,
S — property,
T — quality of the summary.

Yager assumes that the value of a summarized attribute is a crisp number. Amount determination Q is a linguistic variable, for example: "very few", "many", "almost each", and defines how many objects have a given property. The subject of the summary P is determined as an object described with the values from its record. The quality of the summary T is a real number from the interval $[0, 1]$ and can be interpreted as the level of truth (confidence) for a given summary.

Example 1. Very few employees have high salary [0.97].
Here: very few — amount determination Q,
employees — subject of the summary P,
high salary — property S,
[0.97] — quality of the summary T.

In [18], Yager's method is shown for situations when two properties R and S need to be considered:

$$Q\ P\ are\ (have)\ R\ and\ S\ [T], \tag{2}$$

$$and \quad Q\ R\ P\ are\ (have)\ S\ [T]. \tag{3}$$

Example 2. Very few employees are young and have high salary [0.97], Very few young employees have high salary [0.97].

Performing summarization with respect to some query requires a proper definition of linguistic variables determining Q, R and S, as well as computation of T.

Example 3. Six companies employ the following number of people: $D = \{d_i\} = \{14, 45, 110, 12, 31, 50\}$, $(i = 1, 2, \ldots, 6)$. Let the summarization be performed as the answer to the query:

How many companies are small ?

Here we have: S — small, and P — company.

Now, the amount determination, Q, the quality of the summary, T, and the suitable fuzzy sets for this linguistic variable T, are looked for. First, the membership function describing the size of a company (small company) is defined. This measure is applied to each record of database. In the next stage, all values of the membership function are counted and, then, divided by the number of records — the average value R is obtained. Then, the linguistic variable is defined. On the basis of this definition and the value of R, Q and T are determined.

If the membership function describing the size (small) of the company is defined as follows

$$\mu_{small}(d_i) = \begin{cases} 1 & & d_i < 20 \\ \frac{55-x}{35} & for & 20 \leq d_i \leq 55 \\ 0 & & d_i > 55 \end{cases} \tag{4}$$

then one obtains

$$\mu_{small}(d_1) = 1.0 \quad \mu_{small}(d_2) = 0.29 \quad \mu_{small}(d_3) = 0.0$$
$$\mu_{small}(d_4) = 1.0 \quad \mu_{small}(d_5) = 0.69 \quad \mu_{small}(d_6) = 0.14$$

The sum of these values is

$$Y = 1.0 + 0.29 + 1.0 + 0.69 + 0.14 = 3.12$$

Using the average value

$$R = \frac{Y}{\text{numer of small companies}} = \frac{3.12}{6} = 0.52$$

and the definition of the linguistic variable $X = \{few, many, almost\ all\}$ with membership functions as shown in Fig. 1, one obtains the following summarization results

Many companies are small [0.8].
Almost all companies are small [0.37].

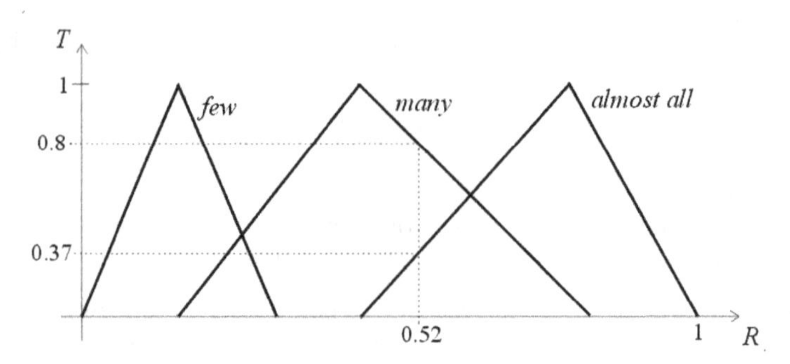

Fig. 1. Linguistic variable X

3 Consideration of textual records

Let us consider a simple database where information about size and condition of a number of companies isgiven (see Table 1 where: Id — identification number, Ne — number of employees). From the database given in Table 1

Table 1. Sample set of information entities with textual records

Id	Ne	Financial condition of the company
1.	7	Condition of company — quite good.
2.	12	Condition of company — sufficiently good.
3.	73	Condition of company — rather good.
4.	4	Condition of company — fairly good.
5.	53	General condition good.
6.	35	Company in good condition.
7.	47	General condition of company — OK.
8.	27	General condition of company — rather good.
9.	1	General condition of company — seriously bad.
10.	40	General condition good.

some different information in an abridged form can be obtained, i.e. answers to diverse queries can be obtained. Examples are:

Query 1: How many of the small companies are in good condition ?

Query 2: How many of the companies being in good condition are small ?

Note that each information entity related to particular company consists of records of both numerical and textual character. Data of numerical type are generally easy to compare and the way of Yagers' summarization is well-known. Dealing with textual records one needs to use a method which enables their comparison, more precisely - computation of their similarity degree. This statement is in accordance with human intuition in which two sentences may be similar to some extent. Qualitative comparison of sentences is based on similarity of words, which can be performed in many ways to call the n-gram matching [3] or its generalized version [13]—[16] as more sophisticated examples.

Let us consider the *Query no 1* and use 'GENERAL CONDITION — GOOD' as reference. Let the records describing the condition of companies be compared with this reference description. Obviously, similar content can be expressed in many different ways even when standardized information entities are used, as in the case of the considered set of textual records. Therefore, application of thesaurus relating synonymous or very similar terms to each other is useful. In the considered example, the list of synonyms to the word 'good' may include the following words and abbreviations excellent, fine, OK, superior, well. Consequently, the record no 7 will obtain its equivalent form given in Table 2, and the comparisons of records may be then performed with the use of any formula based on string matching. The values obtained with the use of the generalized n-gram method are shown in Table 2 (see formulae (a, b) in the Appendix). Next, the procedures for generation of summaries that

Table 2. Similarity between the reference and sentences

Id	Reference pattern	Financial condition of the company	Similarity level for comparison of the two sentences
1.		Condition of company — quite good.	0.45
2.		Condition of company — sufficiently good.	0.45
3.		Condition of company — rather good.	0.48
4.		Condition of company — fairly good.	0.46
5.		General condition good.	1.00
6.	General condition	Company in good condition.	0.53
7.	— good.	General condition of company — OK.	0.64
8.		General condition of company — rather good.	0.57
9.		General condition of company — seriously bad.	0.30
10.		General condition good.	1.00

are to provide answer to the *Query no 1* are described. In order to perform the transformation needed to produce an answer to the query, one must define membership function, which determines what size a given company should be so that it belongs to the set of small companies.
In the case under consideration the membership function takes the form:

$$\mu_{small}(x) = \begin{cases} 0.20x & 0 < x \leq 5 \\ 1 & 5 < x \leq 25 \\ -0.04x + 2.00 \quad for & 25 < x \leq 50 \\ 0 & x > 50 \end{cases} \tag{5}$$

where x is the number of employees. The values taken by the membership function are given in Table 3. Once membership of the companies with respect to their condition and size has been computed the summarization of the database with respect to the first query can be performed. Let

$$\mu_{S_r}(S_{db}) = \mu(S_r, S_{db}) \tag{6}$$

Table 3. Values of membership function with respect to size of companies

Id	Ne	Membership value wrt size
1.	7	1.00
2.	12	1.00
3.	73	0.00
4.	4	0.80
5.	53	0.00
6.	35	0.60
7.	47	0.12
8.	27	0.92
9.	1	0.00
10.	40	0.04

be the membership function describing similarity between any sentence (S_{db}) from the database and the reference sentence (S_r) — values are given in Table 2. Then:

$$R = \frac{\sum_{i=1}^{10} \mu_{S_r}(S_{db_i}) \cdot \mu_{size}(x_i)}{\sum_{i=1}^{10} \mu_{size}(x_i)} \tag{7}$$

According to (7) one obtains

$$R = \frac{1 \cdot 0.45 + 1 \cdot 0.45 + 0.8 \cdot 0.46 + 0.6 \cdot 0.53 + 0.12 \cdot 0.64 + 0.92 \cdot 0.57 + 0.04 \cdot 1}{1.00 + 1.00 + 0.80 + 0.60 + 0.12 + 0.92 + 0.04}$$

$$= \frac{2.15}{4.48} = 0.48$$

Let the linguistic variable X be:

$$X = \{\textit{almost nothing}, \textit{few}, \textit{medium}, \textit{many}, \textit{almost all}\}.$$

By relating the values taken by the linguistic variable X to the quality of the summary T (cf. Fig. 2) one obtains the complete summary with respect to *Query no 1* in the form:

Many small companies are in good condition [0.07].
Medium small companies are in good condition [0.9].

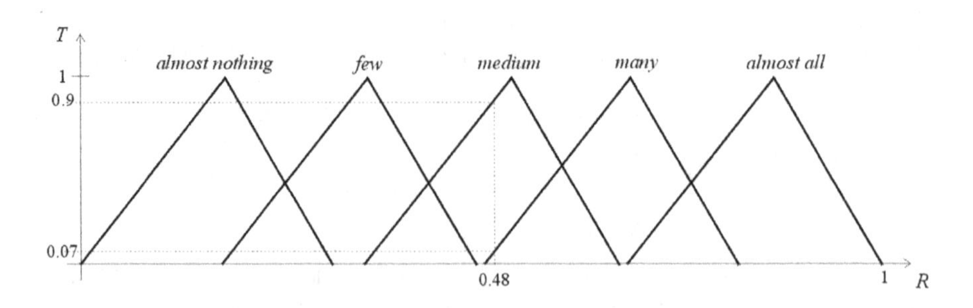

Fig. 2. Linguistic variable X

An analogous procedure can be performed for the second query in order to find out how many of the companies being in good condition are small.

4 Variations and extensions

4.1 Crisp selection of records

Here only the companies determined as small or those whose size is acceptably similar to "smallness" (strictly defined) are considered. The level of acceptable similarity is obviously an arbitrary choice of the user, e.g. the threshold 0.5 can be proposed. Consequently, the companies that do not meet the criterion of "smallness" are rejected and not involved in the summarization (Table 4).

Table 4. Values of membership function for the size of companies bigger than or equal to 0.5

Id	Ne	Financial condition of the company	Membership value wrt size
1.	7	Condition of company — quite good.	1.00
2.	12	Condition of company — sufficiently good.	1.00
3.			
4.	4	Condition of company — fairly good.	0.80
5.			
6.	35	Company in good condition.	0.60
7.			
8.	27	General condition of company — rather good.	0.92
9.			
10.			

Of course, such an arbitrary, hard selection is not motivated when the fuzzy sets theory is used, but it may reduce the computational effort when a huge number of records are dealt with.

4.2 Fuzzy rule-based records selection

The records selection is based on the idea described in [4]. Here, linguistic variables for each of the considered features of object (e.g. size, condition, etc.) are defined. Then, linguistic variables determining the level of similarity (identical, different, etc.) between objects are also defined. For summarization, only those records for which the fuzzy rule describing the users' requirements is relevant are used.

4.3 Consideration of non-similarity

Non-similarity of sentences can be an additional element considered by the generation of summarization. Roughly speaking, this approach is based on the fuzzy theory introduced by Atanassov [1, 2] where

$$0 \leq \mu_A(x) + \nu_A(x) \leq 1 \ \forall_{x \in X} \tag{8}$$

with
$\mu_A : X \rightarrow [0,1]$ — membership function,
$\nu_A : X \rightarrow [0,1]$ — non-membership function.

The problem lies in the separate (though correct) definition of non-similarity of words and sentences. As a result of this approach one obtains the summary in the following general form

$Q\ P\ are\ (have)\ S\ [T]\ -\ [H]$

or, in a verbal version

Q of the objects being subject (P) of the summary have property S, and correctness of this statement is of degree T with hesitancy margin H.

For example:

Few small companies are in good condition [0.40] — hesitancy degree [0.09].

It is questionable whether we really obtain additional knowledge about the data base or only improve the quality of summarization. Moreover, the computational cost increases significantly.

4.4 Fuzzy sets type-2

Fuzzy sets type-2 [10, 11, 21] can also contribute to the summarization approach. Here, the simplest way of their use is reported.

Thanks to fuzzy sets type-2, a greater number of experts (and their knowledge) can be included in the summarization process. Each expert defines (according to their intuition) their own membership function describing the considered features of summarization.

For example, when considering the query: "How many companies are small ?" each expert chooses their own membership function that determines a given feature, i.e. the measure of the size of a company. If a query refers to a verbal feature (e.g. "How many companies are in good financial condition ?") the membership function of a given feature shows the similarity between sentences from the database and a reference sentence. Let us assume that there are only two experts that define membership function with respect to the size of the company. Let the first expert uses formula (5), and the second expert the following one:

$$\mu_{small}(x) = \begin{cases} 0.10x & & 0 < x \leq 10 \\ 1 & & 10 < x \leq 40 \\ -0.05x + 3.00 & for & 40 < x \leq 60 \\ 0.00 & & x > 60 \end{cases} \tag{9}$$

In Table 5 values of membership function for each of the two experts are shown.

Table 5. Values of membership function with respect to size of companies for two experts

Id	Ne	Membership value wrt size for the first expert	Membership value wrt size for the second expert
1.	7	1.00	0.70
2.	12	1.00	1.00
3.	73	0.00	0.00
4.	4	0.80	0.40
5.	53	0.00	0.15
6.	35	0.60	1.00
7.	47	0.12	0.65
8.	27	0.92	1.00
9.	1	0.00	0.00
10.	40	0.04	1.00

We define also the level of confidence for the experts, for example:

$$\mu = \begin{cases} 0.4 \ for\ the\ first\ expert \\ 0.6 \ for\ the\ second\ expert \end{cases} \tag{10}$$

Next, the records are considered separately for each expert, and one obtains:

$$R_1 = \frac{1.00+1.00+0.80+0.60+0.12+0.92+0.04}{10} = 0.45 \quad for\ the\ first\ expert.$$
$$R_2 = \frac{0.70+1.00+0.40+0.15+1.00+0.65+1.00+1.00}{10} = 0.59\ for\ the\ second\ expert.$$

Consequently, one obtains the fuzzy set of the type-2 for which the primary membership determines level of consistency of the considered record with the reference record while the secondary membership reflects the level of confidence for the expert.

$$\tilde{\mu} = \{0.45/0.6, 0.59/0.4\}$$

Next, the fuzzy set of the type-2 is transformed into the set type-1, i.e.

$$R = 0.45 \cdot 0.6 + 0.59 \cdot 0.4 = 0.51$$

and finally synthesis of summarization is performed.

5 Final remarks

To be easily managed by people, information should be condensed and given in a natural language. Today, however, the number and size of documents stored in computers are in fact too large to be handled and interpreted effectively by humans. Consequently, distillation of the most important information from a source (or sources) to produce an abridged version for a particular user (or users) and task is becoming increasingly important from the practical point of view. For this process, called *summarization*, the aid of computer is necessary (paradoxically, it is because of the explosion of information sources, their size and variability).

The paper presents a new approach to summarization of databases containing both numerical and textual records. The method is based on Yager's proposition developed for databases being collections of numerical data. The main novelty of the present contribution lies in the application of fuzzy sets to the measuring of similarity between textual records and given sentences. The use of fuzzy sets theory makes it possible to show consistency of the results with human intuition. The significant advantage of the method is its flexibility with respect to a given query.

References

1. Atanassov K (1986) Intuitionistic Fuzzy Sets. Fuzzy Sets and Systems, **20**:87–96
2. Atanassov K (1999) Intuitionistic Fuzzy Sets; Theory and Applications. Heidelberg, New York, Physica-Verlag, A Springer-Verlag Company
3. Bandemer H, Gottwald S (1995) Fuzzy sets, Fuzzy Logic, Fuzzy Methods with Applications. John Wiley and Sons
4. Bunke H, Fabregas X, Kandel A (2001) Rule-based Fuzzy Object Similarity. Mathware & Soft Computing, **8**:113–128
5. Kacprzyk J, Zadrozny S (1999) On Interactive Linguistic Summarization of Databases via a Fuzzy-logic-based Querying Add-on to Microsoft Access. In: Reusch B (eds) Computational Inteligence. Springer-Verlag, Heidelberg:462–472
6. Kacprzyk J, Zadrozny S (2000) Computing with words: towards a new generation of linguistic quering and summarization of databases. In: Sincak P, Vascak J (eds) Quo vadis computational intelligence ?, Physica Verlag, Heidelberg / New York:144–175
7. Kacprzyk J, Zadrozny S (2001) On Linguistic Approaches in Flexible Querying and Mining of Association Rules. In: Larsen H L, Kacprzyk J, Zadrozny S, Andreasen T and Christiansen H (eds) Flexible Query Answering Systems. Recent Advances. Physica-Verlag (Springer-Verlag), Heidelberg and New York:475–484
8. Kacprzyk J, Zadrozny S (2001) Fuzzy linguistic summaries of databases for an efficient business data analysis and decision support. In: Abramowicz A, Zurada J (eds) Knowledge discovery for business information system. Kluwer Academic Publisher B. V. Boston:129–152
9. Kacprzyk J, Zadrozny S (2001) Fuzzy linguistic summaries via association rules. In: Kandel A, Last M, Bunke H (eds) Data Mining and computational intelligence. Physica-Verlag, Heidelberg / New York:115–139
10. Karnik N N, Mendel J M (1988) An Introduction to Type-2 Fuzzy Logic Systems. University of Southern California, Los Angeles
11. Karnik N N, Mendel J M (1999) Type-2 Fuzzy Logic Systems. IEEE Trans. on Fuzzy Systems, **7**, no **6**:643–658.
12. Mani I and Maybury M T (1999) Advances in Automatic Text Summarization. The MIT Press, Cambridge, Massachusetts, USA

13. Niewiadomski A (2000) Appliance of Fuzzy Relations for Text Documents Comparing. Proceedings of the 5th Conference on Neural Networks and Soft Computing, Zakopane, Poland:347–352.
14. Niewiadomski A, Szczepaniak P S (2002) Fuzzy Similarity in E-Commerce Domains. In: Segovia J, Szczepaniak P S, Niedzwiedzinski M (eds) E-Commerce and Intelligent Methods. Physica-Verlag, A Springer-Verlag Company, Heidelberg, New York
15. Szczepaniak P S, Niewiadomski A (2003) Internet Search Based on Text Intuitionistic Fuzzy Similarity. In: Szczepaniak P S, Segovia J, Kacprzyk J, Zadeh L (eds) Intelligent Exploration of the Web. Physica-Verlag, A Springer-Verlag Company, Heidelberg, New York
16. Szczepaniak P S, Niewiadomski A (2003) Clustering of documents on the basis of text fuzzy similarity. In: Abramowicz W (eds) Knowledge-based information retrieval and filtering from the Web. Kluwer Academic Publ., Boston, New York, Dordrecht, London:219–230.
17. Yager R R (1995) Linguistic summaries as a tool for databases discovery. Workshop on Fuzzy Databases System and Information Retrieval, Yokohama, Japan
18. Yager R R (1990) On Linguistic Summaries of Data. 3rd Int. Conference on Information Processing and Management of Uncertainty in Knowledge-Based Systems, Paris, France
19. Zadeh L A (1965) Fuzzy Sets. Information and Control, **8**:338–353.
20. Zadeh L A (1971) Toward a Theory of Fuzzy Systems. In: Aspects of Network and System Theory. Kalman R. E. and De Claris N. (Eds.): Holt, Rinehart and Winston, New York, USA
21. Zadeh L A (1983) The concept of linguistic variable and its application for approximate reasoning. Information Science, **8**:149–184.

A Appendix. Fuzzy relation for text comparison — a generalized n-gram matching

The fuzzy relation

$$RW = \{(< w_1, w_2 >, \mu_{RW}(w_1, w_2)) : w_1, w_2 \in \mathbf{W}\},$$

on $\mathbf{W}$ — the set of all words within the universe of discourse, for example a considered language or dictionary — is proposed in [13, 14, 15] as a useful instrument for comparison of words.

The proposed form of the membership function $\mu_{RW} : \mathbf{W} \times \mathbf{W} \to [0, 1]$ is

$$\mu_{RW}(w_1, w_2) = \frac{2}{N^2 + N} \sum_{i=1}^{N(w_1)} \sum_{j=1}^{N(w_1)-i+1} h(i, j) \tag{11}$$

where:
$N(w_1)$, $N(w_2)$ — the number of letters in words w_1, w_2, respectively;
$N = max\{N(w_1), N(w_2)\}$ — the maximal length of the considered words;
$h(i, j)$ — the value of the binary function, i.e. $h(i, j) = 1$ if a subsequence,

containing i letters of word w_1 and beginning from the j-th position in w_1, appears at least once in word w_2; otherwise $h(i,j) = 0$;
$h(i,j) = 0$ also if $i > N(w_2)$ or $i > N(w_1)$.
Note that $0.5(N^2+N)$ is the number of possible subsequences to be considered.

Example:

Compare two words: $w_1 =$ *centre* and $w_2 =$ *center.*
Here $N(w_1) = 6$ and $N(w_2) = 6$, thus $N = max\{N(w_1), N(w_2)\} = 6$.

$$\mu_{RW}(w_1, w_2) = \mu_{RW}(w_2, w_1) = \frac{6+3+2+1}{21} \equiv 0.57.$$

because in w_2 there are:

- 6 one-element subsequences of w_1 (*c*, *e*, *n*, *t*, *r*, *e*);
- 3 two-element subsequences of w_1 (*ce*, *en*, *nt*);
- 2 three-element subsequences of w_1 (*cen*, *ent*);
- 1 four-element subsequence of w_1 (*cent*) — this is the longest subsequence of w_1 that can be found in w_2.

Note that the fuzzy relation *RW* is reflexive: $\mu_{RW}(w, w) = 1$ for any word w; but in general it is not symmetrical. This inconvenience can be easily avoided as follows:

$$\mu_{RW}(w_1, w_2) = min\{\mu_{RW}(w_1, w_2),\ \mu_{RW}(w_2, w_1)\}$$

Moreover, *RW* reflects the following human intuition:

- the bigger the difference in length of two words is, the more different they are;
- the more common the letters contained in two words, the more similar the words are.

However, the value of the membership function contains no information on the sense or semantics of the arguments.

Basing on measure (11) it is possible to introduce the similarity measure for sentences or even documents. Here, the sentence or document is considered to be a set (not a sequence) of words. The respective formula is of the form [13, 14, 15]

$$\mu_{RD}(d_1, d_2) = \frac{1}{N} \sum_{i=1}^{N(d_1)} max_{j \in \{1,\ldots,N(d_2)\}}\ \mu_{RW}(w_i, w_j) \tag{12}$$

where:
d_1, d_2 — the documents to be compared;
$N(d_1)$, $N(d_2)$ — the number of words in document d_1 and d_2 (i.e. the length of d_1 and d_2, respectively);
$N = max\{N(w_1), N(w_2)\}$;
μ_{RW} — the similarity measure for words defined by (11).

A Web-knowledge-based Clustering Model for Gene Expression Data Analysis

Na Tang and V. Rao Vemuri

Computer Science Department, University of California, Davis, CA 95616 USA

Abstract. Current microarray technology provides ways to obtain time series expression data for studying a wide range of biological systems. However, the expression data tends to contain considerable noise, which as a result may deteriorate the clustering quality. We propose a web-knowledge-based clustering method to incorporate the knowledge of gene-gene relations into the clustering procedure. Our method first obtains the biological roles of each gene through a web mining process, next groups genes based on their biological roles and the Gene Ontology, and last applies a semi-supervised clustering model where the supervision is provided by the detected gene groups. Under the guidance of the knowledge, the clustering procedure is able to cope with data noise. We evaluate our method on a publicly available data set of human fibroblast response to serum. The experimental results demonstrate improved quality of clustering compared to the clustering methods without any prior knowledge.

1 Introduction

Current DNA microarray technology provides ways to conduct large-scale experiments in a wide range of biological systems. Many problems such as biological interpretation, disease development and drug discovery can thus be further studied by analyzing the data generated from the experiments. The microarray data consists of *expression levels* of many genes over a set of consecutive time points, also referred as time series (or time course) gene expression data. The expression data allow scientists to examine the gene expression changes over time and obtain more discoveries regarding to the time course.

Clustering genes into groups with similar behavior is one of the key processes for time series gene expression data analysis, which provides a way to examine the different patterns of gene modules and study unknown genes based on known genes of the same group. A number of existing approaches are available to cluster time series gene expression data such as HAC [5, 13, 9], k-means [12], SVD and HMM [10]. However, these approaches construct models merely from the gene expression data, in which considerable data noise might be present due to the experiment design and may deteriorate the clustering quality.

Many believe that genes in the same cluster have similar biological roles [5, 13]. Here, a biological role is formally described as the biological process associated with a gene. Their results [5, 13] also illustrated that this knowledge about genes can be inferred from the clustering results. For example, gene KITLG

Mark Last et al. (Eds.): Advances in Web Intelligence and Data Mining (SCI) **23**, 233-242 (2006)
www.springerlink.com

is involved in the "cell proliferation" process. If another gene is in the same cluster as KITLG, then that gene is also likely to be involved in "cell proliferation". Inspired from this fact, we and other [3, 6, 1] believe that if one can find the biological processes associated with the genes and the relations among the processes, then this type of prior knowledge can be used to guide the clustering process in order to generate more meaningful clusters.

Gene Ontology (GO, http://www.geneontology.org/) provides standard terminology for biological processes and constructs a hierarchical structure of these biological processes. Figure 1 shows a fragment of the ontology structure defined by GO. If the biological processes of some genes can be determined, the relationship among these genes can be detected based on the relationship among their biological processes defined by GO. For example, if both gene A and gene B are involved in the "regulation of cell proliferation", they are likely to be in one group. And, this gene-gene relationships can be further incorporated into clustering models. Thus we design a web-knowledge-based clustering model to retrieve the biological processes of genes from the web and further help clustering.

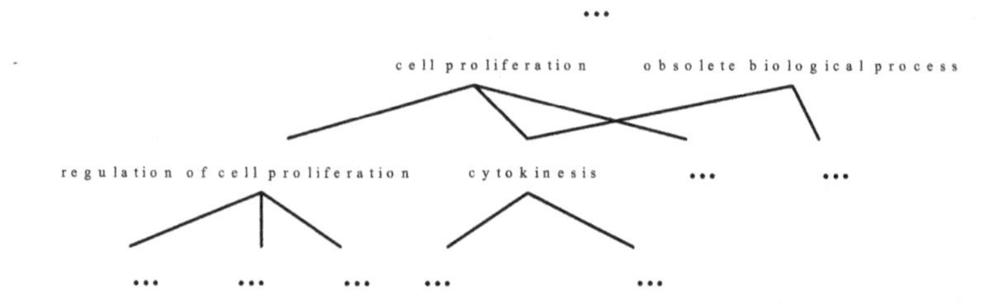

Fig. 1. A fragment of BPO in GO.

Our model (shown in Figure 2) starts with a web knowledge discovery process mining genes' biological processes from web gene databases and specialized web search engines. The gene-gene relationships are then detected by examining relationships among genes' biological processes based on GO. Finally, it applies a semi-supervised clustering model where the supervision is provided by the detected gene-gene relationships. We evaluate our approach on a time series data set of human fibroblast response to serum provided by [7]. The results show that our knowledge-based clustering model generates clusters of better quality compared to the original clustering model without any prior knowledge.

Some papers such as [8] aims at predicting biological processes for unknown genes, and thus classifies time series gene expression data based on GO annotations. Our work aims to analyze any type of gene expression data, and the biological processes are only used as the supplementary knowledge to improve clustering for further research. Some other knowledge-guided clustering methods [3, 6, 1] are available. Our approach is different at the following two aspects 1) we acquire knowledge from PubMed articles instead of solely relying on the knowledge from the gene information databases. Compared to the latter, the former

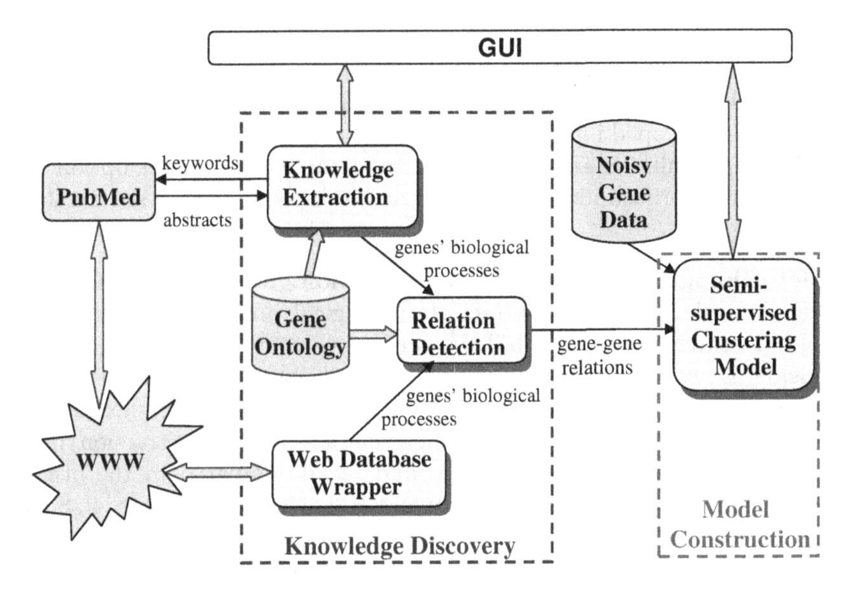

Fig. 2. Implementation architecture of web-knowledge-based clustering model for gene expression data.

serves as a complementary source and may contain more up-to-date information about genes; 2) In our framework, the supervision on clustering combines both constraints and distance learning.

2 Web-knowledge-based Clustering Model

Our web-knowledge-based clustering model, involves two processes: a web knowledge discovery process and a semi-supervised clustering process. The gene-gene relations are obtained from the extracted knowledge through the first process, and further serves as an input for the clustering process.

2.1 Web Knowledge Discovery of Gene-gene relations

Two types of web sources are available to extract the biological process of known genes : web gene databases and web biological documents.

Extract Biological Processes from Web Gene Databases Many web gene databases provide GO annotations for known genes, i.e., they list the biological processes and other properties of known genes in GO terms. One can query on gene names or gene symbols to obtain their biological processes. In addition, such gene information is usually given in a fixed format, which eases the automatic biological process extraction. An example of such a web database is Entrez Gene[1],

[1] http://www.ncbi.nlm.nih.gov/entrez/query.fcgi?db=gene

which is used as the source for finding GO annotations. A web page wrapper is built to extract biological process information from Entrez Gene. For simplicity, hand-crafted rules are used for the wrapper of Entrez Gene via observation. Applying a well-developed information extraction system such as WHISK [11] with a number of training examples to build the wrapper is another option, which can make our approach more general for any web gene information databases.

Extract Biological Process from Web Biological Documents Besides the web gene databases, we extract the biological processes of genes through specialized search engines. PubMed[2], an Entrez search engine on biomedical documents, is used here. Gene symbols serve as keywords for searching, and the abstracts of the search results are analyzed for biological process extraction.

We select the sentence co-occurrence method as our extraction method, because it achieves balanced precision and recall compared to the other two methods, namely, sentence classification and abstract co-occurrence [4]. Sentence classification gives the highest precision but the lowest recall and abstract co-occurrence gives the highest recall but the lowest precision. We also seek ways to improve the precision of the sentence co-occurrence method. For example, stemming and word distance are used to obtain higher extraction accuracy.

We analyze each sentence of the text documents. For each biological process p in GO, we determine if the sentence s contains p for a gene g in GO as follows: 1) If s contains g, goto 2); otherwise, return FALSE; 2) A stop list of words such as "of", "to", "the", which does not indicate any meaning, are removed from s as well as p; 3) When p contains more than one terms, stemming (a method to convert a term into its root) is applied to both s and p. 4) If s contains all the terms in p and any two consecutive terms in p appear in s with less than three other terms in between, return TRUE; otherwise, return FALSE.

If an extraction of a biological process p for gene g is confirmed in a sentence s, both g and the terms appearing in p are highlighted in s for further analysis. An example of an extraction is shown as follows: *"DDB2, while participating in DNA repair, functions as a negative regulator of apoptosis, and may therefore have a pivotal role in regulating immune response and cancer-therapeutic efficacy"*, we extract several biological processes including "DNA repair", "negative regulation of apoptosis" and "immune response" for gene DDB2. All these terms are highlighted. The highlighted area makes users easy to determine if an extraction is correct or not. The correct extractions are selected and combined with the biological processes obtained from Entrez Gene.

Detect Functional Groups from Web Knowledge Based on the biological processes extracted from the previous two steps (2.1.1 and 2.1.2), we detect functional gene groups according to their biological processes. Gene g *strongly* belongs to the group of biological process p if it is associated with p based on the extracted knowledge. Gene g *weakly* belongs to the group of p if the biological

[2] http://www.ncbi.nlm.nih.gov/entrez/query.fcgi?db=PubMed

process p' associated with g is a parent, child or sibling of p. For example, KITLG is a strong member of group "cell proliferation" because its biological processes contain "cell proliferation", while CCND1 is a weak member of this group because its biological process "cytokinesis" is a child of "cell proliferation" according to GO. The distinction between the *strong* members and the *weak* members for a group provides a way to calculate weighted cluster centers.

2.2 Knowledge-based Clustering

Suppose n_g functional groups are detected from the extracted knowledge, the challenge of utilizing these n_g groups for the clustering model remains. We choose the semi-supervised K-means method [2] to incorporate the knowledge with the clustering procedure, because it is a well-designed algorithm combining both constraint-based supervision and distance-based supervision provided by the given knowledge. The semi-supervised method improves the standard K-means clustering by incorporating this supervision into the initialization process and the distance measure based on a probabilistic framework, which are explained in detail below.

Initialization Instead of randomly initializing the clustering centroids, we estimate the initial cluster centroids from the detected n_f functional groups. Since many specific biological processes are usually extracted from the web, the number of all existing functional groups n_f tends to be quite large. Thus, it is usually the case that $n_f > K$, where K is the number of desired clusters. We use a weighted first-farthest traversal algorithm to select K functional groups that are farthest distributed and with considerable group size. Then the cluster centroids are initialized with the weighted means of these K groups:

$$\frac{1}{w_1 \cdot |F_k\_s| + w_2 \cdot |F_k\_w|}(w_1 \sum_{i \in F_k\_s} g_i + w_2 \sum_{j \in F_k\_w} g_j),$$

where F_k_s and F_k_w are the sets of strong members and weak members respectively in the kth selected functional group F_k. The constants w_1 and w_2 satisfy $w_1 > w_2$. The symbols g_i and g_j refer to genes. The formula shows that g_i is a strong member of F_k while g_j is a weak member. The condition $w_1 > w_2$ makes the centroid of F_k biased toward the strong members.

Constraint-sensitive distance measure The constraints induced by the extracted knowledge are enforced into the clustering procedure. The semi-supervised K-means modifies the distance measure so that the assignments conflicting with the provided knowledge are penalized. In this paper, if genes in the same functional group are assigned to different clusters, the distance measure is modified to penalize this violation (so called violation of must-link constraints). Suppose D_{ik} is the distance of a gene g_i from the cluster centroid of C_k. The standard

K-means assigns gene g_i to cluster C_k with the minimum D_{ik} for any C_k. Instead, we assign each gene g_i to C_k to minimize the distortion NEW_D_{ik}, which is defined as:

$$NEW\_D_{ik} = D_{ik} + \sum_{j \in F_k} penalty(g_i, g_j) \cdot D_{ij},$$

where the penalty function is:

$$penalty(g_i, C_k) = \begin{cases} p, & \text{if } i \in F_k \ \&\& \ g_j \notin C_k \\ 0, & otherwise \end{cases}$$

Here, F_k is the kth functional groups that we used for initialization and C_k is the cluster corresponding to F_k. The iterated conditional modes (ICM) applied in [2] is also used in this paper to find the optimal assignment based on the distance measure.

In this paper we only penalize the violation for the must-link constrains but do not consider cannot-link constraints, while both are penalized in the document clustering application in [2]. This is because our gene functional groups might be overlapped to some extent, in which case genes might still have similar biological roles even if they are in different functional groups. Therefore, the cannot-link constraints are not applied in this gene application.

Adaptive distance learning Instead of using static distance measure, a parameterized distance measure is used to incorporate the user-specified constraints and data variance. The modification is exactly the same as [2], so we skip the details here. In essence, the adaptive distance learning brings similar genes closer and pushes dissimilar genes further apart.

As a whole, combined with these three improvements, the gene expression data clustering via semi-supervised K-means is summarized in the chart (next page).

3 Experimental Results

We evaluated our model on a time series gene expression data set (fibroblast response to serum provided by [7]). This data set contains the expression changes of 517 genes corresponding to 497 unique genes during the first 24 h of the serum response in serum-starved human fibroblasts. The expression changes are given as the ratio of the expression level at the given time point to the expression level in serum-starved fibroblasts.

First, we obtained the standard gene symbols for the corresponding gene names in the data from Entrez Gene. For example, "SEPP1" is the gene symbol for the gene name "H.sapiens mRNA for selenoprotein P".

Second, we extracted the GO annotations of biological processes for the given genes via the method in Section 2.1.1. Then we extracted biological processes from biomedical articles via the process in Section 2.2.2. There were totally 1081

Clustering Gene Expression Data via Semi-supervised K-means

Input: Set of gene expressions $\{g_i\}_{i=1}^{N}$, functional groups $\{F_k\}_{k=1}^{n_f}$ containing both strong and weak members, desired number of clusters *K*.
Output: Disjoint *K*-partitioning of $\{g_i\}_{i=1}^{N}$

1. Select *K* farthest distributed groups from the n_f functional groups via the weighted farthest-first traversal algorithm.
2. For each $k \in \{1,...K\}$, initialize the centroid of cluster C_k with $\frac{1}{w_1 \cdot |F_k\_s| + w_2 \cdot |F_k\_w|}(w_1 \sum_{i \in F_k\_s} g_i + w_2 \sum_{j \in F_k\_w} g_j)$.
3. For each $i \in \{1,...N\}$, calculate the parameterized distance from gene *i* to cluster C_k, i.e., $New\_D_{ik}^{A}$. If gene *i* is closest to its own cluster, do nothing; otherwise, move it into the closest cluster.
4. Re-estimate each cluster centroid with $c_k = \frac{1}{|C_k|}\sum_{g_i \in C_k} g_i$;
   Update parameter matrix *A*.
5. Repeat 3& 4 until no genes moving from one cluster to another.

extractions and 596 were correct, which gave a precision rate of 55.1%. Users were responsible to select the correct extractions. This task of selection was not difficult with the highlights of the gene symbols and biological processes. The functional groups were then detected from these two sources of web knowledge based on GO (Section 2.1.3). A total number of 188 groups were detected.

Third, we set the desired number of clusters K as 4, 5, 6 and 7 respectively. The weighted farthest-first traversal algorithm selected K functional groups. Then we started the semi-supervised clustering. During the initialization of the clustering, the constants (w_1, w_2) were set to be (1, 0.5) in this study, which satisfied $w_1 > w_2$ so that strong members dominated the functional group that they belong to.

The parallel coordinate scheme was used to present the clustering results, where different lines stand for different genes and different colors for different clusters. Figure 3 shows the clustering results of the standard K-means and our method with the case $K = 6$. It indicates that although the main patterns of the up-regulated genes are discovered in both methods, the standard K-means fails to distinguish two different patterns of the down-regulated genes that the knowledge-based clustering succeeds to separate (the light-blue cluster and the red cluster in Figure 3 (b)).

We also investigated the biological meaning of partial clustering results for these two methods with the case $K = 6$. The first sixty genes were examined

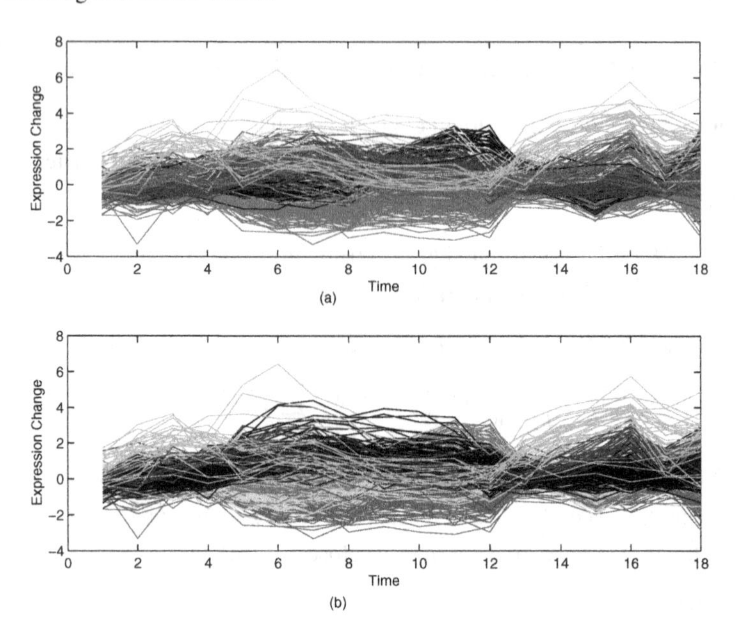

Fig. 3. Clustering results when $K = 6$: (a) K-means; (b) semi-supervised K-means

and the clusters assigned to them by using standard K-means are compared with those by using our method. Three clusters (say C_1, C_2 and C_3) were identified with standard K-means and two clusters (C_1 and C_2) were identified with our method for these sixty genes. While fifty-three of the genes were equally(identically) assigned to C_1 and C_2 by using these two methods, seven genes were assigned into C_3 by using standard K-means and they were still in C_1 and C_2 by using our method. Among these seven different classified genes, five of them were unknown genes, which had no biological information. The other two known genes are CPTI and LUM, members of C_1 and C_2 respectively with our method but members of C_3 with standard K-means. We further examined the biological process of these two genes. LUM is involved in *visual perception* and the members in C_2 (e.g. CYP1B1 and FBN1) are also involved in *visual perception*. This proves that LUM is likely to be a member of C_2 as our method clustered instead of a member of C_3 as standard K-means clustered. Similarly, CPTI is involved in *amino acid metabolism*, which belongs to *cellular metabolism*, while other members of C_1 (e.g. SEPP1 and PIN1) also have the biological processes (*response to oxidative stress* for SEPP1 and *protein folding*) that also belong to *cellular metabolism*. Thus, CPTI is likely to be a member of C_1 also as our method clustered.

To mathematically evaluate the clustering results, two metrics, namely homogeneity (H) and separation (S) are introduced:

$$H = \frac{1}{n}\sum_{i=1}^{n} dist(d_i, center_k)$$

$$S = \frac{1}{\sum_{i \neq j} |C_i| \cdot |C_j|} \sum_{i \neq j} |C_i| \cdot |C_j| \cdot dist(center_i, center_j)$$

The metric H is calculated as the average distance between each data point and the center of the cluster it belongs to. The metric S is calculated as the weighted average distance between cluster centers. The metric H reflects the compactness of the clusters while S reflects the overall distance between clusters. Either a decrement in H or an increment in S brings the better quality of clusters. Table 1 shows that when the knowledge based clustering results in a bigger S (i.e. more separated clusters), there is only a small increment in H (K=4,5,7). In this case, the impact of increasing in the separation (i.e., S) exceeds the impact of a increment in H, which makes the overall clustering result better. When the the knowledge based clustering results in a smaller S, the decrement in H is big enough to contribute to better clustering.

Table 1. Homogeneity and Separation: K-means vs. semi-supervised K-means.

Method/experiments	K=4		K=5		K=6		K=7	
	H	S	H	S	H	S	H	S
K-means	2.5840	5.0161	2.4194	5.0207	2.6830	5.7975	2.2767	4.8822
Semi-supervised K-means	2.6583	5.8019	2.5596	5.7697	2.3901	5.0289	2.2767	5.0922
Increment(+)/Decrement(-)	+2.88%	+15.7%	+5.8%	+14.9%	-4.8%	-10.9%	+1.07%	+3.03%

Therefore, both the observations and the H/S metrics verify that our knowledge-based clustering outperforms the standard K-means.

4 Conclusions and Discussions

This paper presented a general framework of web-knowledge-based clustering for gene expression data. The web was used as the source for gathering useful information to guide clustering. Biological processes were extracted from both the web gene databases and search engines on biomedical documents. Functional groups of genes were then detected from the extracted biological processes based on GO. Finally, the semi-supervised K-means was applied to incorporate the knowledge into the clustering model. The experimental results showed that our knowledge-based clustering model outperformed the clustering model without any knowledge.

A number of interesting problems are left for future research: 1) For time series data, other clustering methods like [10], which considers the dependencies among the time points, may be more suitable than K-means. This type of clustering models are not examined in this paper but will be an interesting issue if we can combine the web knowledge into this type of models. 2) Only abstracts from PubMed are examined for biological process extraction. If full papers are analyzed, we may obtain more useful knowledge. 3) As we mentioned in Section

2.2.2, the functional groups are somewhat overlapped with each other. More work need to be done to deal with overlapping issue to generate more reliable constraints-sensitive distance measure. 4) So far, user need to specify the desired number of clusters to perform clustering. The method to automatic determine the number of clusters in [10] can be also used in this paper.

Acknowledgments: Work reported in this paper is supported in part by the AFOSR grant FA9559- -4-1-0159.

References

1. B. Adryan and R. Schuh. Gene-ontology-based clustering of gene expression data. *Bioinformatics*, 20(16).
2. S. Basu, M. Bilenko, and R. J. Mooney. A probabilistic framework for semi-supervised clustering. In *the Tenth ACM SIGKDD International Conference on Knowledge Discovery and Data Mining*, pages 59–68, Seattle, WA, August 2004.
3. J. Cheng, J. Martin, M. Cline, T. Awad, and M. A. Siani-Rose. Gene expression profiling analysis augmented by mathematically transformed gene ontology. In *International Conference on Intelligent Systems in Molecular Biology ISMB 2002*, August 2002.
4. J.-H. Chiang and H.-C. Yu. Meke: discovering the functions of gene products from biomedical literature via sentence alignment. *Bioinfomatics*, 19(11):1417–1422, 2003.
5. M. B. Eisen, P. T. Spellman, P. O. Brown, and D. Botstein. Cluster analysis and display of genome-wide expression patterns. *Proceedings of Natural Academy Science*, 95(25):14863–14868, December 1998.
6. D. Hanisch, A. Zien, R. Zimmer, and T. Lengauer. Co-clustering of biological networks and gene expression data. *Bioinformatics*, 18(Suppl 1).
7. V. R. Iyer, M. B. Eisen, D. T. Ross, G. Schuler, T. Moore, J. C. F. Lee, J. M. Trent, L. M. Staudt, J. H. Jr., M. S. Boguski, and et al. The transcriptional program in the response of human fibroblasts to serum. *Science*, 283(1):83–87, January 1999.
8. A. Lægreid, T. R. Hvidsten, H. Midelfart, J. Komorowski, and A. K. Sandvik. Predicting gene ontology biological process from temporal gene expression patterns. *Genome Res*, 13:965–979, 2003.
9. G. J. Nau, J. F. L. Richmond, A. Schlesinger, E. G. Jennings, E. S. Lander, and R. A. Young. Human macrophage activation programs induced by bacterial pathogens. *Proceedings of Natural Academy of Sciences of the U. S. A.*, 99(3):1503–1508, February 2002.
10. A. Schliep, A. Schonhuth, and C. Steinhoff. Using hidden markov models to analyze gene expression time course data. *Bioinformatics*, 19:I264–I272, 2003.
11. S. Soderland. Learning information extraction rules for semi-structured and free text. *Machine Learning*, 34(1-3):233–272, 1999.
12. S. Tavazoie, J. Hughes, M. Campbell, R. Cho, and G. Church. Systematic determination of genetic network architecture, 1999.
13. M. L. Whitfield, G. Sherlock, A. J. Saldanha, J. I. Murray, C. A. Ball, K. E. Alexander, J. C. Matese, C. M. Perou, M. M. Hurt, P. O. Brown, and D. Botstein. Identification of genes periodically expressed in the human cell cycle and their expression in tumors. *Molecular Biology of the Cell*, 13(6):1977–2000, June 2002.

Estimations of Similarity in Formal Concept Analysis of Data with Graded Attributes*

Radim Bělohlávek and Vilém Vychodil

Dept. Computer Science, Palacký University Tomkova 40, CZ-779 00, Olomouc, Czech Republic, {radim.belohlavek, vilem.vychodil}@upol.cz

Summary. We study similarity in formal concept analysis of data tables with graded attributes. We focus on similarity related to formal concepts and concept lattices, i.e. the outputs of formal concept analysis. We present several formulas for estimation of similarity of outputs in terms of similarity of inputs. The results answer some problems which arose in previous investigation as well as some natural questions concerning similarity in conceptual data analysis. The derived formulas enable us to compute an estimation of similarity of concept lattices much faster than one can compute their exact similarity. We omit proofs due to lack of space.

Key words: formal concept analysis, fuzzy logic, similarity, concept lattice, hedge

1 Introduction and problem setting

Formal concept analysis (FCA) is a method for analysis of tabular data describing objects and their attributes [8, 9]. There are two basic outputs of FCA, namely, a concept lattice and attribute implications. A concept lattice is a set of all clusters (called formal concepts) extracted from data, hierarchically ordered by subconcept-superconcept relation. Attribute implications are particular expressions describing certain attribute dependencies. Efficient algorithms are known to compute a concept lattice and a non-redundant set of attribute implications which entail all attribute implications true in data.

In the basic setting, attributes are assumed to be bivalent, i.e. either a given attribute y applies to a given object x (indicated by 1 in the data table) or not (indicated by 0). Very often, attributes are graded (fuzzy) rather than bivalent, i.e. an attribute y applies to an object x to a certain degree. FCA of data tables with fuzzy attributes was studied by several authors, we refer to [7] for the first approach, and to [12] and [1]–[5] for the approach we are using in the present paper.

* Supported by grant No. 201/05/0079 of the Czech Science Foundation, by institutional support, research plan MSM 6198959214, and by Kontakt 1–2006–33.

Mark Last et al. (Eds.): Advances in Web Intelligence and Data Mining (SCI) **23**, 243-252 (2006)
www.springerlink.com

The present paper brings up several results related to similarity in FCA of data with fuzzy attributes. Basically, our study is motivated by the following questions: Do similar input data lead to similar outputs of FCA (formal concepts, concept lattices, attribute implications)? Can we obtain estimations of the similarities in question? Can we utilize the similarities to reduce the amount of (input or output) data by putting together similar pieces of data? A study of these problems also tells us about a sensitivity of FCA to exact degrees (in the input data, in the attribute implications, etc.) which is an important issue in fuzzy modeling by itself.

Our paper is a continuation of results from [1] where we studied several issues related to similarity in FCA including a method of parameterized factorization of concept lattices by similarity for which an efficient algorithm was found in [4]. We present results on similarity for concept lattices with hedges [5]. Hedges are parameters controlling the size of concept lattices. Among others, we present formulas for estimation of similarity of concept lattices with hedges in terms of similarity of hedges. Computing the estimations is much faster than computing the exact similarities. Because of the limited scope, we omit proofs and leave them to an extended version of this paper.

2 Preliminaries

We use sets of truth degrees equipped with operations (logical connectives) which form complete residuated lattices, i.e. algebras $\mathbf{L} = \langle L, \wedge, \vee, \otimes, \rightarrow, 0, 1\rangle$ such that $\langle L, \wedge, \vee, 0, 1\rangle$ is a complete lattice with 0 and 1 being the least and greatest element of L, respectively; $\langle L, \otimes, 1\rangle$ is a commutative monoid (i.e. $\otimes$ is commutative, associative, and $a \otimes 1 = 1 \otimes a = a$ for each $a \in L$); $\otimes$ and $\rightarrow$ satisfy so-called adjointness property, i.e. $a \otimes b \leq c$ iff $a \leq b \rightarrow c$, for each $a, b, c \in L$. A truth-stressing hedge (shortly, a hedge) [10, 11] on $\mathbf{L}$ is a unary operation $^* : L \rightarrow L$ satisfying (i) $1^* = 1$, (ii) $a^* \leq a$, (iii) $(a \rightarrow b)^* \leq a^* \rightarrow b^*$, (iv) $a^{**} = a^*$, for all $a, b \in L$. Elements a of L are called truth degrees. $\otimes$ and $\rightarrow$ are (truth functions of) "fuzzy conjunction" and "fuzzy implication". Hedge * is a (truth function of) logical connective "very true" and properties (i)–(iv) have natural interpretations, see [10, 11].

A common choice of $\mathbf{L}$ is a structure with $L = [0, 1]$ (unit interval), $\wedge$ and $\vee$ being minimum and maximum, $\otimes$ being a left-continuous t-norm with the corresponding $\rightarrow$. Three most important pairs of adjoint operations on the unit interval are: Łukasiewicz ($a \otimes b = \max(a + b - 1, 0)$, $a \rightarrow b = \min(1 - a + b, 1)$), Gödel: ($a \otimes b = \min(a, b)$, $a \rightarrow b = 1$ if $a \leq b$, $a \rightarrow b = b$ else), Goguen (product): ($a \otimes b = a \cdot b$, $a \rightarrow b = 1$ if $a \leq b$, $a \rightarrow b = \frac{b}{a}$ else). In applications, we usually need a finite linearly ordered $\mathbf{L}$. For instance, one can put $L = \{a_0 = 0, a_1, \ldots, a_n = 1\} \subseteq [0, 1]$ ($a_0 < \cdots < a_n$) with $\otimes$ given by $a_k \otimes a_l = a_{\max(k+l-n,0)}$ and the corresponding $\rightarrow$ given by $a_k \rightarrow a_l = a_{\min(n-k+l,n)}$. Such an $\mathbf{L}$ is called a finite Łukasiewicz chain.

Two boundary cases of (truth-stressing) hedges are (i) identity, i.e. $a^* = a$ ($a \in L$); (ii) globalization: $1^* = 1$, $a^* = 0$ ($a < 1$). Note that a special case of a complete residuated lattice with a hedge is a two-element Boolean algebra of classical (bivalent) logic.

Having $\mathbf{L}$, we define usual notions [2, 10]: an $\mathbf{L}$-set (fuzzy set) A in universe U is a mapping $A\colon U \to L$, $A(u)$ being interpreted as "the degree to which u belongs to A". Let $\mathbf{L}^U$ denote the collection of all $\mathbf{L}$-sets in U. The operations with $\mathbf{L}$-sets are defined componentwise. For instance, the intersection of $\mathbf{L}$-sets $A, B \in \mathbf{L}^U$ is an $\mathbf{L}$-set $A \cap B$ in U such that $(A \cap B)(u) = A(u) \wedge B(u)$ for each $u \in U$, etc. Binary $\mathbf{L}$-relations (binary fuzzy relations) between X and Y can be thought of as $\mathbf{L}$-sets in the universe $X \times Y$.

Given $A, B \in \mathbf{L}^U$, we define a subsethood degree

$$S(A, B) = \textstyle\bigwedge_{u \in U} \big(A(u) \to B(u)\big), \tag{1}$$

which generalizes the classical subsethood relation $\subseteq$ (note that unlike $\subseteq$, S is a binary $\mathbf{L}$-relation on $\mathbf{L}^U$). Described verbally, $S(A, B)$ represents a degree to which A is a subset of B. In particular, we write $A \subseteq B$ iff $S(A, B) = 1$. As a consequence, $A \subseteq B$ iff $A(u) \leq B(u)$ for each $u \in U$. Given $A, B \in \mathbf{L}^U$, we define an equality degree

$$A \approx B = \textstyle\bigwedge_{u \in U} \big(A(u) \leftrightarrow B(u)\big). \tag{2}$$

It is easily seen that $A \approx B = S(A, B) \wedge S(B, A)$.

A fuzzy relation E in U is called reflexive if for each $u \in U$ we have $E(u, u) = 1$; symmetric if for each $u, v \in U$ we have $E(u, v) = E(v, u)$; transitive if for each $u, v, w \in U$ we have $E(u, v) \otimes E(v, w) \leq E(u, w)$. A fuzzy equivalence [13] in U is a fuzzy relation in U which is reflexive, symmetric, and transitive; a fuzzy equivalence E in U for which $E(u, v) = 1$ implies $u = v$ is called a fuzzy equality. We often denote a fuzzy equivalence by $\approx$ and use an infix notation, i.e. we write $(u \approx v)$ instead of $\approx(u, v)$. If a set U is equipped with a fuzzy equality $\approx$ in U, a fuzzy relation $\preceq$ in U is called a fuzzy order in $\langle U, \approx \rangle$ [2, 13] if $\preceq$ is reflexive, transitive, and antisymmetric w.r.t. $\approx$, i.e. for each $u, v \in U$ we have $(u \preceq v) \wedge (v \preceq u) \leq (u \approx v)$, and if $\preceq$ is compatible with $\approx$, i.e. $(u_1 \preceq v_1) \otimes (u_1 \approx u_2) \otimes (v_1 \approx v_2) \leq (u_2 \preceq v_2)$, see [2] for details.

3 Similarity in concept lattices with hedges

3.1 Concept lattices with hedges

Let X and Y be sets of objects and attributes, respectively, $I \in \mathbf{L}^{X \times Y}$ be a fuzzy relation between X and Y with $I(x, y)$ being interpreted as a degree to which object $x \in X$ has attribute $y \in Y$. The triplet $\langle X, Y, I \rangle$ is called a data table with fuzzy attributes.

Let $*_X$ and $*_Y$ be hedges. For **L**-sets $A \in \mathbf{L}^X$ (**L**-set of objects), $B \in \mathbf{L}^Y$ (**L**-set of attributes) we define **L**-sets $A^\uparrow \in \mathbf{L}^Y$ (**L**-set of attributes), $B^\downarrow \in \mathbf{L}^X$ (**L**-set of objects) by $A^\uparrow(y) = \bigwedge_{x \in X}(A(x)^{*_X} \to I(x,y))$, and $B^\downarrow(x) = \bigwedge_{y \in Y}(B(y)^{*_Y} \to I(x,y))$. We put $\mathcal{B}(X^{*_X}, Y^{*_Y}, I) = \{\langle A, B\rangle \in \mathbf{L}^X \times \mathbf{L}^Y \mid A^\uparrow = B, B^\downarrow = A\}$. For $\langle A_1, B_1\rangle, \langle A_2, B_2\rangle \in \mathcal{B}(X^{*_X}, Y^{*_Y}, I)$, put $\langle A_1, B_1\rangle \leq \langle A_2, B_2\rangle$ iff $A_1 \subseteq A_2$ (or, iff $B_2 \subseteq B_1$; both ways are equivalent). Operators $^\downarrow, ^\uparrow$ form a Galois connection with hedges [5]. $\langle \mathcal{B}(X^{*_X}, Y^{*_Y}, I), \leq\rangle$ is called a (fuzzy) concept lattice with hedges ^{*_X} and ^{*_Y} induced by $\langle X, Y, I\rangle$ [5]. For $*_Y = \mathrm{id}_L$ (identity), we write only $\mathcal{B}(X^{*_X}, Y, I)$. Elements $\langle A, B\rangle$ of $\mathcal{B}(X^{*_X}, Y^{*_Y}, I)$ are naturally interpreted as concepts (clusters) hidden in the input data represented by I. Namely, $A^\uparrow = B$ and $B^\downarrow = A$ say that B is the collection of all attributes shared by all objects from A, and A is the collection of all objects sharing all attributes from B. A and B are called the extent and the intent of the concept $\langle A, B\rangle$, respectively, and represent the collection of all objects and all attributes covered by $\langle A, B\rangle$. $\leq$ models a subconcept-superconcept hierarchy.

For each $\langle X, Y, I\rangle$ we consider a set $\mathrm{Ext}(X^{*_X}, Y^{*_Y}, I) \subseteq \mathbf{L}^X$ of all extents and a set $\mathrm{Int}(X^{*_X}, Y^{*_Y}, I) \subseteq \mathbf{L}^Y$ of all intents of concepts of $\mathcal{B}(X^{*_X}, Y^{*_Y}, I)$, i.e.

$$\mathrm{Ext}(X^{*_X}, Y^{*_Y}, I) = \{A \in \mathbf{L}^X \mid \langle A, B\rangle \in \mathcal{B}(X^{*_X}, Y^{*_Y}, I) \text{ for some } B \in \mathbf{L}^Y\},$$

$$\mathrm{Int}(X^{*_X}, Y^{*_Y}, I) = \{B \in \mathbf{L}^Y \mid \langle A, B\rangle \in \mathcal{B}(X^{*_X}, Y^{*_Y}, I) \text{ for some } A \in \mathbf{L}^X\}.$$

For details, we refer to [2, 3, 5].

3.2 Similarity of concept lattices with hedges

The aim of this section is to study relationships between $\mathcal{B}(X^{*_1}, Y^{*_3}, I)$ and $\mathcal{B}(X^{*_2}, Y^{*_4}, I)$, i.e. between sets of clusters extracted from a data table $\langle X, Y, I\rangle$ which differ in hedges (^{*_1} and ^{*_3} in case of $\mathcal{B}(X^{*_1}, Y^{*_3}, I)$, and ^{*_2} and ^{*_4} in case of $\mathcal{B}(X^{*_2}, Y^{*_4}, I)$). The reason for this study is the following. Hedges ^{*_X} and ^{*_Y} serve as parameters. Their primary role is to control the size of $\mathcal{B}(X^{*_X}, Y^{*_Y}, I)$, i.e. to control the number $|\mathcal{B}(X^{*_X}, Y^{*_Y}, I)|$ of clusters extracted from data table $\langle X, Y, I\rangle$. Basic theoretical and experimental results were presented in [5]. In particular, it was demonstrated in [5] that tuning the hedges leads to a smooth change of the size of concept lattices, and that stronger hedges lead to a smaller number of extracted formal concepts, see [5] and Section 4. Preliminary theoretical results were also obtained in [5]. For instance, it was shown that if both ^{*_1} and ^{*_2} are identities and if ^{*_3} is stronger than ^{*_4}, i.e. $a^{*_3} \leq a^{*_4}$ for each $a \in L$, then $\mathcal{B}(X^{*_1}, Y^{*_3}, I)$ is a subset of $\mathcal{B}(X^{*_2}, Y^{*_4}, I)$.

First, we need to propose a tractable definition of a degree to which $\mathcal{B}(X^{*_1}, Y^{*_3}, I)$ is similar to $\mathcal{B}(X^{*_2}, Y^{*_4}, I)$. The definitions follow. We start by a degree $\mathcal{M}_1 \preceq \mathcal{M}_2$ to which a system $\mathcal{M}_1$ of fuzzy sets is contained in a system $\mathcal{M}_2$ of fuzzy sets. A degree $\mathcal{M}_1 \approx \mathcal{M}_2$ to which $\mathcal{M}_1$ and $\mathcal{M}_2$ are similar is then defined as a "conjunction" of $\mathcal{M}_1 \preceq \mathcal{M}_2$ and $\mathcal{M}_2 \preceq \mathcal{M}_1$.

Definition 1. *For systems* $\mathcal{M}_1, \mathcal{M}_2 \subseteq \mathbf{L}^U$ *of fuzzy sets in* U *define*

$$\mathcal{M}_1 \preceq \mathcal{M}_2 = \textstyle\bigwedge_{A_1 \in \mathcal{M}_1} \bigvee_{A_2 \in \mathcal{M}_2} A_1 \approx A_2,$$
$$\mathcal{M}_1 \approx \mathcal{M}_2 = (\mathcal{M}_1 \preceq \mathcal{M}_2) \wedge (\mathcal{M}_2 \preceq \mathcal{M}_1).$$

Remark 1. (1) $A_1 \approx A_2$ is a degree of equality defined by (2). Note that we may take another suitable definition of $A_1 \approx A_2$. In this paper, however, we do not consider other options.

(2) It can be seen that $\mathcal{M}_1 \preceq \mathcal{M}_2$ is a truth degree of "for each $A_1 \in \mathcal{M}_1$ there is $A_2 \in \mathcal{M}_2$ such that A_1 and A_2 are similar."

The following is another definition which formalizes the same idea.

Definition 2. *For systems* $\mathcal{M}_1, \mathcal{M}_2 \subseteq \mathbf{L}^U$ *of fuzzy sets in* U *define*

$$\mathcal{M}_1 \preceq_c \mathcal{M}_2 = \bigvee\{c \in L \mid \text{ for each } A_1 \in \mathcal{M}_1 \text{ there is } A_2 \in \mathcal{M}_2 : c \leq (A_1 \approx A_2)\},$$
$$\mathcal{M}_1 \approx_c \mathcal{M}_2 = (\mathcal{M}_1 \preceq_c \mathcal{M}_2) \wedge (\mathcal{M}_2 \preceq_c \mathcal{M}_1).$$

Remark 2. Thus, $\mathcal{M}_1 \preceq_c \mathcal{M}_2$ can be seen as the largest degree c such that for each $A_1 \in \mathcal{M}_1$ there is $A_2 \in \mathcal{M}_2$ with $c \leq (A_1 \approx A_2)$.

The following lemma summarizes basic properties of the above concepts.

Lemma 1. $\approx$ *and* $\approx_c$ *are fuzzy equalities in* $2^{\mathbf{L}^U}$. $\preceq$ *and* $\preceq_c$ *are fuzzy orders in* $\langle 2^{\mathbf{L}^U}, \approx\rangle$ *and* $\langle 2^{\mathbf{L}^U}, \approx_c\rangle$. *We have* $(\mathcal{M}_1 \preceq_c \mathcal{M}_2) \leq (\mathcal{M}_1 \preceq \mathcal{M}_2)$ *and* $(\mathcal{M}_1 \approx_c \mathcal{M}_2) \leq (\mathcal{M}_1 \approx \mathcal{M}_2)$. *If* $\mathbf{L}$ *is a finite chain then* $\preceq$ *coincides with* $\preceq_c$ *and* $\approx$ *coincides with* $\approx_c$.

Using the above definition, we are going to define degrees of containment and similarity between concept lattices with hedges. Because of the limited scope we present only definitions and results based on $\preceq$ and $\approx$. For the sake of brevity, we denote

$$\mathcal{B}_{1,3} = \mathcal{B}(X^{*_1}, Y^{*_3}, I) \quad \text{and} \quad \mathcal{B}_{2,4} = \mathcal{B}(X^{*_2}, Y^{*_4}, I).$$

Definition 3. *Put*

$$\mathcal{B}_{1,3} \preceq_{\text{Ext}} \mathcal{B}_{2,4} = \text{Ext}(X^{*_1}, Y^{*_3}, I) \preceq \text{Ext}(X^{*_2}, Y^{*_4}, I),$$
$$\mathcal{B}_{1,3} \preceq_{\text{Int}} \mathcal{B}_{2,4} = \text{Int}(X^{*_1}, Y^{*_3}, I) \preceq \text{Int}(X^{*_2}, Y^{*_4}, I),$$
$$\mathcal{B}_{1,3} \approx_{\text{Ext}} \mathcal{B}_{2,4} = \text{Ext}(X^{*_1}, Y^{*_3}, I) \approx \text{Ext}(X^{*_2}, Y^{*_4}, I),$$
$$\mathcal{B}_{1,3} \approx_{\text{Int}} \mathcal{B}_{2,4} = \text{Int}(X^{*_1}, Y^{*_3}, I) \approx \text{Int}(X^{*_2}, Y^{*_4}, I).$$

Remark 3. (1) Thus, $\mathcal{B}_{1,3} \preceq_{\text{Ext}} \mathcal{B}_{2,4}$ is a degree to which the system of extents of $\mathcal{B}_{1,3}$ is contained in the system of extents of $\mathcal{B}_{2,4}$ in the sense of Definition 2; analogously for $\mathcal{B}_{1,3} \preceq_{\text{Int}} \mathcal{B}_{2,4}$, $\mathcal{B}_{1,3} \approx_{\text{Ext}} \mathcal{B}_{2,4}$, and $\mathcal{B}_{1,3} \approx_{\text{Int}} \mathcal{B}_{2,4}$. Note that in general, $\mathcal{B}_{1,3} \preceq_{\text{Ext}} \mathcal{B}_{2,4}$ may differ from $\mathcal{B}_{1,3} \preceq_{\text{Int}} \mathcal{B}_{2,4}$ and the choice is up to the user (whether he/she is interested in clusters of objects or attributes). One might of course take also $\mathcal{B}_{1,3} \preceq \mathcal{B}_{2,4} = (\mathcal{B}_{1,3} \preceq_{\text{Ext}} \mathcal{B}_{2,4}) \wedge (\mathcal{B}_{1,3} \preceq_{\text{Int}} \mathcal{B}_{2,4})$.

(2) Note that we have $\mathcal{B}_{1,3} \approx_{\text{Ext}} \mathcal{B}_{2,4} = (\mathcal{B}_{1,3} \preceq_{\text{Ext}} \mathcal{B}_{2,4}) \wedge (\mathcal{B}_{2,4} \preceq_{\text{Ext}} \mathcal{B}_{1,3})$ and $\mathcal{B}_{1,3} \approx_{\text{Int}} \mathcal{B}_{2,4} = (\mathcal{B}_{1,3} \preceq_{\text{Int}} \mathcal{B}_{2,4}) \wedge (\mathcal{B}_{2,4} \preceq_{\text{Int}} \mathcal{B}_{1,3})$.

We are going to present formulas for estimation of lower bounds of the degrees introduced in Definition 3. This is particularly interesting if we start with $\mathcal{B}(X^{*_1}, Y^{*_3}, I)$, find out that $\mathcal{B}(X^{*_1}, Y^{*_3}, I)$ is too large, replace hedges ^{*_1} and ^{*_3} by (stronger) hedges ^{*_2} and ^{*_4}, and consider $\mathcal{B}(X^{*_2}, Y^{*_4}, I)$ instead of $\mathcal{B}(X^{*_1}, Y^{*_3}, I)$. Then one wants to know a degree to which $\mathcal{B}(X^{*_2}, Y^{*_4}, I)$ is contained in (similar to) $\mathcal{B}(X^{*_1}, Y^{*_3}, I)$. We need the following definition describing relationships between hedges.

Definition 4. *For hedges $*_1$ and $*_2$ on $\mathbf{L}$ put*

$$(*_1 \preceq *_2) = \textstyle\bigwedge_{a \in L}(a^{*_1} \rightarrow a^{*_2}),$$
$$(*_1 \approx *_2) = \textstyle\bigwedge_{a \in L}(a^{*_1} \leftrightarrow a^{*_2}).$$

Remark 4. (1) Since $a \leftrightarrow b$ (degree of equivalence of a and b) can be seen as a degree to which degrees a and b are similar, $*_1 \approx *_2$ can be interpreted as a degree to which hedges $*_1$ and $*_2$ are similar (yield similar results). Analogously, $*_1 \preceq *_2$ can be interpreted as a degree to which $*_1$ is stronger than $*_2$.

(2) Note that $(*_1 \approx *_2) = (*_1 \preceq *_2) \wedge (*_2 \preceq *_1)$.

Lemma 2. $\approx$ *is a fuzzy equality relation on the set of all truth stressers on* $\mathbf{L}$*;* $\preceq$ *is a fuzzy order on the set of all truth stressers on* $\mathbf{L}$ *equipped with* $\approx$.

Degrees $*_1 \preceq *_2$, $*_1 \approx *_2$ enable us to deduce some natural properties of hedges $*_1$ and $*_2$. As an example, for a hedge $*$ on $\mathbf{L}$ denote

$$\text{fix}(*) = \{a \in L \mid a^* = a\}$$

the set of all fixpoints of $*$. For $K_1, K_2 \subseteq L$, put

$$(K_1 \preceq K_2) = \textstyle\bigwedge_{a \in K_1} \bigvee_{b \in K_2}(a \leftrightarrow b),$$
$$(K_1 \approx K_2) = (K_1 \preceq K_2) \wedge (K_2 \preceq K_1).$$

$K_1 \preceq K_2$ ($K_1 \approx K_2$) is a degree to which K_1 is contained in (similar to) K_2. The following lemma shows that stronger hedges have smaller sets of fixpoints and that similar hedges have similar sets of fixpoints.

Lemma 3. $(*_1 \preceq *_2) \leq (\text{fix}(*_1) \preceq \text{fix}(*_2))$, $(*_1 \approx *_2) \leq (\text{fix}(*_1) \approx \text{fix}(*_2))$.

Before presenting main results of this section, we need some auxiliary claims. For $i = 1, 2, 3, 4$, denote by $^{\uparrow_i}$ and $^{\downarrow_i}$ the mappings induced by a hedge ^{*_i}, i.e. $A^{\uparrow_i}(y) = \bigwedge_{x \in X}(A^{*_i}(x) \rightarrow I(x, y))$ and $B^{\downarrow_i}(x) = \bigwedge_{y \in Y}(B^{*_i}(y) \rightarrow I(x, y))$. Then we have

Lemma 4. *For* $A \in \mathbf{L}^X$, $(*_1 \preceq *_3) \leq S(A^{\uparrow_3}, A^{\uparrow_1})$ *and* $(*_1 \approx *_3) \leq (A^{\uparrow_1} \approx A^{\uparrow_3})$*; dually for* $B \in \mathbf{L}^Y$ *and* ${}^{\downarrow_2}$ *and* ${}^{\downarrow_4}$.

The main results providing estimations of degrees of containment and degrees of similarity of concept lattices with hedges follow.

Theorem 1. *We have the following estimation formulas:*

$$\begin{aligned}
(*_1 \preceq *_2) \otimes (*_3 \approx *_4) &\leq \mathcal{B}_{1,3} \preceq_{\text{Ext}} \mathcal{B}_{2,4},\\
(*_3 \preceq *_4) \otimes (*_1 \approx *_2) &\leq \mathcal{B}_{1,3} \preceq_{\text{Int}} \mathcal{B}_{2,4},\\
(*_1 \approx *_2) \otimes (*_3 \approx *_4) &\leq \mathcal{B}_{1,3} \approx_{\text{Ext}} \mathcal{B}_{2,4},\\
(*_1 \approx *_2) \otimes (*_3 \approx *_4) &\leq \mathcal{B}_{1,3} \approx_{\text{Int}} \mathcal{B}_{2,4}.
\end{aligned}$$

The next two theorems provide formulas for the case when the hedges by attributes are equal and, moreover, are identities.

Theorem 2. *We have the following estimation formulas:*

$$\begin{aligned}
(*_1 \preceq *_2) &\leq \mathcal{B}(X^{*_1}, Y^{*_Y}, I) \preceq_{\text{Ext}} \mathcal{B}(X^{*_2}, Y^{*_Y}, I),\\
(*_1 \approx *_2) &\leq \mathcal{B}(X^{*_1}, Y^{*_Y}, I) \preceq_{\text{Int}} \mathcal{B}(X^{*_2}, Y^{*_Y}, I),\\
(*_1 \approx *_2) &\leq \mathcal{B}(X^{*_1}, Y^{*_Y}, I) \approx_{\text{Ext}} \mathcal{B}(X^{*_2}, Y^{*_Y}, I),\\
(*_1 \approx *_2) &\leq \mathcal{B}(X^{*_1}, Y^{*_Y}, I) \approx_{\text{Int}} \mathcal{B}(X^{*_2}, Y^{*_Y}, I).
\end{aligned}$$

Theorem 3. *For* $*_Y$ *being identity on* L *we have*

$$(*_1 \preceq *_2) \leq \mathcal{B}(X^{*_1}, Y^{*_Y}, I) \preceq_{\text{Int}} \mathcal{B}(X^{*_2}, Y^{*_Y}, I).$$

Remark 5. Note that for ^{*_Y} being identity we not only have the previous better estimation but one can show that for $\langle A, B\rangle \in \mathcal{B}(X^{*_1}, Y^{*_Y}, I)$ there is $\langle C, D\rangle \in \mathcal{B}(X^{*_2}, Y^{*_Y}, I)$ such that both $(*_1 \preceq *_2) \leq (A \approx C)$ and $(*_1 \preceq *_2) \leq (B \approx D)$. Note also that the case when ^{*_X} is identity (dual situation to ^{*_Y} being identity) is important in studying fuzzy attribute implications.

We postpone further ramifications of the previous results to an extended version of this paper.

4 Example and remarks

This section presents an illustrative example of similarities of concept lattices with hedges. Let $\mathbf{L}$ be a finite Łukasiewicz chain with $L = \{0, 0.25, 0.5, 0.75, 1\}$. There are five hedges on $\mathbf{L}$; they are depicted in Fig. 1: the left-most hedge (denoted by $*_1$) is globalization on L, the right-most one (denoted by $*_5$) is identity on L; for $*_2$ we have $0^{*_2} = 0.25^{*_2} = 0$, $0.5^{*_2} = 0.75^{*_2} = 0.5$, $1^{*_2} = 1$, etc. Table 1 (left) contains the fuzzy order $\preceq$ on hedges: a table

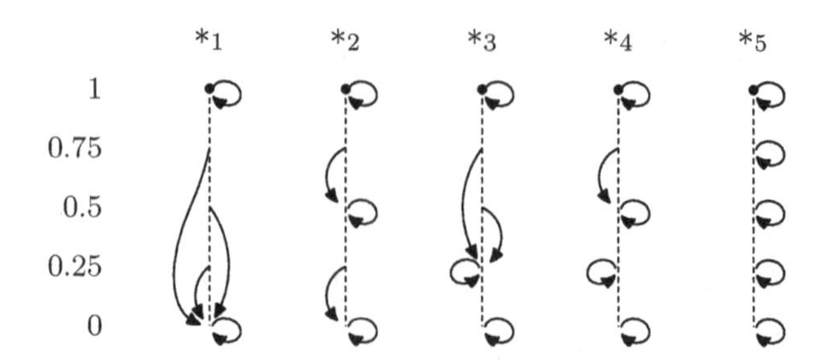

Fig. 1. Truth stressers of five-element Łukasiewicz chain

Table 1. Fuzzy order and fuzzy equality on truth-stressers

$\preceq$	$*_1$	$*_2$	$*_3$	$*_4$	$*_5$
$*_1$	1	1	1	1	1
$*_2$	0.5	1	0.75	1	1
$*_3$	0.75	0.75	1	1	1
$*_4$	0.5	0.75	0.75	1	1
$*_5$	0.25	0.75	0.5	0.75	1

$\approx$	$*_1$	$*_2$	$*_3$	$*_4$	$*_5$
$*_1$	1	0.5	0.75	0.5	0.25
$*_2$	0.5	1	0.75	0.75	0.75
$*_3$	0.75	0.75	1	0.75	0.5
$*_4$	0.5	0.75	0.75	1	0.75
$*_5$	0.25	0.75	0.5	0.75	1

entry corresponding to a row $*_i$ and a column $*_j$ contains degree $*_i \preceq *_j$. Table 1 (right) displays the fuzzy equality $\approx$ on hedges.

Consider a data table $\langle X, Y, I\rangle$ given by table in Fig. 2 (left). The set X of object consists of "Mercury", "Venus", ..., set Y contains four attributes: size of the planet (small / large), distance from the sun (far / near). Since we have five hedges on $\mathbf{L}$, the input data table $\langle X, Y, I\rangle$ induces 25 (possibly different) output concept lattices with hedges $\mathcal{B}(X^{*_1}, Y^{*_1}, I), \ldots, \mathcal{B}(X^{*_5}, Y^{*_5}, I)$. For brevity, each concept lattice $\mathcal{B}(X^{*_i}, Y^{*_j}, I)$ will be denoted by $\mathcal{B}_{i,j}$. All the concept lattices are depicted in Fig. 2 (right) ($\mathcal{B}_{i,j}$ lies on the intersection of row $*_i$ and column $*_j$). Each concept lattice is depicted by its Hasse diagram. The nodes of the diagram represent the clusters (concepts) in the data $\langle X, Y, I\rangle$; the edges represent the partial ordering of the clusters (subconcept-superconcept hierarchy), see Section 3.1. Note that $\mathcal{B}_{1,1}$ (both hedges are globalization) contains 8 clusters (concepts) while $\mathcal{B}_{5,5}$ (both hedges are identity) consists of 216 clusters.

Looking at the concept lattices in Fig. 2, we might say that, e.g., $\mathcal{B}_{4,5}$, $\mathcal{B}_{5,4}$, and $\mathcal{B}_{5,5}$ are (very) similar while $\mathcal{B}_{1,1}$ is quite different (much simpler) from each of $\mathcal{B}_{4,5}$, $\mathcal{B}_{5,4}$, and $\mathcal{B}_{5,5}$. This intuitive observation agrees with degrees of similarity of these fuzzy concept lattices. Indeed, Table 2 (left) contains degrees of similarity of intents, i.e. a table entry on the intersection of row $\mathcal{B}_{i,k}$ and column $\mathcal{B}_{j,l}$ contains degree $\mathcal{B}_{i,k} \approx_{\mathrm{Int}} \mathcal{B}_{j,l}$. We have, e.g., $\mathcal{B}_{4,5} \approx_{\mathrm{Int}} \mathcal{B}_{5,5} = 0.75$ while $\mathcal{B}_{1,1} \approx_{\mathrm{Int}} \mathcal{B}_{5,5} = 0.25$. Table 2 (right) contains estimations of degrees of similarity of intents: a table entry on the intersection of row $\mathcal{B}_{i,k}$ and column $\mathcal{B}_{j,l}$ contains truth degree $(*_i \approx *_j) \otimes (*_k \approx *_l)$. Observe that in some cases, estimations given by Table 2 (right) are equal to the values of

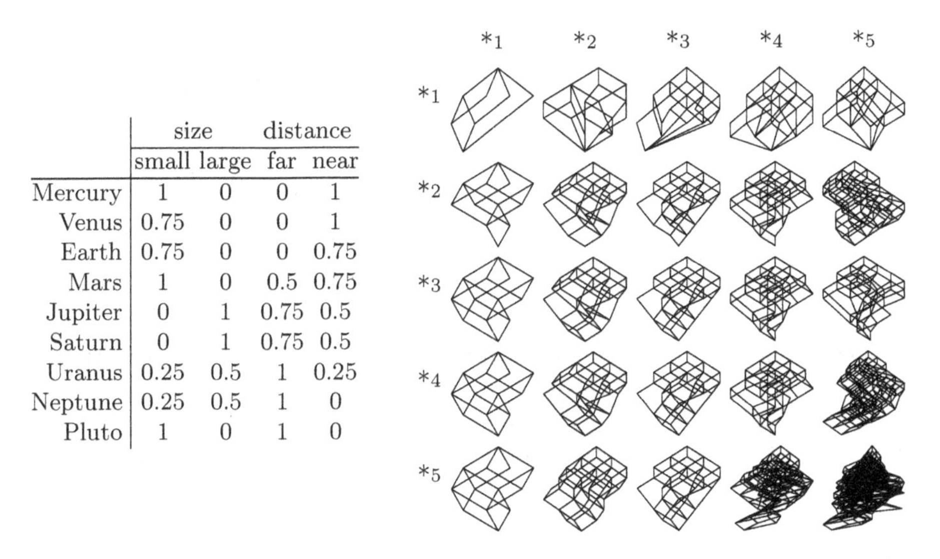

	size		distance	
	small	large	far	near
Mercury	1	0	0	1
Venus	0.75	0	0	1
Earth	0.75	0	0	0.75
Mars	1	0	0.5	0.75
Jupiter	0	1	0.75	0.5
Saturn	0	1	0.75	0.5
Uranus	0.25	0.5	1	0.25
Neptune	0.25	0.5	1	0
Pluto	1	0	1	0

Fig. 2. Data table with fuzzy attributes (left); concept lattices generated from the data table by all combinations of truth stressers $*_X$ and $*_Y$ from Fig. 1 (right).

similarity over intents, in some cases, however, the estimation is strictly lower. On the other hand, the cost of computing values $(*_i \approx *_j) \otimes (*_k \approx *_l)$ is much smaller than the cost of computing $\mathcal{B}_{i,k} \approx_{\mathrm{Int}} \mathcal{B}_{j,l}$ especially in case of large input data. Table 3 depicts fuzzy order over extents, i.e. $\mathcal{B}_{i,k} \preceq_{\mathrm{Ext}} \mathcal{B}_{j,l}$, and its estimation, i.e. $(*_i \preceq *_j) \otimes (*_k \approx *_l)$. Note that the estimations of Theorems 2 and 3 provide closer approximations of the estimated degrees (details in the extended version).

Table 2. Similarity over intents and its estimation

$\approx_{\mathrm{Int}}$	$\mathcal{B}_{1,1}$	$\mathcal{B}_{3,3}$	$\mathcal{B}_{4,5}$	$\mathcal{B}_{5,4}$	$\mathcal{B}_{5,5}$
$\mathcal{B}_{1,1}$	1	0.5	0.5	0.5	0.25
$\mathcal{B}_{3,3}$	0.5	1	0.5	0.5	0.5
$\mathcal{B}_{4,5}$	0.5	0.5	1	0.75	0.75
$\mathcal{B}_{5,4}$	0.5	0.5	0.75	1	0.75
$\mathcal{B}_{5,5}$	0.25	0.5	0.75	0.75	1

est.	$\mathcal{B}_{1,1}$	$\mathcal{B}_{3,3}$	$\mathcal{B}_{4,5}$	$\mathcal{B}_{5,4}$	$\mathcal{B}_{5,5}$
$\mathcal{B}_{1,1}$	1	0.5	0	0	0
$\mathcal{B}_{3,3}$	0.5	1	0.25	0.25	0
$\mathcal{B}_{4,5}$	0	0.25	1	0.5	0.75
$\mathcal{B}_{5,4}$	0	0.25	0.5	1	0.75
$\mathcal{B}_{5,5}$	0	0	0.75	0.75	1

5 Future research

Future research as well as topics which did not fit the limited extent of this paper include the following: factorization of concept lattices with hedges and collections of attribute implications by putting together similar concepts and

Table 3. Fuzzy order over extents and its estimation

$\preceq_{\text{Ext}}$	$\mathcal{B}_{1,1}$	$\mathcal{B}_{3,3}$	$\mathcal{B}_{4,5}$	$\mathcal{B}_{5,4}$	$\mathcal{B}_{5,5}$
$\mathcal{B}_{1,1}$	1	1	1	1	1
$\mathcal{B}_{3,3}$	0.5	1	0.75	1	1
$\mathcal{B}_{4,5}$	0.5	0.75	1	0.75	1
$\mathcal{B}_{5,4}$	0.5	0.75	0.75	1	1
$\mathcal{B}_{5,5}$	0.25	0.75	0.75	0.75	1

est.	$\mathcal{B}_{1,1}$	$\mathcal{B}_{3,3}$	$\mathcal{B}_{4,5}$	$\mathcal{B}_{5,4}$	$\mathcal{B}_{5,5}$
$\mathcal{B}_{1,1}$	1	0.75	0.25	0.5	0.25
$\mathcal{B}_{3,3}$	0.5	1	0.5	0.75	0.5
$\mathcal{B}_{4,5}$	0	0.25	1	0.75	1
$\mathcal{B}_{5,4}$	0	0.25	0.5	1	0.75
$\mathcal{B}_{5,5}$	0	0	0.75	0.75	1

similar implications; results concerning similarity and validity of attribute implications; similarity of theories consisting of attribute implications [6] (do similar data tables have similar non-redundant bases of attribute implications? etc.); similarity results based on other measures of similarity of fuzzy sets.

References

1. Bělohlávek R.: Similarity relations in concept lattices. *J. Logic Comput.* 10(6):823–845, 2000.
2. Bělohlávek R.: *Fuzzy Relational Systems: Foundations and Principles.* Kluwer, Academic/Plenum Publishers, New York, 2002.
3. Bělohlávek R.: Concept lattices and order in fuzzy logic. *Ann. Pure Appl. Logic* **128**(2004), 277–298.
4. Bělohlávek R., Dvořák J., Outrata J.: Fast factorization by similarity in formal concept analysis of data with fuzzy attributes (submitted). Preliminary version in Proc. CLA 2004, pp. 47–57.
5. Bělohlávek R., Vychodil V.: Reducing the size of fuzzy concept lattices by hedges. Proc. FUZZ-IEEE, pp. 663–668, Reno, Nevada, 2005.
6. Bělohlávek R., Vychodil V.: Fuzzy attribute logic: attribute implications, their validity, entailment, and non-redundant basis. Proc. IFSA 2005, Vol. I, pp. 622–627.
7. Burusco A., Fuentes-Gonzáles R.: The study of the L-fuzzy concept lattice. *Mathware & Soft Computing*, **3**(1994), 209–218.
8. Carpineto C., Romano G.: *Concept Data Analysis. Theory and Applications.* J. Wiley, 2004.
9. Ganter B., Wille R.: *Formal Concept Analysis. Mathematical Foundations.* Springer, Berlin, 1999.
10. Hájek P.: *Metamathematics of Fuzzy Logic.* Kluwer, Dordrecht, 1998.
11. Hájek P.: On very true. *Fuzzy Sets and Systems* **124**(2001), 329–333.
12. Pollandt S.: *Fuzzy Begriffe.* Springer-Verlag, Berlin/Heidelberg, 1997.
13. Zadeh L. A.: Similarity relations and fuzzy orderings. *Information Sciences* **3**(1971), 159–176.

Kernels for the Relevance Vector Machine - An Empirical Study

David Ben-Shimon and Armin Shmilovici

Dept. of Information Systems Engineering, Ben-Gurion University, P.O.B 653, 84105 Beer-Sheva, Israel {dudibs,armin}@bgumail.bgu.ac.il

Abstract. The Relevance Vector Machine (RVM) is a generalized linear model that can use kernel functions as basis functions. Experiments with the Matérn kernel indicate that the kernel choice has a significant impact on the sparsity of the solution. Furthermore, not every kernel is suitable for the RVM. Our experiments indicate that the Matérn kernel of order 3 is a good initial choice for many types of data.
Keywords: Machine Learning, Relevance Vector Machine, Kernel Regression, Matérn Kernel

1 Introduction

Suppose that N noisy observations of an unknown function $f : R^d \to R$ are available:

$$Y_i = f(X_i) + \varepsilon_i \tag{1}$$

Suppose that f can be expressed in a form of some infinite expansion:

$$f(x) = \sum_{j=1}^{\infty} \theta_j^* g_j(x) \tag{2}$$

where $\{g_j(x)\}_0^\infty$ is an unknown family of basis functions. In the regression problem estimating f reduces to estimation of a suitable truncation of the vector of all parameters $\Theta = (\theta_0^* ... \theta_n^*)^T$, using the observations $\{X_i, Y_i\}_{i=1}^N$ and (2) is called a generalized linear model (GLM). To limit the number of parameters such that the coefficients θ_j^* decrease in a certain way as $j \to \infty$ some smoothness or regularity assumptions have to be stated about f. Generally speaking, smoothness conditions require that the unknown function f belongs to a particular restricted functional class. Otherwise, convergence can be arbitrary slow[1].

One commonly used implementation of (2) is the Parzen-Rosenblatt density estimator defined as:

Mark Last et al. (Eds.): Advances in Web Intelligence and Data Mining (SCI) **23**, 253-263 (2006)
www.springerlink.com

$$\hat{f}_N(x) = \frac{1}{Nh_N^d} \sum_{i=1}^{N} k\left(\frac{x_i - x}{h_N}\right) \tag{3}$$

where the positive number h_N is called the bandwidth or scaling factor and the function k is called a kernel. A kernel function is a positive definite function [2] which decreases very fast outside the window $[x - h_N, x + h_N]$, thus, the estimator (3) is a moving average of the observations belonging to that window. The accuracy of the approximation (3) depends on how densely observation points fill the input space. Efficient uniform error bounds are available for kernel estimators when the function f is further restricted to the class of functions bounded by a polynomial of (unknown) orders [1]. Noisy observations introduce error in the estimation of the regression coefficients θ_j^*. The total mean square error of the estimates will be the sum of the stochastic part (because of the noise) and of the bias due to the approximation error. Thus, the optimal choice for the regression problem will depend more on the characteristics of the kernel function and less on the characteristics of the unknown f.

The use of kernels has received considerable attention in machine learning [2]. The kernel matrix is symmetric and positive definite matrix, thus, it can be defined as some kind of similarity between pairs of data points such as $k(x, x') = \left\langle \Psi(x), \Psi(x') \right\rangle$. The transform $\Psi : X \to H$ from X, the input space to H, is often used to embed the training data into a high dimensional feature space. The assumption is that it could be easier to obtain the solution for a specific problem in the feature space. Using this technique, there is no restriction on the dimensionality of the data, since usually the number of examples N is much larger than the number of dimensions d.

The freedom in the choice of the mapping Ψ enables us to design a large variety of similarity measures adapted to the given problem [2], [10]. In practice, most applications of kernel methods just use the Gaussian kernel $k(x, x_i) = exp^{-||x-x_i||^2/2\sigma^2}$ where the $||||$ operator indicates the distance between the any two points, and σ is the width parameter of the Gaussian. The Gaussian kernel is also called the universal kernel. It is an infinitely smooth function, thus, may not be the best choice for noisy datasets.

The Matérn kernel [2], [3] is unique because it has an extra parameter v to explicitly control the smoothness of the kernel. The Matérn function of order v belongs to the class of functions bounded by a polynomial of order v. One formulation of the Matérn kernel takes the following form:

$$M(x, x_i) = \frac{2(\frac{\sqrt{v}}{\sigma}||x - x_i||)^v}{\Gamma(v)} K_v(2\frac{\sqrt{v}}{\sigma}||x - x_i||) \tag{4}$$

where $\Gamma(v)$ is the gamma function and $K_v(x)$ is the modified Bessel function of the second kind[1] of order v, and σ in this case is the width scaling parameter of the Matérn function.

When $v \to \infty$ the Matérn kernel degenerates to the Gaussian kernel and when v=0.5 it degenerates to the exponential kernel . Thus, the Matérn kernel is able to define a wide range of kernel functions. Figure 1 illustrates the Matérn kernel with different degrees of smoothness and its ability to behave as the Gaussian and the exponential kernels.

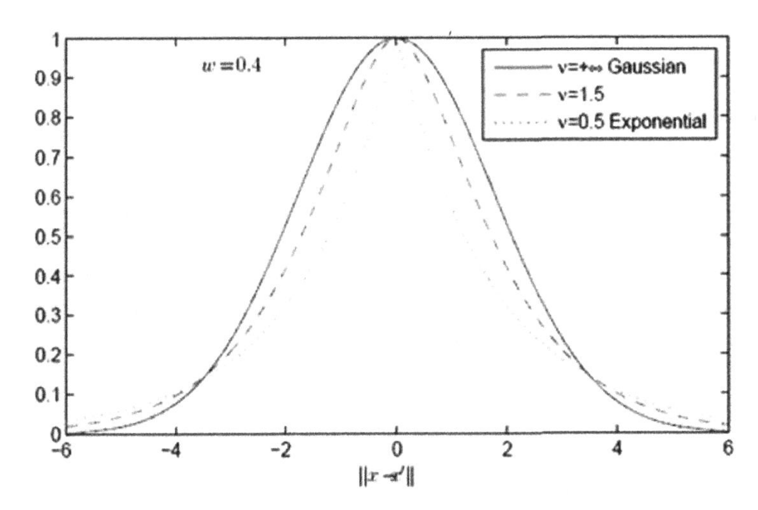

Fig. 1. The Matérn Kernel; w denotes the standard deviation (the width) and v is the smoothness parameter

Sparsity is generally considered a desirable feature of a machine learning algorithm. Sparse algorithms prefer a simple solution. In the context of GLM, as sparse solution will have a small number of non-zero coefficients. The Relevance Vector Machine (RVM) is a method for training a GLM such as (2). In the literature, it was presented as a method for sparse kernel regression.

$$y(\mathbf{x}, \mathbf{w}) = \sum_{i=1}^{N} w_i \cdot k(x, x_i) + \epsilon \tag{5}$$

$k(\mathbf{x}, x_i)$ is a bi-variate kernel function centered on each one of the N training data points x_i, $\mathbf{w} = [w_i...w_N]^T$ is a vector of regression coefficients, and ϵ is the noise. This means that it will select a subset, often a small subset, of the kernel functions in the final model.

Though it is stated that the RVM can use any basis functions [4], the examples in the literature apply only the Gaussian Kernel. The novelty in this

[1] $K_v(z) = \frac{\Gamma(v+\frac{1}{2})(2z)^v}{\sqrt{\pi}} \int_0^{\infty} \frac{\cos(t)}{(t^2+z^2)^{v+\frac{1}{2}}} dt$

paper is the first investigation of the RVM for kernels other than the Gaussian, and specifically the Matérn kernel. We study for the first time the effect of the smoothness of the kernel function on the convergence and sparseness of the RVM solution for datasets with various attributes such as non linearity and noise.

The rest of this paper is as follows: section 2 introduces the RVM algorithm; section 3 presents experiments with different kernels; and section 4 concludes with a discussion.

2 The RVM Algorithm

2.1 The Regression RVM

Consider a dataset of input-target pairs $\{X_i, t_i\}_{i=1}^{N}$. Each target t_i is assumed Normally distributed with mean $y(x_i)$ and uniform variance σ^2 of the noise ϵ so $p(\mathbf{t}|\mathbf{x}) = N(t|y(\mathbf{x}), \sigma^2)$. The targets are also assumed *jointly* Normal distributed as $N(\mu, \Sigma)$, where (μ, Σ) are the unknowns to be determined by the algorithm. The conditional probability of the targets given the parameters and the data can now be expressed as (6).

$$p(\mathbf{t}|\mathbf{w}, \sigma^2) = (2\pi\sigma^2)^{\frac{N}{2}} exp\left\{-\frac{1}{2\sigma^2}||\mathbf{t} - \mathbf{\Phi w}||\right\} \tag{6}$$

where the data is hidden in the NxN kernel function matrix $\mathbf{\Phi}$ representing all the pairs $\Phi_{i,j} = k(x_i, x_j), i, j \in [1...N]$.($\mathbf{\Phi}$ could be extended to include a possible bias term).

The goal of the RVM is to accurately predict the target function, while retaining as few basis functions as possible in (5). Sparseness is achieved via the framework of sparse Bayesian learning and the introduction of an additional vector of hyper parameters α_i that controls the width of a Normal prior distribution over the precision of each element of w_i.

$$p(w_i|\alpha_i) = \sqrt{\frac{\alpha_i}{2\pi}} exp(1 - \frac{1}{2}\alpha_j w_j^2) \tag{7}$$

A large parameter α_i indicates a prior distribution sharply peaked around zero. For a sufficiently large α_i, the basis function is deemed irrelevant and w_i is set to zero, maximizing the posterior probability of the parameters' model (7). As an analogy to the Support Vector Machine [5], the non-zero elements of $\mathbf{w}$ are called Relevance Values, and their corresponding data-points are called Relevance Vectors (RVs).

The solution is derived via the following iterative type II maximization of the marginal likelihood $p(\mathbf{t}|\boldsymbol{\alpha}, \sigma^2)$ with respect to $\boldsymbol{\alpha}$ and σ^2.

$$\alpha_i^{new} = \frac{1 - \alpha_i \Sigma_{ii}}{\mu_i^2} \tag{8}$$

$$(\sigma^2)^{new} = \frac{||\mathbf{t} - \mathbf{\Phi}\boldsymbol{\mu}||^2}{N - \Sigma_{i=1}^{N}(1 - \alpha_i \Sigma_{ii})} \tag{9}$$

The unknowns $(\boldsymbol{\mu}, \mathbf{\Sigma})$ are computed as:

$$\mathbf{\Sigma} = (\mathbf{\Phi^T B \Phi + A})^{-1} \tag{10}$$

$$\boldsymbol{\mu} = \mathbf{\Sigma \Phi^T B t} \tag{11}$$

where $\mathbf{B} \equiv \sigma^{-2}\mathbf{I}_{NxN}$. The basic RVM algorithm cycles between (8),(9),(10),(11) reducing the dimensionality of the problem when any α_i larger than a preset threshold. The algorithm stops when the likelihood $p(\mathbf{t}|\boldsymbol{\alpha}, \sigma^2)$ ceases to increase. Further information about the algorithm, as well as priors for $(\boldsymbol{\alpha}, \sigma^2)$ is presented in [4].

2.2 The Classification RVM

In the classification problem each target t_i is Binary: $t_i \in \{0, 1\}$. The model (5) is assumed to be noise-free. That is $\sigma^2 \equiv 0$. Note that equation (5) can not produce a binary function by itself without an additional rounding to the closest value $\{0, 1\}$. The sigmoid function $\rho(y) = 1/(1 + e^{-y})$ is used to generalize the linear model. The main idea of the sigmoid function is to make an approximation of the regression case to the two-class classification problem. With the sigmoid link function we can adopt the Bernoulli distribution $P(t|x)$ and rewrite the likelihood as:

$$p(t|w) = \prod_{i=1}^{N} \rho\left(\Phi_i^T w\right)^{t_i} \left(1 - \rho(\Phi_i^T w)\right)^{1-t_i} \tag{12}$$

In the classification RVM framework, we need to find two solutions to two different coupled problems, an optimization problem and a regression RVM problem [4]. The multi-class problem is solved with an assembly of binary classifiers. The classification RVM will not be considered in this paper

2.3 Attributes of the RVM

The RVM is an approximate Bayesian method, thus it can generate not only a predicted values, but also the probability distribution of the values [6]. For further details of the basic RVM look in [4]. For a discussion of convergence and sparseness look in [7].

The matrix inversion operation in (10), which requires $O(N^3)$ operations is the computationally intensive part of the algorithm. The matrices $\mathbf{\Phi}$ and $\mathbf{\Sigma}$ are full rank, thus require initially $O(N^2)$ space complexity. Furthermore, it is common that the inversion of a large matrix becomes ill-conditioned after several cycles even for positive definite matrices unless the parameters of the kernel function are optimized. These problems limit the practicality of the basic RVM algorithm for moderately sized problems. Fortunately, practical approaches were developed for reducing the runtime complexity to $O(N^2)$ [8].

An important step in GLM learning is to find a feature space - a projection of the data on highly dimensional space - where the data is linear for regression problems and linearly separable for classification problems. The choice of projection (the kernel function) is important for the accuracy and the convergence of the RVM. Note that the RVM typically produces very sparse solutions compared to the SVM, when its kernel is as the SVM kernel [4].

3 Comparative Experiments

3.1 The Matérn Kernel

The purpose of the following experiments is to check the sensitivity of the RVM to the kernel choice, the smoothness of the kernel function and various attributes of the dataset. In the Matérn kernel it is possible to control the smoothness of the kernel function. Thus, the following types of the kernel functions were considered: Gaussian kernel, Matérn of orders $v = 1, 2, 3, 4$ (respectively Matérn1, Matérn2, Matérn3 and Matérn4). Higher orders Matérn are not that different than the Gaussian. Moreover, we also test the finitely supported kernel function.

Instead of taking a set of unrelated benchmark data sets, we used the *Pumadyn* family of datasets[2]. These are eight synthetic datasets generated from a realistic simulation of the dynamics of the Puma 560 robotic manipulator. The regression problem is to predict the angular acceleration of one of the robotic links. Each dataset has a unique combination of three attributes: dimensionality (8 or 32 attributes), non-linearity (fairly linear or non-linear), and output noise (moderate or high). Table 1 present the details about the datasets and the split between the training set and the test set. We selected this specific set of benchmark datasets in order to study for the first time the dependency of the RVM solution on both the selected kernel and the attributes of the dataset (such as non-linearity).

We used a MATLAB implementation of the working set RVM from [8]. The Training set was used for the learning phase, and the error was measured on the testing set. Each kernel was simulated for 10 different repetitions and the same randomizations were used for testing each kernel. The width parameter for each kernel was optimized manually via cross-validation experiments on the training set. Table 2 presents the width parameters found for each combination of kernel and dataset.

[2] www.cs.toronto.edu/ delve/data/pumadyn/desc.htmwww.cs.toronto.edu/ delve /data/pumadyn/desc.html

Table 1. Details of the Pumadyn family of datasets

Name	Size	# of Attributes	Level of noise	Level of non linearity	Training set/test set
8fh	8192	8	high	fairly linear	6144/2048
8fm	8192	8	moderate	fairly linear	6144/2048
8nh	8192	8	high	non-linear	6144/2048
8nm	8192	8	moderate	non-linear	6144/2048
32fh	8192	32	high	fairly linear	6144/2048
32fm	8192	32	moderate	fairly linear	6144/2048
32nh	8192	32	high	non-linear	6144/2048
32nm	8192	32	moderate	non-linear	6144/2048

Table 2. The selected width parameters

Name	Matérn1	Matérn2	Matérn3	Matérn4	Gauss
8fh	3	4	8	18	2
8fm	5	4	8	25	2
8nh	7	8	16	25	2
8nm	12	14	18	32	10
32fh	50	50	50	90	25
32fm	20	25	125	175	25
32nh	180	190	135	135	135
32nm	135	140	150	160	60

We used two measures in these experiments, the number of RVs and the accuracy (RMSE). Tables 3 and 4 presents the comparative RMSE and the comparative number of RVs respectively. The standard deviation of each measure is also presented in the tables.

Table 3. Comparative RMSE

Name	Matérn1	Matérn2	Matérn3	Matérn4	Gauss
8fh	3.14 ± 0.04	3.14 ± 0.05	3.16 ± 0.06	3.17 ± 0.04	3.14 ± 0.05
8fm	1.07 ± 0.02	1.05 ± 0.02	1.05 ± 0.02	1.23 ± 0.03	1.05 ± 0.02
8nh	3.24 ± 0.06	3.22 ± 0.04	3.25 ± 0.03	4.25 ± 0.06	3.25 ± 0.04
8nm	1.18 ± 0.02	1.14 ± 0.02	1.20 ± 0.02	3.54 ± 0.03	1.22 ± 0.02
32fh	$.021 \pm 3*10^{-4}$	$.02 \pm 3*10^{-4}$	$.02 \pm 3*10^{-4}$	$.02 \pm 3*10^{-5}$	$.02 \pm 3*10^{-4}$
32fm	$.005 \pm 7*10^{-5}$	$.005 \pm 7*10^{-5}$	$.005 \pm 9*10^{-5}$	$.005 \pm 1*10^{-5}$	$.005 \pm 6*10^{-5}$
32nh	$.034 \pm 7*10^{-4}$	$.033 \pm 5*10^{-4}$	$.033 \pm 4*10^{-4}$	$.033 \pm 3*10^{-5}$	$.033 \pm 5*10^{-4}$
32nm	$.028 \pm 5*10^{-4}$	$.027 \pm 5*10^{-4}$	$.027 \pm 3*10^{-4}$	$.027 \pm 5*10^{-5}$	$.027 \pm 5*10^{-4}$

Table 4. Comparative number of RVs

Name	Matérn1	Matérn2	Matérn3	Matérn4	Gauss
8fh	46.7±5.3	36.5±2.5	34.2±1.7	12.3±1.4	38.5±2.9
8fm	136.1±10.3	79.8±3.2	57.2±2.6	16.1±1	70.6±3.1
8nh	160.7±8.5	93±2.47	45.9±5.8	19.3±1.4	142.4±8
8nm	530.4±85.4	173.3±6	82.8±3.9	17.7±1.3	82.8±2.5
32fh	240.8±37.1	41±11	38.6±3.4	11.3±2.5	79.1±3.7
32fm	265.1±14.3	64±7	34.5±4.3	17±4.6	140±13.4
32nh	268.3±19.5	28±21.5	7.3±1.6	7.3±1.4	9.9±1.7
32nm	324±10	31.1±12.5	16.9±8.6	8.8±2.6	67.7±5.2

Analysis of the results in tables 2, 3, 4 indicates that:

- The Gaussian and Matérn of orders 1,2,3,4 achieved a similar accuracy. Thus, from an accuracy point of view there is no difference among them. The Matérn4 demonstrates a significantly lower accuracy for three of the datasets.
- The Matérn4 achieved the sparsest results (less than 0.3%) for all the data sets. The Matérn3 is also typically sparser than the Gaussian, however, it has a similar accuracy as the Gaussian. It seems that the Matérn4 presents a danger of under-fitting the data.
- Regarding which kernel is more suitable for a given problem (non-linearity, noise) it is hard to decide. While the Matérn4 obtained the sparsest results for all the datasets, the Matérn3 also retained the accuracy, thus, it is recommended.

The main contribution of these experiments is that we found a kernel which is better than the Gaussian - we suggest using the Matérn of order 3.

We can also analyze for the first time the sensitivity of the RVM to different attributes of the dataset, and as expected, a decrease in the noise level (e.g. from puma8fh to puma8fm) results in an increase in the number of RVs, since less features of the function are now masked by the noise.

3.2 The Finitely Supported Matérn Kernel

The typical kernel measures the distance/similarity between any two points in the data. In a finitely supported kernel, we set the corresponding value in the kernel ma-trix to zero whenever the similarity between two points x_i and x_j is below a certain threshold. In a typical dataset, data is distributed among separate clusters in the multidimensional space. Effectively, a data is similar only to data in the same cluster, and will not be considered similar to data from different clusters. Thus, a finitely supported kernel matrix is expected to containing a majority of zero values (a sparse matrix). The advantage in a sparse matrix is that there exist efficient sparse linear algebra and sparse matrix computation techniques [9] that reorder the nonzero values to be around the diagonal of the kernel matrix and invert a sparse matrix in $O(N^2)$- effectively accelerating the RVM.

Simply truncating the kernel below a certain threshold does not result in a positive definite matrix in general. However, any kernel can easily become a compactly supported kernel by multiplying it with the "hyper-triangular" kernel (13).

$$max\left\{\left(1-\frac{||x_i-x_j||}{\sigma}\right)^v, 0\right\} \tag{13}$$

where $\sigma > 0$ is a width parameter (same as the one used in the regular Matérn kernel) and $v \geq (d+1)/2$ in order to insure positive definiteness (d is the dimensionality of the data which generated the kernel matrix).

For the experiments with the finitely supported Matérn kernel, we used only the first four Pumadyn datasets from table 1 which have $d = 8$, thus we choose $v = 5$ (for a higher smoothness than that, the Matérn is quite similar to the Gaussian). We used the same experimental setting as before. Table 5 presents the results of the experiments with the order 5 finitely supported hyper-triangular kernel. The finitely supported Matérn kernel was generated by the multiplication of (13) with (4). Unfortunately, for three of the datasets we failed to find an appropriate kernel width that leads to convergence of the RVM. Maybe the high noise and relative linearity of the first dataset facilitates finding a single appropriate width parameter.

Table 5. Experiments with finitely supported kernels

Name	Matérn5fs				Triangular5			
	width	Time	RV	Rmse	width	Time	RV	Rmse
8fh	50	394± 140	128.4± 13.8	3.17± 0.04	45	363± 160	129± 20.4	3.2± 0.05
8fm		no convergence				no convergence		
8nh		no convergence				no convergence		
8nm		no convergence				no convergence		

Comparing the results in table 5 to the results in tables 2, 3, 4, we see that:

- The hyper-triangular kernel of order 5 behaved fairly similar to the finitely supported Matérn kernel of order 5. When the effective width of the Matérn is larger than the effective width of the Triangular kernel, this could be expected, since in this case the Matérn will have a fairly constant value within the effective support of the Triangular kernel, and the multiplication of the kernels will not differ much from the values of the Triangular kernel.

- The accuracy of the finitely supported kernels is fairly similar to that of the regular kernels, while the number of RVs is much larger.
- A careful analysis of the two kernels that did converge, indicate 0% sparsity for the width parameters selected.

While these experiments are not conclusive, it indicates that finite supported kernels are not a good choice for the RVM - the algorithm does not converge for sparse kernels.

4 Discussion

The problem of selecting the best kernel and tuning its parameters lies in the core of all the kernel based methods. In this paper we investigated for the first time the sensitivity of the RVM to the kernel choice. We found that Matérn of order 3 provides a fair trade-off between sparsity and error. Considering that cubic splines (splines of order 3) are well known to provide good approximations for many types of functions, this is not surprising.

It turns out - unlike conjectured by [4] - that not every kernel is suitable for the RVM. Surprisingly, when we trained the RVM using the finitely supported Matérn kernels, the results were very poor convergence if any, and a very long training time relatively to the ordinary Matérn kernels. One possible explanation to the poor results could be the existence of regions of zero derivative of the finitely support kernels which causes the RVM to slow down, or stop its convergence. This phenomena merits further investigation, since sparse linear algebra can potentially accelerate the RVM.

We did not manage to answer the question if there is a kernel that is best suited for a given problem such as non-linearity, noise or whether kernels for classification should be different than kernels for regression. However, this is the first time (to the best of our knowledge) that a kernel different than the Gaussian, or specifically the Matérn family, was ever used for the RVM.

References

1. A. Juditsky, H. Hjalmarsson, A. Beneviste, B. Delyon, L. Ljung, J. Sjoberg, Q. Zhang (1995), Nonlinear Black-box Models in System Identification: Mathematical Foundations, Automatica, 31(12), pp. 1725-1750.
2. M.G.Genton (2001). Classes of Kernels for Machine Learning: A Statistics Perspective. Journal of Machine Learning Research 2, pages 299-312.
3. B. Matérn (1960). Spatial Variation. New York, Springer.
4. M.E. Tipping (2001). Sparse Bayesian Learning and the Relevance Vector Machine. Journal of Machine Learning Research 1, 211-244.

5. A. Shmilovici (2005). Support Vector Machines. In O. Maimon and L. Rokach (editors), Data Mining and Knowledge Discovery Handbook: A Complete Guide for Practitioners and Researchers, Springer.
6. C.E. Rasmussen, J. Quinonero-Candela, (2005). Healing the Relevance Vector Machine through Augmentation. Proceedings of the 22nd International Conference on Machine Learning, August 7-11, Bonn, Germany.
7. D. Wipf, J. Palmer, B. Rao (2004). Perspectives on Sparse Bayesian Learning. Advances in Neural Information Processing systems, 16. Cambridge, Massachussettes, MIT Press.
8. D. Ben-Shimon, A. Shmilovici (2006). Accelerating the Relevance Vector Machine via Data Partitioning, Journal of Computing and Decision Sciences, forthcoming.
9. J.R.Gilbert, C. Moler, R. Schreiber (1992). Sparse matrices in MATLAB design and implementation. SIAM Journal on Matrix Analysis, 13(1), pages 333-356.
10. N. Cristianini, J. Shawe-Taylor (2003). An Introduction to Support Vector Machines and other kernel-based learning methods. Cambridge University Press.

A Decision-Tree Framework for Instance-space Decomposition

Shahar Cohen[1], Lior Rokach[2], and Oded Maimon[1]

[1] Department of Industrial Engineering, Tel-Aviv University, Israel
shaharco@post.tau.ac.il,maimon@eng.tau.ac.il
[2] Department of Information Systems Engineering, Ben-Gurion University of the Negev, Israel liorrk@bgu.ac.il

Summary. This paper presents a novel instance-space decomposition framework for decision trees. According to this framework, the original instance-space is decomposed into several subspaces in a parallel-to-axis manner. A different classifier is assigned to each subspace. Subsequently, an unlabelled instance is classified by employing the appropriate classifier based on the subspace where the instance belongs. An experimental study which was conducted in order to compare various implementations of this framework indicates that previously presented implementations can be improved both in terms of accuracy and computation time.

1 Introduction

The multiple-classifier approach may improve the performance of a certain classification method. Both decomposition methodology as well as ensemble methodology applies a multiple-classifier approach for solving a classification task. Nevertheless the main idea of ensemble methodology is to combine a set of classifiers, each of which solves the same original task, in order to obtain a more accurate and reliable result than using a single classifier [4]. In a typical ensemble setting, each classifier is trained on data taken or re-sampled from a common dataset. On the other hand, the purpose of decomposition methodology is to break down a complex problem into several manageable problems, and to let each classifier solve a different task, i.e., individual classifiers cannot provide a solution to the original task.

Instance-space decomposition is a specific decomposition approach in which the original instance-space is decomposed into several subspaces. For each subspace, a different classifier is generated. Subsequently an unlabelled instance can be classified by selecting the appropriate classifier according to the subspace to which the instance belongs.

Several researchers proposed methods for combining, selecting and weighting the classification results of a set of given classifiers [5]. These methods are

Mark Last et al. (Eds.): Advances in Web Intelligence and Data Mining (SCI) **23**, 265-274 (2006)
www.springerlink.com

based on evaluating the level to which a certain classifier is appropriate for a certain region in the instance-space.

NBTree is an instance space decomposition method that induces a decision tree and a naïve-Bayes hybrid classifier [3]. Naïve-Bayes, which is a classification algorithm based on Bayes' theorem and a naïve independence assumption, is very efficient in terms of its processing time. To induce an NBTree, the instance space is recursively partitioned according to attributes values. The result of the recursive partitioning is a decision tree whose terminal nodes are naïve-Bayes classifiers. Since subjecting a terminal node to a naïve-Bayes classifier means that the hybrid classifier may classify two instances from a single hyper-rectangle region into distinct classes, the NBTree is more flexible than a pure decision tree. In order to decide when to stop the growth of the tree, NBTree compares two alternatives in terms of error estimation - partitioning into a hyper-rectangle regions and inducing a single naïve-Bayes classifier. The error estimation is calculated by cross-validation, which significantly increases the overall processing time. Although NBTree applies a naïve-Bayes classifier to decision tree terminal nodes, classification algorithms other than naïve-Bayes are also applicable. However, the cross-validation estimations make the NBTree hybrid computationally expensive for more time-consuming algorithms such as neural networks.

Although different researchers have addressed the issue of instance space decomposition, there is no research that suggests an automatic procedure for mutually exclusive instance space decompositions, which can be employed for any given classification algorithm and in a computationally efficient way. This paper introduces a decision-tree framework for instance-space decomposition (DFID). This framework grows a decision tree and applies a certain induction algorithm to its terminal nodes. While growing the tree, DFID can use several splitting criteria. Nevertheless a new splitting criterion that combines gain ratio measure with a subset grouping procedure is suggested, and its contribution to the problem is discussed. The DFID implementation with the new splitting criterion is called CPOM (Contrasted POpulations Miner).

The proposed algorithm can be used for developing lookahead based algorithms for induction of decision trees. Lookahead based algorithms attempts to predict the profitability of a split at a node by estimating its effect on deeper decedents of the node [6, 2]. For this purpose, one can use the DFID while employing a decision tree induction algorithm as the inner inducer.

2 Decision-tree Framework for Instance-space Decomposition (DFID)

Given a training set S drawn from a distribution D over the labelled instance space U, and an inducer I, the aim' of the instance-space decomposition problem is to find an optimal decomposition W such that a unique classifier is derived for each of the regions and the sum of generalization errors over the induced classifiers $M_i = I(S \cap \psi_i)$ will be minimized over the distribution D. Beside the error measure, this paper considers also the time complexity measure, and the number of leaf nodes in the composite classifier.

This paper focuses on hierarchical partitioning that can be represented by a univariate decision tree. DFID is a heuristic framework for instance space decomposition. An NBTree [3] can be seen as a specific implementation of DFID. DFID can be also used to implement lookahead based algorithms for induction of decision trees. CPOM, described below, is another DFID implementation.

As Figure 1 shows, the proposed DFID procedure receives a training set S and an inducer I. The framework flow begins by checking for stopping criteria. If a stopping criterion is met, then a single classifier is induced from the training set by applying I to S, and the procedure terminates. As long as no stopping criterion is met, the training set is recursively partitioned. The split is then optionally subjected to validation. The aim of the validation is to ensure that the split under consideration is indeed beneficial. (If a splitting rule is found to be invalid, then another split is generated until there are no more possible splits, or until a valid split was found (whichever comes first). If there are no more possible splits, I is applied to S, and the procedure terminates. As soon as a valid split is found, S is split accordingly. Subsequently, each subset recursively goes through DFID flow in a similar fashion (Stopping and splitting criteria as well as validation techniques are discussed below).

Distinct implementations of DFID may differ in all or some of the procedures, which implement three components – stopping criteria, splitting criteria and splitting validation. Possible implementations of these three components are now discussed.

2.1 Stopping Criteria

Stopping criteria are represented by the "Consider Partition" condition block in Figure 1. It should be noted however, that this condition block is not the only scenario in which DFID recurrence stops. Another, more natural though passive scenario, for stopping recurrence, is the lack of any valid splitting rule.

NBTrees [3] uses a simple stopping criterion according to which a split is not considered when the dataset consists of 30 instances or less. Splitting too few instances will not affect the final accuracy much yet will lead, on the other hand, to a complex composite classifier. Moreover, since each sub classifier is required to generalize instances in its region, it must be trained on samples of sufficient size. Kohavi's stopping criterion can be revised to a rule that never considers further splits in subsets consisting of $\beta||S||$ instances or less, where $0 < \beta < 1$ is a proportion and $|S|$ is the number of instances in original training set S.

Another heuristic stopping criteria is to never consider splitting if a single sub classifier can accurately describe the current region. Practically, the rule can be checked by comparing an accuracy estimation of the sub classifier to a pre-defined threshold; the procedure terminates if the estimate exceeds the threshold. The motivation for this condition is that if a single sub classifier is good enough, there is probably no point in partition considerations.

2.2 Split Validation

Since splitting criteria, are heuristic, it may be beneficial to regard the splitting rule as a recommendation that should be validated. However in the DFID framework, validation is optional.

Kohavi validates a split by estimating the reduction in error gained by the split and comparing it to a predefined threshold of 5%. In an NBTree, it is enough to examine only a single split in order to conclude that there is no valid split if the one examined is invalid. This follows since in an NBTree, the attribute selected for the split is the one that maximizes a utility measure which is strictly increasing with the reduction in error. If a split in accordance with a selected attribute cannot reduce the accuracy in more than 5%, then a split according to another attribute will also fail to do so.

This work suggests a new split validation which is nested in a new split criterion. More details are provided below but speaking in very general terms, a splitting rule is regarded as invalid if it splits the space into a single region.

2.3 Splitting Criteria

A core DFID issue is how to generate splitting rules which result from splitting criteria. It should be noted at the outset that any splitting criterion that is used to grow a pure decision tree, is also applicable to the DFID framework [8].

Kohavi [3] suggested a new splitting criterion which is to select the attribute with the highest utility. Kohavi defined utility as the 5-fold cross-validation accuracy estimate of using a naïve-Bayes algorithm for classifying regions generated by a split. The regions are partitions of the initial subspace according to a particular attribute values.

This paper suggests a novel splitting criterion called grouped gain ratio. Grouped gain ratio is based on gain ratio splitting criterion [7] followed by a subset grouping heuristic.

The gain ratio splitting criterion, selects a single attribute. The space may be partition according to values of the selected attribute [7]. An alternative approach however is to group values of the selected attribute. A subset grouping heuristics to perform this grouping is now described.

The grouping heuristic is based on the sub sets corresponding to the sub spaces. A partition of the instance space is beneficial to the classification task if several classifiers, one for each subspace, are more accurate than a single classifier for the entire space. For simplicity, consider the case of two distinct subsets representing two distinct sub spaces and described by two distinct classifiers. If the two classifiers are similar to each other, the distinction between the two populations is not beneficial, since a single classifier can describe both subsets. The term similar is abstract, but it is clear that if the two classifiers are identical then the distinction between the two subsets is pointless. The opposite conjecture also makes sense, i.e. if the two classifiers are very different from each other then there is probably a point in the distinction.

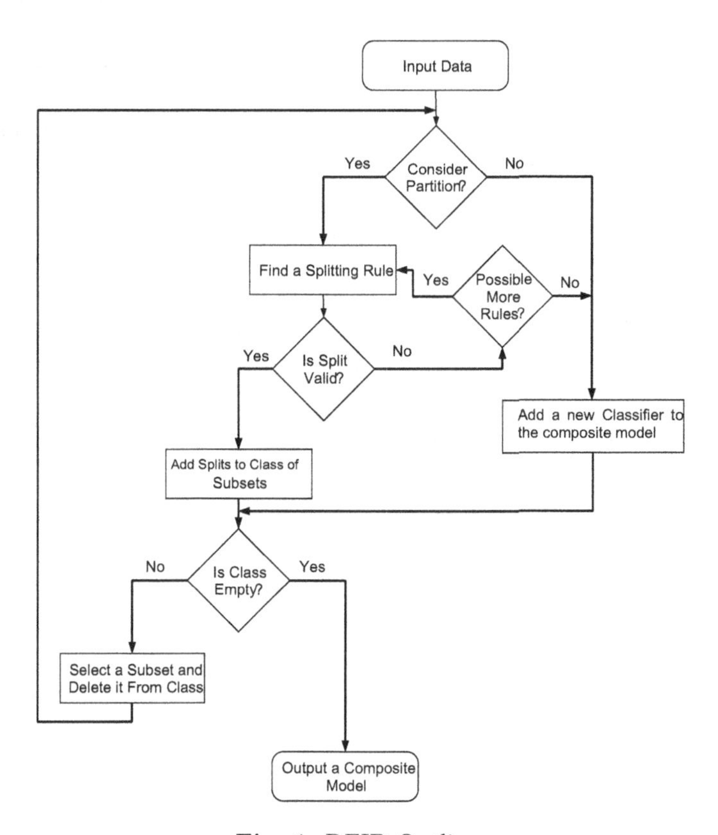

Fig. 1. DFID Outline

The intuition regarding to classifier comparisons raises the question of what is similar, and how to compare classifiers. This paper uses a comparison heuristic termed Cross Inspection. The heuristic is outlined in Figure 2. This procedure, based on two samples and two inducers as input parameters, starts by randomly partitioning each sample into a training set and a test set. Then two classifiers are induced one from each training set. If the error of the first classifier over the first test set is significantly different from the error of the first classifier over the second test set, or vice versa, then the two classifiers are regarded as different. The hypothesis in cross inspection is tested by statistically estimating the difference in proportions of two binomial random variables [1].

The comparison of classifiers by cross inspection considers only two distinct classifiers. However, there is a need to compare more than two classifiers at a time if the selected attribute has more that two possible values. For example if it is believed that graduate students from different schools behave differently, one may seek the appropriate partition according to the school of the graduate. The domain of 'school' attribute might have multiple values, all of which will have to be compared simultaneously. The simultaneous comparison may be the basis for partition recommendation. In the graduate student's example, a successful partition will group similar schools into a single group, where different schools will be in different groups.

Since an exhaustive search over possible partitions is unacceptable in terms of complexity, grouped gain ratio uses a greedy grouping procedure instead. Grouped gain ratio is outlined in Figure 3. The heuristic does not explicitly guarantee that any two classifiers in a group are equivalent, but equivalence is assumed to be a transitive relation. The greedy grouping procedure is a simple clustering method and other clustering methods like, graph coloring, may also be suitable here.

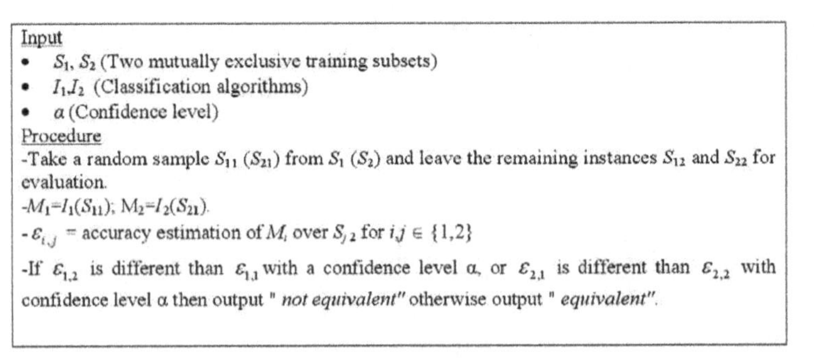

Input

- S_1, S_2 (Two mutually exclusive training subsets)
- I_1,I_2 (Classification algorithms)
- α (Confidence level)

Procedure

-Take a random sample S_{11} (S_{21}) from S_1 (S_2) and leave the remaining instances S_{12} and S_{22} for evaluation.

-M_1=I_1(S_{11}); M_2=I_2(S_{21}).

-$\varepsilon_{i,j}$ = accuracy estimation of M_i over S_{j2} for $i,j \in \{1,2\}$

-If $\varepsilon_{1,2}$ is different than $\varepsilon_{1,1}$ with a confidence level α, or $\varepsilon_{2,1}$ is different than $\varepsilon_{2,2}$ with confidence level α then output " *not equivalent*" otherwise output " *equivalent*".

Fig. 2. Cross Inspection Procedure Outline

2.4 CPOM Algorithm

CPOM is a DFID implementation which splits instance spaces according to grouped gain ratio. It should be noted that grouped gain ratio is inherently equipped with a split validation mechanism (i.e. if a grouped gain ratio groups all sub spaces into a single space, then the split recommended by gain ratio criterion is invalid). CPOM does not use any explicit split validation method.

CPOM uses the two stopping rules. First CPOM compares the number of instances in the subset under consideration with a predefined ratio of the number of instances in the initial dataset. If the subset is too small, it stops.

Input:
- $S_1, S_2, \ldots, S_q$ (subsets of training data)
- $M_1, M_2, \ldots, M_q$ (classifiers corresponding to the subsets)

Procedure:

-For each pair of subsets mark 'equivalent' or 'not equivalent'. Using the cross-inspection procedure

-Create L – a list of subsets of training data. For each member of L, compute the number of instances that are described by classifiers equivalent to the member. Sort L descending by these corresponding numbers.

-While L is not empty:

- Declare a new cluster consisting of all the subsets that are equivalent to the first member of L.
- Remove from L all subsets that belong to the new cluster (including the first member).

Fig. 3. Grouped Gain Ratio Heuristic

Secondly CPOM compares the accuracy estimation of a single classifier to a pre defined threshold. It stops if the accuracy estimate exceeds the threshold.

3 Experiment Study

Following Kohavi [3], all DFID implementations evaluated in this paper use naïve Bayes classification algorithm for classifying tree terminal nodes. The evaluation took place using datasets chosen manually from the well-known UCI Machine learning repository. The chosen datasets vary in relation to a number of characteristics, such as the number of target classes, instances and of explaining attributes.

The datasets passed a simple preprocessing stage. In this stage, missing values were replaced by a distinctive value, and numeric attributes were made discrete by dividing their original range into ten equal-size intervals (or one per observed value, whichever was least). Accuracy results may have been improved by using a more robust way to treat missing values [7]. Accuracy measures were derived using 10-fold cross-validation except in Letter and Nurse datasets, where it was assumed that there are enough instances to use a leave 1/3 out validation.

Each dataset was tested using three DFID implementations – NBTree, CPOM with grouped gain ratio as a splitting criterion, and CPOM with simple gain ratio splitting criteria (referred to as CPOM with no grouping). NBTree implementation was done according to the report in [3]. In CPOM implementation, maximal accuracy to consider a split was chosen to be 95%, and a minimal data subset size was chosen to be one fifth of the initial data set size.

Table 1 describes accuracy estimations of the three implementations over UCI repository datasets. For Letter and Nurse datasets, a point estimate is derived, where in other datasets, a 95% confidence interval is construct based on a two-tailed t-distribution. The results in Table 1 indicates that the grouping procedure tends to improve the accuracy of the composite classifier compared to a composite classifier that is solely based on gain ratio splitting rule. As a matter of fact, in none of the sets does gain ratio significantly outperform grouped gain ratio. Thus, it is likely to assume that the grouping heuristic is a beneficial one, i.e. subspaces that are described by similar classifiers should be grouped together.

Table 2 presents the number of leaf nodes in the composite classifier obtained and the number of inter-classifiers needed in order to obtain this classifier. The number of leaves is a way for assessing classifier comprehensibility, and the number of inter-classifiers is a mean for assessing algorithm processing time. The table shows that CPOM tends to build trees with fewer leaves. Trees with fewer leaves are considered to be desirable since they are easier for humans to understand.

Table 2 also shows that CPOM required significantly fewer inter-classifiers than NBTrees in all of the examined datasets. This result is partly due to the more compact trees built by CPOM, but it also due to the slow splitting rule used by NBTree. In order to chose the attribute with the highest utility, NBTree, estimates accuracies of all splitting possibilities, where each possibility is evaluated using a 5-fold cross validation. The number of classifiers needed for this splitting rule becomes a significant burden when the number of attributes and splits increases. CPOM, on the other hand, does not use cross validation and needs to build sub-classifier only for the attributes selected by gain ratio splitting criterion. Table 2 implies that in terms of time consumption, CPOM shows a complete superiority over NBTrees. Due to this superiority, it is likely to expect the generalization of CPOM to classifiers, other than naïve-Bayes, to be more feasible in terms of processing time than generalizations of NBTrees.

Finally CPOM shows overall accuracy results that are better than overall NBTrees results. Checking accuracy estimations in Table 1, it can be verified that CPOM significantly outperforms NBTrees in 7 of the datasets, while NBTrees shows significant better results in only 3 datasets.

4 Summary and Future Research

This paper introduced an implementation of decision-tree based framework for instance space decomposition called CPOM. This implementation employs a new splitting rule, termed grouped gain ratio. In the grouped gain ratio, an attribute is first selected according to gain ratio criterion. A greedy grouping heuristic then groups together similar sub spaces.

The algorithm recursively splits an instance space according to the values of explained attributes until stopping rules are met. Subsequently, for each region formed by the recursive splits (referred to as partition), a unique classifier is induced using some classification algorithm.

With datasets manually selected from the well-known UCI Machine Learning repository, CPOM was superior both compared to NBTrees and the procedure this paper refers to CPOM with no grouping. Specifically, CPOM was

Table 1. Accuracy estimations of the three implementations over UCI's datasets

Dataset	CPOM	CPOM no Grouping	NBTree
Andor	86.23±1.72	87.01±1.70	87.20±6.75
B.Can	97.42±1.16	93.84±1.86	96.56±1.46
Hyes	75.75±9.16	64.39±9.43	67.42±7.11
Led	60.90±5.12	61.81±7.22	60.90±4.38
Mush	99.37±0.17	99.28±0.23	99.95±0.07
Nurse	93.07	93.01	96.59
Sonar	76.44±7.62	70.19±9.09	62.98±7.68
Soyb.	93.41±1.49	93.99±1.03	92.53±1.86
TTT	76.51±1.87	78.70±1.43	75.67±5.82
Vote	92.64±2.50	91.03±3.32	92.06±4.51
Wine	96.62±3.35	91.01±4.24	90.44±5.69
Zoo	95.04±5.05	95.04±5.05	91.08±6.25
Letter	77.70	77.70	80.06
Aust	86.52±3.05	84.34±2.92	84.78±2.64
Car	93.92±0.74	89.64±0.96	85.30±0.92
Monk1	97.58±3.97	70.16±8.49	91.12±4.16

Table 2. No. of Leaves and No. of Classifiers of CPOM vs. NBTree

Data	**NBTree**		**CPOM**	
	Leaves	**# Models**	**Leaves**	**# Models**
Andor	28	6730	8	14
B.Can	28	2250	2	162
Hyes	10	225	4	54
Led	7	1920	8	28
Mush	2634	235480	10	262
Nurse	759	45500	10	104
Sonar	19	9000	2	772
Soyb.	44	11500	18	122
TTT	1	135	7	18
Vote	35	4080	2	8
Wine	10	1950	1	368
Zoo	8	540	5	36
Letter	226	43520	27	1650
Aust	59	6930	5	554
Car	1	105	8	90
Mon1	5	340	7	72

overall more accurate results, with less execution time and in a more comprehensive representation. Moreover, the grouping heuristic significantly improved the accuracy results compared to CPOM with no grouping.

Grouped gain ratio can be tailored as a splitting criterion for pure decision trees. As to future research, it is suggested to compare and evaluate this possibility in regard to other splitting criteria.

References

1. Dietterich, T.G. (1998). Approximate statistical tests for comparing supervised classification learning algorithms. Neural Computation, 10 (7).
2. Esmeir S. and Markovitch S. (2004). Lookahead-based Algorithms for Anytime Induction of Decision Trees. In Proceedings of The Twenty-First International Conference on Machine Learning, Banff, Alberta, Canada. Morgan Kaufmann, pp. 257–264.
3. Kohavi R. (1996). Scaling Up the Accuracy of Naive-Bayes Classifiers: a Decision-Tree Hybrid, Proceedings of the Second International Conference on Knowledge Discovery and Data Mining, pp. 202–207.
4. Maimon O. and Rokach L. (2005) Decomposition Methodology for Knowledge Discovery and Data Mining: Theory and Applications Series in Machine Perception and Artificial Intelligence - Vol. 61 World Scientific Publishing.
5. Mertz, C. J. (1999). Using Correspondence Analysis to Combine Classifiers. Machine Learning 36(1), 33-58.
6. Murthy, S. and Salzberg, S. (1995), Lookahead and pathology in decision tree induction, in C. S. Mellish, ed., Proceedings of the 14th International Joint Conference on Artificial Intelligence, Morgan Kaufmann, pp. 1025–1031
7. Quinlan, J.R. (1993). C4.5: Programs for Machine Learning, Morgan Kaufmann.
8. Rokach L., Maimon O. (2005): Top-down induction of decision trees classifiers - a survey. IEEE Transactions on Systems, Man, and Cybernetics, Part C 35(4): 476–487.

On prokaryotes' clustering based on curvature distribution

L.Kozobay-Avraham[1], A. Bolshoy[2], and Z. Volkovich[3]

[1] Institute of Evolution, Haifa University, Haifa, Israel. limor@research.haifa.ac.il
[2] Institute of Evolution, Haifa University, Haifa, Israel. bolshoy@research.haifa.ac.il
[3] Software Engineering Department, ORT Braude Academic College, Karmiel 21982, Israel.
Affiliate Professor. Department of Mathematics and Statistics. The University of Maryland, Baltimore County, USA. vlvolkov@ort.org.il

1 Abstract

Massive determination of complete genomes sequences has led to development of different tools for genome comparisons. Our approach is to compare genomes according to typical genomic distributions of a mathematical function that reflects a certain biological function. In this study we used comprehensive genome analysis of DNA curvature distributions before starts and after ends of prokaryotic genes to evaluate the assistance of mathematical and statistical procedures. Due to an extensive amount of data we were able to define the factors influencing the curvature distribution in promoter and terminator regions. Two clustering methods, K-means and PAM were applied and produced very similar clusterings that reflect genomic attributes and environmental conditions of species' habitat.

2 Introduction

The term DNA curvature refers to a characteristic of DNA fragments, which are bent without application of any external forces. This property also named intrinsic curvature or sequence-dependent DNA curvature. The presence of curved DNA was established by biological experiments since the early 80-ies ([15], [22], [3], [7]). Based on experimental results, some computational models, including our model ([2], [18]) were developed to predict the magnitude of DNA curvature with high reliability. The genes in prokaryotes usually preceded by curved DNA sequence, which presumably enhances gene expression

Mark Last et al. (Eds.): Advances in Web Intelligence and Data Mining (SCI) **23**, 275-284 (2006)
www.springerlink.com

([16], [21], [12]). Also evidences on curved DNA located at the end of prokaryotes' genes, in transcription termination sites, are recently found ([8], [12]). Characterization of the factors influencing presence of curved DNA along the sites of gene regulation can greatly improve our current understanding of the regulation processes and improve promoter and terminator prediction algorithms.

Cluster analysis and other statistical tests were performed on the genomic data. For every genome, we predicted DNA curvature distributions, before start of genes (promoter regions) and after end of genes (terminator regions), by using our CURVATURE program [18]. Randomized sequences were constructed and curvature excess profiles in the standard deviation units were calculated for each genome. While in our previous papers [11], [12] the curvature excess profiles were used for intuitive classification of the prokaryotic genomes, in the current manuscript we take advantage of the Euclidian distances calculating between the genomes for the further cluster analysis. The aim of the study is to discover the underlying genomic and environmental factors causing the obtained partitions. Correlation analysis between clusters based on curvature profiles of promoter regions and of terminator regions was also performed.

3 Clustering methods

Various clustering problems are found in many areas of the bionformatics. Let $\{x_1, ..., x_m\}$ be a set of vectors in a subset X of the n-dimensional Euclidean space R^n. Given a natural number $k > 1$ we are looking for a set of points $C = \{c_j\}$, $j = 1, ..., k$ solves the optimization problem

$$\min_{c_j} R(c_1, ..., c_k) = \sum_{j=1}^{k} \sum_{z \in X} \min_{c_j} \|z - c_j\|^2 \tag{1.1}$$

The sets $\pi_j = \{x \in X|\ min(d(x, c_t), t = 1, ..., k) = j)$ are named clusters and the points $\{c_1, \ldots, c_k\}$ are named clusters' centroids. $\|\cdot\|$ is the standard Euclidean distance. Two of the most widespread clusters methods are k-means and k-medoids (PAM) iterative procedures.The k-means method (see [4]) produces an approximate solution to this optimization problem in an iterative way:

1. Selection of an initial partition.
2. Minimization: calculate the mean (centroid) of each cluster's points.
3. Classification: assign each element to the nearest current centroid.
4. Repeat until the partition is stable, that is, until the centroids no longer change.

However, k-means often leads to a partition consisting of so-called "non-optimal stable clusters". The k-means approach can be considered as a simplification of the well known Expectation Maximization (EM) algorithm (see, for example [5]).

PAM (Partitioning Around Medoids)(see, for example [10]) was developed to find the k most representative objects (medoids) that represent k clusters such that non-selected objects are clustered with the medoid to which it is the most similar. The total distance between non-medoid objects and their representative medoid may be reduced by swapping one of the medoids with one of the objects iteratively. Obviously, it is time consuming even for the moderate number of objects and small number of medoids. The PAM approach looks more robust and efficient than the k-means algorithm and is implemented in clustering packages R and S-Plus.

Two following indexes of the cluster stabiliry are used often. Given a partition $\Pi = \{\pi_1, \ldots, \pi_k\}$, $2 \leq k$, we denote by B_k and W_k the dispersion matrices of between and within group sums of squares (see, for example defin-ion in [14]).

1. The Krzanowski and Lai index [13] is defined by means of the following relationships

$$diff_k = (k-1)^{2/n} tr(W_{k-1}) - k^{2/n} tr(W_k)$$

$$KL_k = |diff_k| / |diff_{k+1}|.$$

   The estimated number of clusters is the maximal value of the index KL_k.
2. Sugar and James [19] proposed an information theoretic approach for an estimation of the true number of clusters. A transformation power t is predetermined (A typical value is $t = n/2$.) The estimated number of clusters maximazes the value of the index

$$J_k = \left(tr(W_k)^{-t} - tr(W_{k-1})^{-t} \right).$$

4 Biological Methods

4.1 Genomic sequences and their attributes

For further analysis we took 205 complete prokaryotic genomes from the GenBank, the public genome library of the National Center Biotechnology. The following genomic characteristics were gathered from the genomic annotations and from the literature: optimal growth temperature, genome size, A+T composition, and taxonomic description.

Optimal growth temperature. The organisms belong to four temperature groups, as defined in literature: (1) psychophiles - organisms that defined by their ability to grow at 0°C and below; (2) mesophiles - organisms that grow best between 10° and 30°C; (3) thermophiles - organisms that grow best in hot conditions, between 30° and 50°C; (4) hyperthermophiles - organisms that thrive in extremely hot environments - that is, hotter than around 60°C with the optimal temperatures that are between 80°C and 110°C.

Taxonomy. Prokaryotes fall into one of two groups, Archaebacteria (ancient forms thought to have evolved separately from other bacteria) and Eubacteria. Archaebacteria, sometimes called the Archaea, emerged at least 3.5 billion years ago and live in environments that resemble conditions existing when the earth was young. Many hyperthermophiles are archaeas. Our genomic database consist of 23 Archaea and 182 Bacteria. Among the Archaea genomes 19 genomes are thermophiles or hyperthermophiles; and 4 genomes are mesophiles.

Genome sizes. Prokaryotes have relatively small genomes: from very short genomes with lengths less than 1Mb up to 5-9 Mb. A size of a genome is a relevant factor because the smallest genome-sized prokaryotic species, the obligate endocellular parasites, when compared to their free-living relatives, have preferentially lost many regulatory elements, including factors, presumably due to the rather stable environment inside host cells, which renders extensive gene regulation useless ([1], [6], [17]). DNA curvature excess is related to gene regulation ([11]).

A+T composition. This feature is relevant to our analysis because magnitude of DNA curvature is related to it. It was shown in many experiments that strongly curved DNA fragments as a rule possess rather high A+T content.

4.2 Curvature calculation

Our CURVATURE program [18] is based on the stepwise calculation of geometric transformations according to the set of the previously estimated parameters (([9], [2]). The program calculates a DNA map of a given sequence with an input arc-size parameter. A curvature value at a sequence-position i corresponds to a curvature of the arc approximating to the predicted DNA path, when the arc approximates a path segment of the length equal to a program parameter arc size with a center of the segment in the position i. The DNA curvature was measured in DNA curvature units (cu) introduced by Trifonov and Ulanovsky ([20]). For example, a segment of 125 bp of length with a shape close to a half–circle has a curvature value of about 0.34 cu. Such strongly curved pieces appear infrequently in genomic sequences.

4.3 Construction of curvature-excess profile

The whole genome sequence was used as an input for the program and a map of curvature distribution using a given window size of 125 base pairs (bp) along the whole sequences was produced. To construct gene-start or gene-end genomic profiles relevant pieces of produced DNA curvature genome maps were averaged.

Averaging. Gene positions were taken from genome annotations that accompany every genome sequence in GenBank. In our study of curvature distribution around the starts or ends of genes, we only processed genes flanked by intergenic regions longer than 125 nucleotides. The reason for this choice is that shorter intergenic regions can hardly include regulatory signals. Hundreds of genes of a genome i were processed to obtain a genomic profile (g_i). The standard errors (s_i) were estimated by bootstrap method using 1,000 runs.

Preparation of randomized genomes. We constructed control genomes for testing the significance of the results and comparing properties of natural and artificial genomes. The construction procedure consisted of three steps: a) a genome was cut in separate genic and intergenic pieces at every start and end gene junction; b) each piece was separately reshuffled preserving dinucleotide composition, and c) all the pieces were reassembled in the original order. A randomized DNA curvature map was obtained by averaging maps of 10 shuffled genomes. To construct randomized profiles relevant pieces of produced randomized genome maps were averaged (r_i).

Curvature excess calculation. Curvature excess related to a genome i is an apparent deviation between genomic profile (g_i) and randomized profile (r_i). Excess curvature value at position k is measured in standard deviation units (s_i) and calculated as follows: $ce_{ik} = (g_{ik} - r_{ik})/(s_{ik})$.

5 Results and Discussion

5.1 Clustering

We performed cluster analyses using the K-means and the PAM methods, based on the Euclidian distances between curvature excess profiles before and after the genes. The partitions are provided by means of the two leading principal components. This approach can be motivated from the point of view of the covered part of the total dispersion. In the considered cases these fractions typically are about 80 percents for the first component and about 90 percents for two leading components together. As the experiments show partitions obtained by means of the two mentioned above cluster algorithms K-means and PAM. The graphs of the two mentioned above indexes demonstrate that both of the indexes precisely indicate the true number of clusters as 3. In all further text the number of clusters is assumed to be 3.

5.2 Correlation between clusters and genomic characteristics

In order to define factors bringing together genomes to one cluster according to the curvature distribution we examined characteristics of the genomes containing every cluster. Cluster 1, the smallest cluster containing genomes

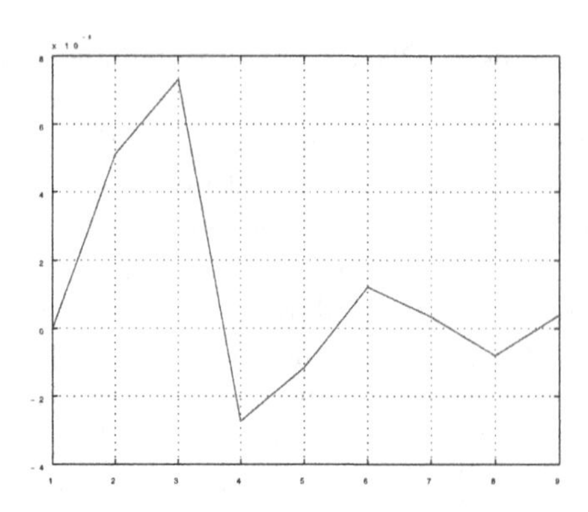

Fig. 1.1. Graph of the Sugar and James index

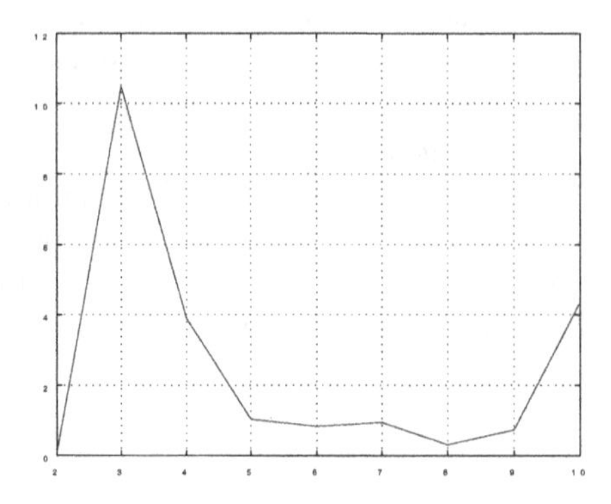

Fig. 1.2. Graph of the Krzanowski and Lai index

with the highest curvature excess values in promoter regions is rather homogeneous. The cluster contains exclusively mesophilic prokaryotes that have genome sizes larger than 1.4 Mb and high A+T composition. PAM method gives rather similar results: 92% genomes in the smallest cluster are also big A+T-rich mesophilic genomes.

Temperature\ cluster	psychophiles	mesophiles	thermophiles	hyperthermophiles
1	0	48	0	0
2	4	67	4	4
3	0	57	8	13

Table 1. Crosstabulation between temperature classification and clustering based on curvature excess in upstream regions.

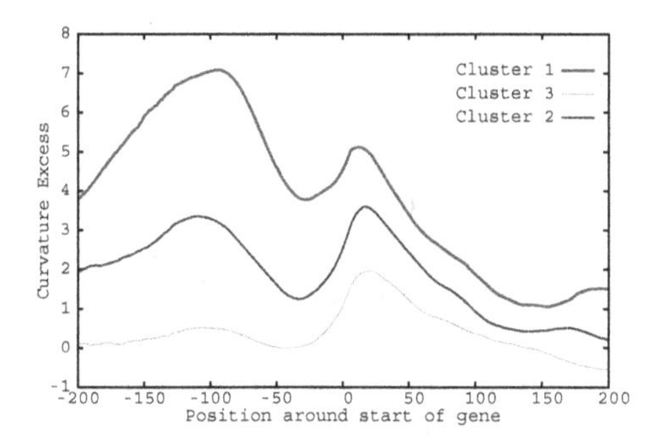

Fig. 1.3. Genomic profiles based on curvature excess distributions in the neighborhood of the starts of genes. Three clusters were obtained using K-means algorithms and the graphs present centroid profiles related to each of three clusters. The highest profile is related to the cluster 1, and the lowest profile corresponds to the cluster 3.

This table presents the results of clustering obtained by this algorithm applied to curvature excess in promoter regions crosstabulated with temperature classification. FM correlation coefficient is equal to 0.48 for the correlation between the temperature and curvature classifications. As we mentioned in the section Biological Methods, subsection Taxonomy, the majority of the processed Archaea are hyperthermophiles and vice versa. Naturally, the correlation between the two classifications, taxonomy and K-means clustering based on curvature promoter profiles is similar to the abovementioned correlation with the growth temperature. However, we observe that clustering of mesophilic Archaea is similar to clustering of mesophilic Bacteria, while clustering of thermophilic Eubacteria is as of Archaebacteria.

Genome size and A+T composition were also found to have an influence on curvature distribution [11]. Lengths of the currently available prokaryotic genomes range in size from 490,885 to 9,105,828 bp. With respect to the genomes size, we arbitrarily divided the genomes into two groups with threshold of 1.4 Mbp. In order to verify our intuitively chosen threshold of 1, 400,000 bp, we tested the correlation between size and curvature excess in promoter regions. Actually, we performed three correlation tests: 1) genome size was taken as sequential variable; 2) genome size was divided according to the median size, and 3) according to our arbitrary threshold of 1.4 Mbp. The results showed a significant correlation between genome size as sequentially variable and curvature excess in promoter regions. Performing T-tests, once according to the median (2.4 Mbp) and secondly according to the arbitrary value (1.4 Mbp), we found that the differences between the mean value of the groups were higher when the threshold of 1.4 Mbp was used.

With respect to A+T composition, we divided the genomes into four groups: 0 – 30%, 30 – 47%, 47 – 65% and 65% – 100%. The A+T composition of the genome is a mean value. However, A+T composition of prokaryotic intergenic regions is higher than of the appropriate genic sequences in almost all prokaryotic genomes. Therefore, the threshold of 47% A+T, between the second and the third group, was chosen to include genomes with A+T composition over 50% in intergenic regions. Correlation coefficient analysis between clusters and A+T composition was also performed. The results showed that cluster 1 does not contain any genomes of 'low' A+T composition (below 47%), also called 'GC-rich' genomes. This cluster mainly consists of the genomes with A+T content in a range of 47 – 65%. Surprisingly only 3 genomes (out of 35) from the forth group, genomes with the very high A+T content represented in this cluster. Examination of the remaining 32 'very A+T-rich' genomes revealed that most of them are 'small' genomes.

5.3 Correlation between DNA curvature in the neighborhoods of starts and ends of genes

Cluster analysis of the classifications was also performed on curvature excess distribution immediately after the ends of genes (terminator sites). We found that clustering based on curvature profiles makes sense whether profiles are related to regions before genes or after them. In order to find a relationship between results of cluster analyses based on different profiles correlation analysis was performed. The correlation coefficient FM is 0.5.

e_clusters\ s_clusters	1	2	3
1	35	7	0
2	13	50	31
3	0	22	47

Table 2. Crosstabulation between *s_clusters* (clusters obtained using promoter curvature profiles) and *e_clusters* (clusters obtained using curvature distributions after ends of genes).

We can see that in this table the most striking features are the high values on the main diagonal, zeroes and small values in the other cells related to the clusters with high curvature excess.

6 Conclusions

In this study we used DNA curvature excess profiles to reduce a comprehensively big text file (genome) to a numerical vector consisting of 200 real positive numbers smaller than 0.5. Such 205 vectors in a 200-dimensional Euclidian space were used for further data clustering based on two very widespread methods: K-means and PAM. Correlation coefficient tests between these two methods revealed high correlation, either with clusters based on promoter regions or terminator regions. However, examining the genomes in the clusters according to some characteristics, which known to have influence on curvature, K-means algorithm seems slightly more efficient. We found that main factors influencing curvature distribution in promoter regions of the prokaryotes are in the order of importance: 1) optimal growth temperature, 2) genome size, and 3) A+T composition. The absence of excessive curvature in almost all thermophiles and hyperthermophiles brings them altogether to a mutual cluster using clustering based both on promoter or terminator profiles; while majority of the mesophilic 'AT-rich' genomes presented in other clusters. The correlation between clusters based on curvature excess profile of promoters and terminators indicates on relationship in regulation mechanisms between initiation and termination of transcription. So, the clustering provides an indication that environmental factors (mesophilic vs. hyperthermophilic) influence regulation mechanisms. The conservative patterns of DNA curvature distribution across big mesophilic AT-rich eubacterial and archaeal genomes provide evidence that curved DNA is evolutionary preserved and determined by external (environmental) and internal (genomic) factors. The absence of excessive curvature in almost all of the 'small' genomes reflects also some evolutionary selection. Small genomes are usually obligate endocellular parasites that during the evolution adapted to utilize their genome host and consequentially lost nonfunctional sequences, such as regulatory elements.

References

1. S. Andersson, A. Zomorodipour, J. Andersson, T. Sicheritz-Pontent, U. Alsmark, R. Podowski, A. Naslund, A. Eriksson, H. Winkler, and C. Kurland. The genome sequence of Rickettsia prowazekii and the origin of mitochondria. *Nature*, 396:109–110, 1998.
2. A Bolshoy, P. McNamara, R.E. Harrington, and E.N. Trifonov. Curved DNA without A-A: experimental estimation of all 16 DNA wedge angles. *Proc Natl Acad. Sci. U.S.A.*, 88:2312–2316, 1991.
3. S. Diekmann and J.C. Wang. On the sequence determinants and flexibility of the kinetoplast DNA fragment with abnormal gel electrophoretic mobilities. *J. Mol. Biol.*, 186:1–11, 1985.
4. E. Forgy. Cluster analysis of multivariate data: Efficiency vs. interpretability of classifications. *Biometrics*, 21(3):768, 1965.
5. C. Fraley and A.E. Raftery. How many clusters? Which clustering method? Answers via model-based cluster analysis. *The Computer Journal*, 41(8):578–588, 1998.
6. C. Fraser, J. Gocanye, O. White, M. Adams, R. Clayton, R. Fleischmann, D. Bult, A. Kerlavage, G. Sutton, J. Kelly, and et al. The minimal gene complement of Mycoplasma genitalium. *Science*, 270:397–403, 1995.
7. J. Griffith, M. Bleyman, C.A. Rauch, P.A. Kitchin, and P.T. Englund. Visualization of the bent helix in kinetoplast DNA by electron microscopy. *Cell*, 46:717–724, 1986.

8. S. Hosid and A. Bolshoy. New elements of the termination of transcription in prokaryotes. *J. Biomol. Struct. Dyn.*, 22:347–354, 2004.
9. W. Kabsch, C. Sander, and E.N. Trifonov. The ten helical twist angles of B-DNA. *Nucleic Acids Res.*, 10(3):1097–1104, 1982.
10. L. Kaufman and P.J. Rousseeuw. *Finding Groups in Data: An Introduction to Cluster Analysis.* Wiley, New York, 1990.
11. L. Kozobay-Avraham, S. Hosid, and A. Bolshoy. Curvature distribution in prokaryotic genomes. *In Silico Biol.*, 4(3):361–375, 2004.
12. L. Kozobay-Avraham, S. Hosid, and A. Bolshoy. Involvement of DNA curvature in intergenic regions of prokaryotes. *Nucleic Acids Res.*, submited, 2006.
13. W. Krzanowski and Y. Lai. A criterion for determining the number of groups in a dataset using sum of squares clustering. *Biometrics*, 44:23–34, 1985.
14. J. Mardia, K. Kent and J. Bibby. *Multivariate Analysis.* Academic Press, San Diego, 1979.
15. J.C. Marini, S.D. Levene, D.M. Crothers, and P.T. Englund. Bent helical structure in kinetoplast DNA. *Proc. Natl. Acad. Sci. U.S.A.*, 79:7664–7668, 1982.
16. J. Perez-Martin, F. Rojo, and de V. Lorenzo. Promoters responsive to DNA bending: a common theme in prokaryotic gene expression. *Microbiol. Rev.*, 58:268–290, 1994.
17. S. Shigenobu, H. Watanabe, M. Hattori, K. Sakaki, and H. Ishikawa. Genome sequence of the endocellular bacterial symbiont of aphids Buchnera sp. aps. *Nature*, 7:81–86, 2000.
18. E.S. Shpigelman, E.N. Trifonov, and A. Bolshoy. Curvature: software for the analysis of curved DNA. *Comput. Appl. Biosci.*, 9:435–440, 1993.
19. C. Sugar and G. James. Finding the number of clusters in a data set : An information theoretic approach. *J of the American Statistical Association*, 98:750–763, 2003.
20. E. N. Trifonov and L. E. Ulanovsky. Inherently curved DNA and its structural elements. In *Unusual DNA Structures, Wells, R. D. and Harvey, S. C. (eds.)*, pages 173–187. Springer-Verlag, Berlin, 1987.
21. E.N. Trifonov. Curved DNA. *CRC Crit Rev Biochem*, 19:89–106, 1985.
22. H.M. Wu and D.M. Crothers. The locus of sequence-directed and protein-induced DNA bending. *Nature*, 308:509–513, 1984.

Look-Ahead Mechanism Integration in Decision Tree Induction Models

Michael Roizman and Mark Last

Ben-Gurion University of Negev, Department of Information Systems Engineering
Beer-Sheva 84105, Israel

Abstract. Most of decision tree induction algorithms use a greedy splitting criterion. One of the possible solutions to avoid this greediness is looking ahead to make better splits. Look-Ahead has not been used in most decision tree methods primarily because of its high computational complexity and its questionable contribution to predictive accuracy. In this paper we describe a new Look-Ahead approach to induction of decision tree models. We present a computationally efficient algorithm which evaluates quality of subtrees of variable-depth in order to determine the best split attribute out of a set of candidate attributes with a splitting criterion statistically indifferent from the best one.

1 Introduction

Decision trees are a way of representing a series of disjunctive rules that lead to a class or value; they classify instances by sorting them down the tree from the root to some leaf node, which provides the classification of the instance. Decision trees are grown through an iterative splitting of data into disjoint sets, where the goal is to maximize the "distance" between groups at each split.

One of the distinctions between decision tree methods is how they measure this distance. Each split may be introduced as separating the data into new sets which are as different from each other as possible. Trees left to grow without bound take longer to build and become unintelligible, but more importantly they overfit the data. Tree size can be controlled via stopping rules that limit the growth. One common stopping rule is simply to limit the maximum depth to which a tree may grow. Another stopping rule is to establish a lower limit on the number of records in a node and not do splits below this limit.

A common criticism of decision-tree algorithms is that they choose a split using a "greedy" approach, where the decision on which variable to split does not take into account any effect the split might have on future splits. In addition, all splits are made sequentially, so that each split is dependent on its predecessor. As a result the final solution could be very different if a different first split is made. One of the possible solutions to avoid this greediness is looking ahead to make better splits.

Look-Ahead has not been used in decision tree systems primarily because it is very computationally intensive.

Mark Last et al. (Eds.): Advances in Web Intelligence and Data Mining (SCI) **23**, 285-294 (2006)
www.springerlink.com

Look-Ahead search addresses the *horizon effect* by evaluating the quality of tests based on the quality of the subtrees they create to a certain depth. The horizon effect is the risk of a greedy algorithm to stop tree construction prematurely after being trapped in a local optimum [5].

Apparently, reducing the greediness of the splitting measure by looking ahead for future splits in deeper levels should benefit the induction process, but it turned out false in game tree evaluation [15]. Some studies show the same Look-Ahead pathology in decision tree learning [3], [9]. Related works on this topic show that the Look-Ahead can even hurt the predictive accuracy rather than improving it [3], [4].

One of the main results of Murthy and Salzberg (1995) is that "decision tree induction exhibits pathology, in the sense that Lookahead can produce trees that are both larger and less accurate than trees produced without it" [3]. They experimentally showed that Look-Ahead search fails to produce significant benefits; actually it often hurts the tree quality. They suppose that the pathology is a side-effect of how splitting measure is defined - the locally optimum split does not necessarily improve the global tree.

Elomaa and Malinen (2005) also showed that in many cases more extensive search in the hypotheses space produces hypotheses with larger error [5]. They claim that Look-Ahead affects statistical variance and this is the reason for machine learning bias amplification [7]. The variance of a learning algorithm can be ill-effected when the number of available choices grows, or the choices the algorithm makes depend on a smaller fraction of the training sample (for reasoning, see Dietterich and Kong 1995). This appears to be exactly what happens in Look-Ahead, i.e. more consideration is given to a greater number of choices - different attributes to split by - and, as the Look-Ahead trees are grown, in effect we split the sample recursively and consider smaller and smaller fractions of it [4].

The recent studies have not shown significant improvement in accuracy while looking ahead for better splits [4], [11]. Elomaa and Malinen (2005) introduced a new feature selection voting algorithm. The motivation for the algorithm is not to perform actual look-ahead but to reduce the greediness of information gain split, and at the same time to avoid pathology of a more thorough search.

The empirical results of their research are as follows: the limited look-ahead algorithm showed better prediction accuracy with some look-ahead depths in 14 of the 20 data sets examined in total (in some cases the advantage was very slight). In general, the deeper search performed by their algorithm tends to produce larger decision trees. These results somewhat coincide with those of Murphy and Pazzani (1994), who found in their empirical tests that slightly larger trees predict better than the smallest ones [5].

Esmeir and Markovitch (2004) presented a Look-Ahead based anytime algorithm for constructing decision trees. The main argument for this algorithm is that it can make use of additional time in order to generate better decision trees. In *Look-Ahead by Stochastic ID3 (LSID3)* algorithm which they presented, each candidate split is evaluated by summing the estimated size of each subtree. For each subtree there are several estimations, and since each estimation is an upper bound, the minimal one is considered.

They showed that for several real world and synthetic datasets, LSID3 can make use of higher budget of time. In most cases, when LSID3 was allocated few minutes, it produced trees of smaller size and of higher accuracy. The usage a few more time is shown to be worthwhile when the concepts are hard and involve interdependencies between the attributes. In these cases, most of the existing greedy methods fail.

The goal of this paper is nevertheless to seek for a way to reduce the computational intensity of Look-Ahead operations while gaining an accuracy benefit over greedy decision tree induction algorithms.

In this research we use a split quality measure to find all locally optimal binary splits and then grow alternative subtrees by activating a bounded Look-Ahead. The alternative trees are then evaluated and compared to find the one that provides the best prediction performance.

The remainder of this paper is organized as follows. Section 2 outlines the greedy decision-tree induction process. In Section 3 we show splitting measures used in the research, Section 4 presents the proposed Look-Ahead algorithm based on J48 (Java version of C4.5), Section 5 describes the design of evaluation experiments and Section 6 summarizes this paper and presents future research.

2 Decision Tree Learning (C4.5)

We use the usual framework for supervised learning from examples based on the C4.5 algorithm [2]. Let $\mathcal{X}$ be the instance space. Each instance $\chi \in \mathcal{X}$ is described by the values of a set of attributes $A = \{a_i, \ldots, a_k\}$. The domain of each attributes value a_i may be nominal or numerical. In this research we concentrate on numerical attributes. Let $Y = \{\gamma_i, \ldots, \gamma_k\}$ be the set of class labels, where for labeled example $(\chi, \gamma) \in \mathcal{X} \times \mathcal{Y}$, γ is a correct class of instance χ. Suppose we are given a training set S of n labeled examples $\{s_1, \ldots, s_k\}$.

The basic algorithm for top-down tree induction is described as follows. First, all the examples S are assigned to a single *root* node. Then, if S contains examples with different class labels, then it is divided into two subsets according to the *values* of the splitting attribute. The resulting subsets are assigned to two new children of the root node. This process is then recursively used on these new leaves.

At this stage of the research we consider attributes with numerical values, so the simplest method for splitting, called *binarization*, where the real axis is split in two intervals, will be used to split the attribute numeric values. The proposed algorithm will evaluate three different splitting criteria: *Information Gain, Gain Ratio* and *Gini Index*. These criteria are used to determine the splitting attribute at any level of the induced tree. Their definitions are provided in "Splitting Rules" section below.

3 Splitting Rules

Most decision tree induction algorithms use some kind of greedy heuristics known as splitting criteria / rule. In our research, we deal with three classical splitting rules. Several other splitting rules have been proposed in the literature but we give a background only for the relevant three ones used by C4.5, CART, and other decision-tree algorithms..

3.1 Information Gain: Entropy (Numeric values only)

Suppose we want to split a set S. The IG for this binary split, using attribute a_i with the splitting threshold at b is defined as the difference in entropy between the unpartitioned data and the data partitioned by the chosen split

$$H(S) - (H(S_{a_i \leq b}) + H(S_{a_i > b})) \tag{1}$$

where, $H(S) = \sum_{j \in (1,\ldots,m)} Pr(\gamma = \gamma_i) \cdot \log_2 \left(\frac{1}{Pr(\gamma = \gamma_i)} \right)$
and $H(S_{a_i \leq b}) = \sum_{j \in (1,\ldots,m)} Pr(a_i \leq b, \gamma = \gamma_i) \cdot \log_2 \left(\frac{1}{Pr(a_i \leq b, \gamma = \gamma_i)} \right)$
Also, $H(S_{a_i > b})$ is defined similarly [6].

3.2 Gain Ratio

The problem with Information Gain is that it prefers attributes with many values over those with few values. That is why if there is an attribute with very large number of values (e.g. Date), the algorithm that uses Information Gain as splitting criteria will choose this attribute as a root node and, as a result, very broad tree with depth one will be induced. This tree will perfectly classify training set data, but if we take any sample with unseen data, the prediction accuracy of the classifier will be poor. In order to avoid this problem, Split Information measure is incorporated. It is sensitive to how broadly and uniformly the attribute splits the data [6]:

$$SplitInformation(S, A) = -\sum_{i=1}^{c} \frac{|S_i|}{|S|} \log_2 \frac{|S_i|}{|S|}$$

Where S_1 through S_c are the c subsets that are the result of partitioning S by the c-valued attribute. So, the new splitting criterion that uses Split Information is defined as follows [6]:

$$GainRatio(S, A) = \frac{Gain(S, A)}{SplitInformation(S, A)} \tag{2}$$

3.3 Gini Index

Gini index tries to minimize the impurity contained in the training subsets generated after branching decision tree. Its function is defined as follows:

$$I(t) = \sum_{j \neq i} p(j|t)\, p(i|t) = 1 - \sum_{j} p^2(j|t) \qquad (3)$$

Where t is a given node with estimated class probabilities $p(j|t)$ [8].

4 Look-Ahead Mechanism Integration

The proposed algorithm is based on C4.5 [10] as it was indicated in the previous section, but its implementation is based on the Java version of C4.5 - J48 [12]. We call the new algorithm LA-J48 (Look-Ahead J48).

4.1 LA-J48 Attribute Selection Algorithm

Here we present the method for reduction of the greediness of split selection through Look-Ahead with a limited computational effort. The Look-Ahead is applied to every attribute that was chosen as good enough to become a split one. The definition is done by statistical comparison of all attributes to the "best" one, chosen by the best split measure value. Figure 1 shows the logic flow and emphasizes the main stages of the proposed algorithm.

Suppose we build some decision tree T which includes a leaf l that doesn't satisfy stopping criterion. Now we want to decide by which attribute to split the node l. We define S_l as a set of attributes assigned to leaf l. The Split Measure (SM) calculation function runs for every attribute in S_l according to the selected splitting measure (see Section 3). The result of this stage of the algorithm is an array of splitting measures SM_i of the attributes for node l.

At the next step of the algorithm we do the statistical evaluation of all calculated splitting measures. The process is described later in Section 4.2. We choose $BestSM(S_l)$ - the best split attribute for this leaf by greedy splitting measure and find all other attributes with a calculated split measure value not significantly different from $BestSM(S_l)$. The set of these attributes is defined as E_l, where any one of them is a potential candidate for splitting the leaf l. Then, Look-Ahead method is activated for every one of these attributes sequentially (in the order of calculated splitting measure values). Look-Ahead method runs the original J48 with previously chosen split measure, where we suppose that the root node of each subtree is $a_e \in E_l$, $e = 1, \ldots, k$. Applying a stopping criterion (threshold) for Look-Ahead to greedy generation of subtrees for splitting attributes is necessary to avoid overfitting, so the algorithm runs until the stopping criterion is satisfied or a subtree reaches maximum given depth d, $(d = 1, \ldots, 10)$.

The result of this stage is a number of trees, rooted each one at potential split attribute. These trees have their own structure and constitute all possible ways

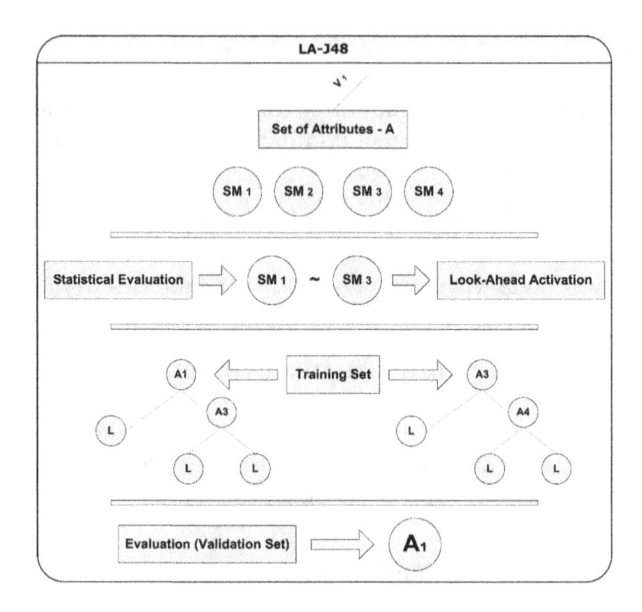

Fig. 1. LA-J48 algorithm logic flow

of global tree induction continuation. Here we choose the splitting attribute. The importance of this stage is obvious and the evaluation of these trees that is described in Section 4.3 defines the behavior of the proposed algorithm and its ability to choose better splitting attribute. It is very important to gather all possible measures of these trees in order to perform better evaluation and to determine the way in which the induction will continue.

After the evaluation of every tree rooted at a_e the "best" tree is chosen and its root a_e becomes the split attribute for leaf l. From here the algorithm divides the data set of node l into subsets on possible values of chosen splitting attribute and continues the induction in a recursive manner.

4.2 Splitting Measure Values Evaluation

At every level of the tree, the calculated splitting measure values have to be evaluated and compared statistically. This is done using different confidence levels (0.9, 0.95, 0.99, 0.995).

The statistical comparison of the splitting measures of different attributes is done by checking the significance of the test statistic. For example, while evaluating the *IG* splitting measure, the *likelihood-ratio* statistic is calculated for every attribute. Then the ratios $Ratio_i$ between the likelihood-ratio of the attribute chosen as the "best" one and likelihood-ratio of all other attributes are calculated.

$$Ratio_i = \frac{BestG^2\left(G_i^2\right)}{G_i^2}$$

Each calculted ratio is the F statistic and its significance is tested according to the F distribution. If the test statistic is not found to be significant, we consider the attribute which likelihood-ratio was a denominator to be the potential split attribute for given node. First, the *Mutual Information* (MI) of each attribute is calculated according to following formula:

$$MI\left(A_{i'};A_i/z\right)=\sum_{j=0}^{M_i-1}\sum_{j'=0}^{M_{i'}-1}P\left(V_{ij};V_{i'j'};z\right)\bullet\log\frac{P\left(V_{i'j'}^{ij}/z\right)}{P\left(V_{i'j'}/z\right)\bullet P\left(V_{ij}/z\right)}$$

Where

$P\left(V_{i'j'}/z\right)$ - an estimated conditional (*a posteriori*) probability of a value j' of the candidate input attribute i', given the node z.

$P\left(V_{ij}/z\right)$ - an estimated conditional (*a posteriori*) probability of a value j of the target attribute i, given the node z.

$P\left(V_{i'j'}^{ij};z\right)$ - an estimated conditional (*a posteriori*) probability of a value j' of the candidate input attribute i' and a value j of the target attribute i, given the node z.

$P\left(V_{ij};V_{i'j'};z\right)$ - an estimated join probability of a value j of the target attribute i, a value j' of the candidate input attribute i' and the node z.

At the next step we calculate the likelihood-ratio statistic of each candidate input attribute and the target attribute, given the node, by:

$$G^2\left(A_i;A_{i/z}\right)=2\bullet(ln2)\bullet E^*(z)\bullet MI\left(A_{i'};A_i/z\right)$$

Where $E^*(z)$ is a number of tuples.

The number of degrees of freedom for the likelihood-ration statistic is calculated by:

$$DF\left(A_{i'};A_i/z\right)=\left(NI_{i'}(z)-1\right)\bullet\left(NT_i(z)-1\right)$$

Where

$NI_{i'}(z)$ - number of values of a candidate input attribute i' at node z.

$NT_i(z)$ - number of values of a target attribute i at node z.

The distribution probability of G^2 is X^2 so the distribution of ratio between two likelihood-ratios is F [13].

The number of degrees of freedom for F-test statistic $F_{i_{(\alpha,v_1,v_2)}}$ is:

$$v_1=\left(NI_{Best'}(z)-1\right)\bullet\left(NT_{Best}(z)-1\right)$$

$$v_2=\left(NI_{i'}(z)-1\right)\bullet\left(NT_i(z)-1\right)$$

After the significance of all F-test statistics is checked, the insignificant ones are considered as potential split attributes for the given node z and the algorithm applies the Look-Ahead method on them.

4.3 Subtrees Evaluation

The goal of this evaluation is examining the similarity of the subtrees, comparison of different measures gathered during Look-Ahead like tree size, number of nodes, building time and etc. In order to decide what attribute to split after the Look-Ahead activation, the accuracy of the induced subtrees will be compared using a validation set. The one with the best accuracy will be chosen and the main induction process will continue while splitting on the root of the "best" subtree. At this stage of the algorithm we compare all possible ways of decision tree future evolution. Depending on different measures of the subtrees we decide how the algorithm will proceed. Also, the collected measures of the subtrees will help us to understand the pathology of Look-Ahead in decision tree induction [4] and to estimate the contribution of LA-J48 in avoiding this pathology.

4.4 LA-J48 Algoritm Pseudo Code

Algorithm 1 Look-Ahead Mechanism Integration

1: $S_l \leftarrow$ A set of attributes assigned to leaf l in some tree T
2: **For Each** *attribute* $\in S_l$ **Do**
3: $\quad SM(attribute_i) \leftarrow$ CalculateSplittingMeasure($attribute_i$)
4: $BestSM(S_l) \leftarrow$ CalculateBestSplittingMeasure(SM)
5: Create array of attributes E_l
5: **For Each** *attribute* $\in S_l$ **Do**
6: $\quad$ **If** IsNotSignificantlyDiff($BestSM, SM_i$)
7: $\quad\quad a_e \leftarrow attribute_e \mid a_e \in E_l$
8: **For Each** $a_e \in E_l$ **Do**
9: $\quad t_l \leftarrow$ BuildTreeJ48(a_e)
10: $\quad$ Calculate measures for t_l: tree size, number of nodes, build time, etc.
11: $\quad Accuracy\ (t_l) \leftarrow$ EvaluateTree(*Validation Set*)
12: $BestTree \leftarrow$ CompareSubTrees(t)
13: Split l by the root of the *BestTree*

5 Design of Evaluation Experiments

The performance evaluation of the obtained decision tree will be done by 10-fold cross validation. 10-fold cross validation will divide the whole dataset into 10 subsets. 10 runs will be done when each run will use another subset as a Test Set. In every run, the remaining 9 parts will be divided into 2 parts (75% and 25%) for Training Set and Validation Set respectively. The classification performance of the classifier will be compared to that of Elomaa and Malinen (2005) and to the J48 (C4.5) algorithm without Look-Ahead. Decision tree induction will be evaluated by different measures like running time of the algorithm as a whole and its parts (Look-Ahead procedure for example), resulting tree size, subtrees similarity, etc. According to these measures different conclusions can be drawn about the usability of the new model, its predictive performance, reliability, types of datasets where the look-ahead induction works better and so on. Based on the empirical results, conclusions will be drawn about the pathology of Look-Ahead methods in decision tree induction as described in Elomaa and Malinen (2005) and its possible effect on the new proposed algorithm.

6 Summary and Future Research

This paper presents a novel approach for Looking Ahead in decision tree induction. The main goal of this method is an improvement of classifier accuracy and reducing the computational intensity of existing Look-Ahead methods. After implementing the algorithm we are going to perform empirical evaluations, to study the pathology described in related studies [3], [4], [9], and to examine its effect on the proposed algorithm. Initial results of our experiments will be presented at the workshop.

7 Acknowledgments

This work was partially supported by the National Institute for Systems Test and Productivity at University of South Florida under the USA Space and Naval Warfare Systems Command Grant No. N00039-01-1-2248.

References

[1] Breslow, L. A. and Aha, D. W. 1997. "Simplifying decision trees: A survey." Knowl. Eng. Rev. 12, 1 (Jan. 1997), 1-40.

[2] J. R. Quinlan. "C4.5: Programs for Machine Learning." Morgan Kaufmann Publishers Inc., 1993.

[3] Murthy, S. and Salzberg, S. (1995). "Lookahead and pathology in decision tree induction, in C. S." Mellish, ed., 'Proceedings of the 14th International Joint Conference on Artificial Intelligence', Morgan Kaufmann, pp. 1025-1031.

[4] T. Elomaa. "On look-ahead and pathology in decision tree learning." Journal of Experimental and Theoretical Artificial Intelligence 17, 1-2 (Jan. 2005) 19-33. With T. Malinen.

[5] T. Elomaa, T. Malinen. "On Lookahead Heuristics in Decision Tree Learning." ISMIS 2003: 445-453.

[6] Mitchell, T.M. "Machine Learning." McGraw-Hill, New York (1997).

[7] T. G. Dietterich and E. B. Kong. "Machine learning bias, statistical bias, and statistical variance of decision tree algorithms." Technical report, Department of Computer Science, Oregon State University, 1995.

[8] Breiman, L., Friedman, J. H., Olshen, R. A., and Stone, C. J. (1984). "Classification and Regression Trees." Wadsworth, Belmont, CA

[9] J. R. Quinlan and R. M. Cameron-Jones. "Oversearching and layered search in empirical learning." In IJCAI'95, pages 1019-1024. Morgan Kaufmann, 1995.

[11] Esmeir, S. and Markovitch, S. 2004. "Lookahead-based algorithms for anytime induction of decision trees." In Proceedings of the Twenty-First international Conference on Machine Learning (Banff, Alberta, Canada, July 04 - 08, 2004). ICML '04, vol. 69. ACM Press, New York, NY, 33.

[12] I. Witten and E. Frank. "Data Mining: Practical Machine Learning Tools and Techniques, 2nd Edition." Morgan Kaufmann, ISBN 0120884070, 20

[13] Oded Z. Maimon , Mark Last, Knowledge Discovery and Data Mining: The Info-Fuzzy Network (Ifn) Methodology, Kluwer Academic Publishers, Norwell, MA, 2001

Feature Selection by Combining Multiple Methods

Lior Rokach[1], Barak Chizi[2], and Oded Maimon[2]

[1] Department of Information Systems Engineering, Ben-Gurion University of the Negev, Israel liorrk@bgu.ac.il
[2] Department of Industrial Engineering, Tel-Aviv University, Israel barakc,maimon@eng.tau.ac.il

Summary. Feature selection is the process of identifying relevant features in the dataset and discarding everything else as irrelevant and redundant. Since feature selection reduces the dimensionality of the data, it enables the learning algorithms to operate more effectively and rapidly. In some cases, classification performance can be improved; in other instances, the obtained classifier is more compact and can be easily interpreted. There is much work done on feature selection methods for creating ensemble of classifiers. Thus, these works examine how feature selection can help ensemble of classifiers to gain diversity. This paper examines a different direction, i.e. whether ensemble methodology can be used for improving feature selection performance. In this paper we present a general framework for creating several feature subsets and then combine them into a single subset. Theoretical and empirical results presented in this paper validate the hypothesis that this approach can help finding a better feature subset.

1 Introduction

Feature selection is a common issue in statistics, pattern recognition and machine learning [8]. The aim of feature selection is to distil the most useful subset of features from a given subset.

There are two main strategies for performing feature selection. The first known as *filter* [6] operates independent of any learning algorithm – undesirable features are filtered out of the data before learning begins. These algorithms use heuristics based on general characteristics of the data to evaluate the merit of feature subsets. The second strategy argues that the bias of a particular induction algorithm should be taken into account when selecting features.

The second strategy, known as *wrapper* [6], uses a learning algorithm along with a statistical re-sampling technique such as cross-validation to select the best feature subset for this specific learning algorithm.

Mark Last et al. (Eds.): Advances in Web Intelligence and Data Mining (SCI) **23**, 295-304 (2006)
www.springerlink.com

A sub-category of filter methods that will be refer to as rankers, are methods that employ some criterion to score each feature and provide a ranking. From this ordering, several feature subsets can be chosen, either manually of setting a threshold.

The main idea of ensemble methodology is to combine a set of models, each of which solves the same original task, in order to obtain a better composite global model, with more accurate and reliable estimates or decisions than can be obtained from using a single model. The idea of building a predictive model by integrating multiple models has been under investigation for a long time.

In the past few years, experimental studies conducted by the machine-learning community show that combining the outputs of multiple classifiers reduces the generalization error [12]. Ensemble methods are very effective, mainly due to the phenomenon that various types of classifiers have different "inductive biases". Indeed, ensemble methods can effectively make use of such diversity to reduce the variance-error [15], without increasing the bias-error. In certain situations, an ensemble can also reduce bias-error, as shown by the theory of large margin classifiers [1].

A common strategy for manipulating the training set is to manipulate the input attribute set. The idea is to simply give each classifier a different projection of the training set. Ensemble feature selection methods [10] extend traditional feature selection methods by looking for a set of feature subsets that will promote disagreement among the base classifiers. Ho [4] has shown that simple random selection of feature subsets may be an effective technique for ensemble feature selection because the lack of accuracy in the ensemble members is compensated for by their diversity. Tsymbal and Puuronen [14] presented a technique for building ensembles of simple Bayes classifiers in random feature subsets.

The hill-climbing ensemble feature selection strategy [2], randomly construct the initial ensemble. Then, an iterative refinement is performed based on hill-climbing search in order to improve the accuracy and diversity of the base classifiers. For all the feature subsets, an attempt is made to switch (include or delete) each feature. If the resulting feature subset produces better performance on the validation set, that change is kept. This process is continued until no further improvements are obtained.

The Genetic Ensemble Feature Selection strategy uses genetic search for ensemble feature selection [10]. It begins with creating an initial population of classifiers where each classifier is generated by randomly selecting a different subset of features. Then, new candidate classifiers are continually produced by using the genetic operators of crossover and mutation on the feature subsets. The final ensemble is composed of the most fitted classifiers.

An approach for constructing an ensemble of classifiers using rough set theory was presented in [5]. The method searches for a set of reducts, which include all the indispensable attributes. A reduct represents the minimal set of attributes which has the same classification power as the entire attribute set.

Oliveira et al. [9] suggest creating a set of feature selection solutions using a genetic algorithm. Then they create a Pareto-optimal front in relation to two different objectives: accuracy on a validation set and number of features. Following that they select the best feature selection solution. Masulli and Rovetta [7] have employed ensemble methodology for feature selection. Nevertheless their method was specifically developed for micro-array data, and no general framework was proposed.

Recently Torkkola and Tuv [13] and Tuv and Torkkola [16] examined the idea of using ensemble classifiers such as decision trees in order to create a better features ranker. They have showed that this ensemble can be very effective in variable ranking for problems with up to a hundred thousand input attributes. Note that this approach uses inducers for obtaining the ensemble. Thus, it concentrates on wrapper feature selectors.

There is much work done on feature selection methods for creating ensemble of classifiers. These works examine how feature selection can help ensemble of classifiers to gain diversity. Nevertheless there is hardly works that examine the other way around, i.e. how can ensemble of feature selectors improve current feature selection results. The aim of this paper is theoretically and experimentally examine whether ensemble feature subsets can be used for improving non-ranker feature selection filters methods.

2 Problem Definition and Theoretical Observations

The problem of feature selection ensemble is that of finding the best feature subset by combining a given set of feature selectors, such that if a specific inducer is run on it, the generated classifier will have the highest possible accuracy. Following Kohavi and John [6] we adopt the definition of optimal feature subset with respect to a particular inducer.

Definition 1. *Given an inducer I, a training set S with input feature set $A = \{a_1, a_2, ..., a_n\}$ and target feature y from a fixed and unknown distribution D over the labeled instance space, the subset $B \subseteq A$ is said to be optimal if the expected generalization error of the induced classifier $I(\pi_{B\cup y}S)$ will be minimized over the distribution D.*

where $\pi_{B\cup y}S$ represents the corresponding projection of S and $I(\pi_{B\cup y}S)$ represent a classifier which was induced by activating the induction method I onto dataset $\pi_{B\cup y}S$.

Definition 2. *Given an inducer I, a training set S with input feature set $A = \{a_1, a_2, ..., a_n\}$ and target feature y from a fixed and unknown distribution D over the labeled instance space, and an optimal subset B, a Feature Selector FS is said to be consistent if it selects an attribute $a_i \in B$ with probability $p > 1/2$ and it selects an attribute $a_j \notin B$ with probability $q < 1/2$.*

Definition 3. *Given a set of feature subsets* $B_1, ..., B_\omega$ *the majority combination of features subsets is a single feature subset that contains any attribute* a_i*such that* $f_c(a_i) > \frac{\omega}{2}$ *where* $f_c(a_i, B_1, ..., B_\omega) = \sum_{j=1}^{\omega} g(a_i, B_j)$ *and*

$$g(a_i, B_j) = \begin{cases} 1 & a_i \in B_j \\ 0 & otherwise \end{cases}$$

3 Independent Algorithmic Framework

Roughly speaking, the feature selectors in the ensemble can be created dependently or independently. In the dependent framework the outcome of a certain feature selector affect the creation of the next feature selector. Alternatively each feature selector is built independently and their results are combined in some fashion. In this paper we concentrate on independent framework. Figure 1 presents the proposed algorithmic framework. This simple framework gets as an input the following arguments:

1. A Training set (S) – A labeled dataset used for feature selectors.
2. A set of feature selection algorithms $\{FS_1, \ldots, FS_\xi\}$ – A feature selection algorithm is an algorithm that obtains a training set and outputs a subset of relevant features. Recall that in this paper we employ non-wrapper and non-ranker feature selectors.
3. Ensemble Size (ω)
4. Ensemble generator (G) – This component is responsible for generating a set of ω pairs of feature selection algorithms and their corresponding training sets. We refer to G as a class that implements a method called "genrateEnsemble".
5. Combiner (C) – The combiner is responsible to create the subsets and combine them into a single subset. We refer to C as a class that implements the method "combine".

The proposed algorithm simply uses the ensemble generator to create a set of pairs of feature selection algorithms and their corresponding training sets. Then it call the combine method in C to execute the feature selection algorithm on its corresponding dataset and then combine the various feature subsets into a single subset.

Require: S, $\{FS_1, \ldots, FS_\xi\}$, G, C
Ensure: A combined feature subset.
1: $\{(S_1, FS_1), \ldots, (S_\omega, FS_\omega)\} \leftarrow$ G.genrateEnsemble $(S, (FS_1, \ldots, FS_\xi), \omega)$
2: Return C.combine $(\{(S_1, FS_1), \ldots, (S_\omega, FS_\omega)\})$

Fig. 1. Pseudo-code of Independent Algorithmic Framework for Feature Selection

3.1 Combining Procedure

We begin by describing two implementations for the combiner component. In the literature there are two types of methods to combine the results of the ensemble members: weighting methods and meta-learning methods. In this paper we concentrate on weighting methods. The weighting methods are best suited for problems where the individual members have comparable success or when we would like to avoid problems associated with added learning (such as over-fitting or long training time).

Simple Weighted Voting

Figure 2 presents an algorithm for selecting a feature subset based on the weighted voting of feature subsets. As this is an implementation of the abstract combiner used in Figure 1, the input of the algorithm is a set of pairs; every pair is built from one features selector and a training set. It executes the feature selector on its associated training set to obtain a feature subset. Then the algorithm employs some weighting method and attaches a weight to every subset. Finally it uses a weighted voting to decide which attribute should be included in the final subset. We considered the following methods for weighting the subsets:

1. **Majority Voting** – In this weighting method the same weight is attached to every subset such that the total weights is 1, i.e. if there are ωsubsets then the weight is simply$1/\omega$. Note that the inclusion of a certain attribute in the final result requires that this attribute will appear in at least $\omega/2$ subsets. This method should have a low false positive rate, because selecting an irrelevant attribute will take place only if at least $\omega/2$ feature selections methods will decide to select this attribute.
2. **"Take-It-All"** – In this weighting method all subsets obtain a weight that is greater than 0.5. This lead to the situation in which any attribute that has been in at least one of the subsets will be included in the final result. This method should have a low false negative rate, because loosing a relevant attribute will take place only if all feature selections methods will decide to filter out this attribute.
3. **"Smaller is Heavier"** – The weight for each selector is defined by its bias to smallest subset. Selectors that tend to provide a small subset will gain more weight than selectors that tend to provide a large subset. This approach is inspired by the fact that the precision rate of selectors tend to decrease as the size of the subset increases. This approach can be used to avoid noise caused by feature selectors that tend to select most of the possible attributes. More specifically the weights are defined as (note that in this case the weights are normalized and sum up to 1):

$$w_i = \frac{|B_i|}{\sum_{j=1}^{\omega} |B_j|} \Bigg/ \sum_{k=1}^{\omega} \frac{|B_k|}{\sum_{j=1}^{\omega} |B_j|} \qquad (1)$$

Require: $\{(S_1, FS_1), \ldots, (S_\omega, FS_\omega)\}$
Ensure: A Combined feature subset
1: **for all** $(S_i, FS_i) \in F$ **do**
2: B_i = FS$_i$.getSelectedFeatures(S_i)
3: **end for**
4: $\{w_1, \ldots, w_\omega\} = getWeight\,(\{B_1, ..., B_\omega\})$
5: $B \leftarrow \emptyset$
6: **for all** $a_j \in A$ **do**
7: totalWeight=0
8: **for** $i = 1$ to ω **do**
9: **if** $a_j \in B_i$ **then**
10: totalWeight $\leftarrow$ totalWeight+W$_i$
11: **end if**
12: **end for**
13: **if** totalWeight$>$ 0.5 **then**
14: B $\leftarrow B \cup a_j$
15: **end if**
16: **end for**
17: Return B

Fig. 2. Pseudo-code of combining procedure

3.2 Feature Ensemble Generator

In order to make the ensemble more effective, there should be some sort of diversity between the feature subsets. Diversity may be obtained through different presentations of the input data or variations in feature selector design.

Multiple Feature Selectors

In this approach we simply use a set of different feature selection algorithms. The basic assumption is that using different algorithms have different inductive biases' and thus they will create different feature subsets.

Bagging

The most well-known independent method is bagging (bootstrap aggregating). In this case each feature selector is executed on a sample of instances taken with replacement from the training set. Usually each sample size is equal to the size of the original training set. Note that since sampling with replacement is used, some of the instances may appear more than once in the same sample and some may not be included at all. So the training samples are different from each other, but are certainly not independent from statistics point of view

4 Experimental Study

In order to illustrate the theoretical results shown above, a comparative experiment has been conducted on benchmark data sets. The following subsections describe the experimental set-up and the obtained results.

4.1 Dataset Used

The selected algorithms were examined on 10 data sets of which have been selected manually from the UCI Machine Learning Repository. The datasets chosen vary across a number of dimensions such as: the number of target classes, the number of instances, the number of input features and their type (nominal, numeric).

4.2 Algorithms Used

Table 1 presents 10 feature ensemble alternatives examined in this experiment. All multiple generators used 5 different feature selection algorithms. All the algorithms use Correlation-based Feature Subset Selection (CFS) as a subset evaluator [3]. CFS Evaluates the worth of a subset of attributes by considering the individual predictive ability of each feature along with the degree of redundancy between them. Subsets of features that are highly correlated with the class while having low intercorrelation are preferred. The algorithms differ by their search method: Best First Search (BFS), Forward Selection Search by using Gain Ratio, Chi-Square, OneR classifier, and Information Gain. The bagging approach was used by employing the CFS with BFS as a search method.

4.3 Evaluation Method

Based on the problem formulation described above, the main goal of the feature selection is to mininize the generalization error of a particular inducer.

J48 algorithm is used as the induction algorithm. J48 is a java version of the well-known C4.5 algorithm [11].

In order to estimate the generalization error 10-fold cross-validation procedure was used. Each dataset was randomly divided into 10 equal parts in order to provide 10 different iterations of feature selection and classification. For each iteration 1/10 of the dataset was used as the test set and 9/10 of the dataset was used as train.

Table 1. Accuracy Results of Various Ensemble Alternatives

Alternative	Generator	Combiner
BMV10	Bagging of size 10	Majority Voting
BMV5	Bagging of size 5	Majority Voting
BSH10	Bagging of size 10	"Smaller is Heavier"
BSH5	Bagging of size 5	"Smaller is Heavier"
BTA10	Bagging of size 10	"Take-It-All"
BTA5	Bagging of size 5	"Take-It-All"
MMV	Multiple	Majority Voting
MSH	Multiple	"Smaller is Heavier"
MTA	Multiple	"Take-It-All"

4.4 Results

Table 2 summarizes the experimental results of the various feature ensemble implementations. It can be seen that the MTA (Multiple Take-It-All) implementation achieved on average the best results. All other methods achieved on average comparable results.

Table 2. Accuracy Results of Various Ensemble Alternatives

Dataset	BMV10	BMV5	BSH10	BSH5	BTA10	BTA5	MMV	MSH	MTA
Arrhythmia	68.13	67.93	69.51	67.87	64.44	65.25	66.89	67.34	68.71
Audiology	72.02	71.49	72.02	71.49	73.35	73.48	71.81	71.89	77.54
Balance	78.18	78.18	78.18	78.18	78.18	78.18	77.61	77.61	78.18
Bridges	57.88	58.43	57.88	57.60	58.18	58.73	58.45	58.17	58.43
Car	77.50	77.50	77.50	77.50	78.05	77.50	77.50	77.50	86.39
Kr-vs-kp	90.34	90.34	90.34	90.34	90.34	90.34	90.34	90.34	90.69
Letter	85.90	85.90	85.90	85.90	85.90	85.90	85.90	85.90	85.97
Pendigits	95.18	95.14	95.03	95.14	95.21	95.21	95.20	95.20	95.63
Soybean	88.95	89.08	88.78	89.08	88.61	88.65	88.52	88.52	88.44
Spambase	92.03	91.94	91.94	91.94	92.44	92.03	92.34	92.27	92.38
Splice	93.30	93.30	93.30	93.30	93.59	93.26	93.30	93.30	93.30
Average	81.76	81.75	81.85	81.67	81.66	81.68	81.62	81.64	83.24

For benchmarking the ensemble approach, all the feature selection algorithms mentioned above were separately experimented on the same datasets. Moreover we examined the result obtained with no feature selection. Table 3 summarizes the comparison of MTA with these algorithms. The superscript "+" indicates that the accuracy rate of MTA was significantly higher than the corresponding algorithm at confidence level of 5%. The "–" superscript indicates the accuracy was significantly lower.

When observing the results, two important observations appear: First, as seen on Table 3, MTA did much better than other methods for feature selection. Applying t-test (paired two sample for means) validates the above

Table 3. Classification Results

Datasets	J48 Without FS	MTA	CFS-BFS	CFS-Rank Search-Gain Ratio	CFS-Rank Search-Chi Square	CFS-Rank Search-OneR	CFS-Rank Search-Information Gain
Arrhythmia	61.84^{+}	68.71	68.19	68.58	66.89	66.04	64.93^{+}
Audiology	75.00	77.53	72.01^{+}	73.27^{+}	71.40^{+}	71.73^{+}	71.89
Balance	74.40^{+}	78.17	77.61^{+}	78.17	77.61^{+}	77.61^{+}	77.61^{+}
Bridges	55.55^{+}	58.43	57.59^{+}	58.19	58.16	58.43	58.16
Car	89.45	86.38	77.50^{+}	77.50^{+}	77.50^{+}	86.38	77.50^{+}
Kr-vs-kp	99.63^{-}	90.68	90.33	71.15^{+}	90.33	90.33	90.33
Letter	85.92	85.97	85.89	85.37^{+}	85.89	85.97	85.89
Pendigits	95.93	95.62	95.04	95.25	95.19	95.62	95.19
Soybean	88.12	88.44	88.57	88.44	88.44	88.44	88.45
Spambase	92.20	92.37	92.00	90.51^{+}	92.37	90.88	92.06
Splice	91.88^{+}	93.29	93.29	89.72^{+}	93.29	93.29	93.29

observation, by providing $p < 0.05$ for all methods vs. MTA. Statistically the empirical results validate the theoretical results shown on section 2. Ensemble method may yield better feature subset when the goal is to improve classification accuracy.

Another important observation from Table 3 indicates that employing MTA before using induction algorithm provides better results than not using MTA. Using t-test (paired two sample for means) validates the above by providing $p < 0.05$ for J48 without any feature selection procedure vs. MTA. When handling noisy data sets which involved irrelevant and redundant information, feature selection can provide better classification accuracy. Nevertheless, this improvement is not guaranteed. Some feature selection techniques might reduce the accuracy in certain datasets (see for instance the Audiology dataset). However in this experimental study, it becomes evident that MTA has almost never reduced the accuracy of the inducer (the only dataset in which MTA has significantly reduced accuracy was Kr-vs-kp). Thus, the MTA can be referred as a more reliable preprocessing step for induction algorithm.

5 Conclusions

This paper examines theoretically and experimentally whether ensemble of feature subsets can be used for improving feature selection performance in terms of the final classification accuracy.

This paper examines several methods for ensemble feature selection. Experiments on benchmark data sets showed that MTA did much better than other methods for feature selection. In addition, the MTA method did not reduce the initial classification accuracy and managed to remove the irrelevant and the redundant attributes.

References

1. Bartlett P. and Shawe-Taylor J., Generalization Performance of Support Vector Machines and Other Pattern Classifiers, In "Advances in Kernel Methods, Support Vector Learning", Bernhard Scholkopf, Christopher J. C. Burges, and Alexander J. Smola (eds.), MIT Press, Cambridge, USA, 1998.
2. Cunningham P., and Carney J., Diversity Versus Quality in Classification Ensembles Based on Feature Selection, In: R. L. de Mántaras and E. Plaza (eds.), Proc. ECML 2000, 11th European Conf. On Machine Learning,Barcelona, Spain, LNCS 1810, Springer, 2000
3. Hall, M. Correlation- based Feature Selection for Machine Learning. University of Waikato, 1999.
4. Ho T. K. The random subspace method for constructing decision forests. IEEE Transactions on Pattern Analysis and Machine Intelligence, 20(8):832–844, 1998.
5. Hu, X., Using Rough Sets Theory and Database Operations to Construct a Good Ensemble of Classifiers for Data Mining Applications. ICDM01. pp. 233-240, 2001.
6. Kohavi R. and John, G. Wrappers for feature subset selection. *Artificial Intelligence*, special issue on relevance, 97(1– 2): 273– 324, 1996
7. Masulli, F. and Rovetta, S. Random Voronoi ensembles for gene selection in DNA microarray data, in Udo Seiffert and Lakhmi C. Jain, editors, Bioinformatics using Computational Intelligence Paradigms, World Scientific Publishing, Singapore, 2003
8. Miller, A. Subset Selection in Regression. Chapman and Hall, New York, 1990.
9. Oliveira L.S., Sabourin R., Bortolozzi F., and Suen C.Y. A Methodology for Feature Selection using Multi-Objective Genetic Algorithms for Handwritten Digit String Recognition, International Journal of Pattern Recognition and Artificial Intelligence, 17(6):903-930, 2003.
10. Opitz, D., Feature Selection for Ensembles, In: Proc. 16th National Conf. on Artificial Intelligence, AAAI, pages 379-384, 1999.
11. Quinlan, J. *C4.5: Programs for machine learning.* Morgan Kaufmann, Los Altos, California, 1993.
12. Quinlan, J. Induction of decision trees. *Machine Learning*, 1: 81– 106, 1996
13. Torkkola, K. and Tuv, E. Variable selection using ensemble methods. *IEEE Intelligent Systems*, 2005, (Vol. 20, No. 6): 68-70.
14. Tsymbal A., and Puuronen S., Ensemble Feature Selection with the Simple Bayesian Classification in Medical Diagnostics, In: Proc. 15thIEEE Symp. on Computer-Based Medical Systems CBMS'2002, IEEE CS Press, 2002.
15. Tumer, K. and Ghosh J., Error Correlation and Error Reduction in Ensemble Classifiers, Connection Science, Special issue on combining artificial neural networks: ensemble approaches, 8 (3-4): 385-404, 1999.
16. Tuv, E. and Torkkola, K. Feature filtering with ensembles using artificial contrasts. In *Proceedings of the SIAM 2005 Int. Workshop on Feature Selection for Data Mining*, Newport Beach, CA, April 23 2005, pp. 69-71.

Clustering and Classification of Textual Documents Based on Fuzzy Quantitative Similarity Measure — a Practical Evaluation

Piotr S. Szczepaniak[1,2]

[1] Institute of Computer Science, Technical University of Lodz, Sterlinga 16/18, 90-217 Lodz, Poland
[2] Systems Research Institute, Polish Academy of Sciences Newelska 6, 01-447 Warsaw, Poland

Summary. Clustering enables more effective information retrieval. In practice, similar approaches are used for ranking and clustering. This paper presents a practical evaluation of a method for clustering of documents which is based on certain textual fuzzy similarity measure. The similarity measure was originally introduced in [12] — cf. also [13], and later used in internet-related applications [14, 15, 18]. Two textual databases [21, 22] of predefined clusters and of diverse level of freedom in the contents of documents were used for experiments that employed some variants of the basic clustering method [19].

Key words: document classification, clustering of textual documents, information retrieval, fuzzy set, fuzzy similarity measure, text comparison

1 Introduction

The internet is becoming an important source of information. Diverse algorithms (models) for ranking Web documents according to the likelihood of relevance assigned by a system to a given user query have been developed. Following the taxonomy proposed in [1] the models may be divided into two groups, namely classic and alternative. The former include vector-space, probabilistic and Boolean models. The latter are remarkably numerous and varied; they cover: fuzzy set models, extended Boolean models, generalized vector space models, neural network models, etc. Since ranking and clustering are closely related, the same or slightly modified approaches are in use.

In textual database searching, the methods used so far for text comparison have been based mainly on the classical identity relation, according to which two given texts are either identical or not (if we neglect some simple preprocessing operations). To define the distance between textual objects is a more

Mark Last et al. (Eds.): Advances in Web Intelligence and Data Mining (SCI) **23**, 305-317 (2006)
www.springerlink.com

complicated task than to determine the distance from a data vector to the prototype; examples like Jaccard, Cosine and Dice Coefficients can frequently be found in literature, e.g. [10, 11]. The similarity between two documents is usually determined on the basis of the number of topical terms that appear in both of the documents under consideration.

Novel methods aim at approximate matching the requirement formulated as a query [9]. Here, the fuzziness in the sense of Zadeh [20] is the attempt to establish more sophisticated method for text similarity analysis [12, 13, 18], and consequently for clustering of retrieved documents [19].

Clustering enables more effective information retrieval when part of a large collection of diverse documents is already dealt with, but the real importance of the problem becomes evident when the largest library, namely the Web, is to be used [1, 17].

2 Information retrieval and clustering

In information retrieval, the standard way is to represent a given set $\boldsymbol{B}$ of documents and a set $\boldsymbol{T}$ of querying/indexing terms by vectors. In the vector space model, each document is represented by a vector whose elements are numbers characterising the *weight* (significance) of term t_j in document d_i. These so-called indexing weights can be computed from frequencies of occurrence of the terms in a given document [6, 7]. The degrees of match between the query and the documents are computed by means of comparison of the vectors that represent the query and the document, respectively.

For document clustering, the same test of matching can be used; the only and simple modification lies in the use of a similarity measure for the document-document comparison.

Defining a document cluster, one may generally adopt the view rooted in [3, 8] and say that
a cluster is a group of documents that are more similar to one another than to the members of any other cluster.

In clustering the problems involved cover: representation of documents, choice of a similarity measure, and a grouping method. For any kind of clustering it is crucial to define the term 'similarity' so that it can be measured in a well-defined way.

Clusters can be well-separated or overlapping, and their number as well as the reference documents (prototypes) may be known or may not exist before the clustering procedure is started.

The objective of the document partition is to create c clusters in the collection (body) $\boldsymbol{B}$ of documents. Let us assume that the number c is known or desired. The hard partition defined by Bezdek [3] can be applied here.
Let $p(\boldsymbol{B})$ denote the power of set $\boldsymbol{B}$. The hard partition of $\boldsymbol{B}$ is a family of sub-collections $\{B_i \,|1 \leq i \leq c\} \subset p(\boldsymbol{B})$ which

- collectively form collection $\boldsymbol{B}$,

- are disjoint,
- are non empty and any single B_i is a full collection.

If the documents of the sub-collection fulfil some similarity requirement, we will speak about the cluster. In the literature, the situation when each document of the collection belongs exactly to one cluster is frequently called a hierarchical clustering. It is easy to notice that the decision about the membership of documents in a cluster is not always simple. It can be said that to a certain extent each of the considered documents belongs to two or more groups — their membership is fuzzy.
Fuzzy clustering applied to different tasks of information retrieval took a remarkable place in last years' publications. The topics that were frequently addressed include matching of queries with documents, representation of user profiles, and document-document matching.

As mentioned above, to perform clustering a suitable representation of documents, a similarity measure, and a grouping method should be selected and implemented. Numerical representation of documents enables application of any hard or fuzzy clustering method. When speaking about fuzzy clustering, the popular fuzzy c-means algorithms should be mentioned [4, 5]. This is a large family of methods that work iteratively and form clusters by minimising an objective function. The clusters are called fuzzy because the membership of the considered documents in given clusters is graded.

The clustering method presented in [19] and evaluated in this paper applies hard partitioning as defined above but it uses a novel method for fuzzy evaluation of similarity between textual documents, cf. also [12, 18].

3 Method under evaluation

The problem at hand is as follows. Given a collection $\boldsymbol{B}$ of N textual documents, find a family of $c < N$ sub-collections $\{B_i \,|1 \leq i \leq c\}$ that group documents being more similar to one another than to members of other clusters or at least more similar to their own cluster centre than to others.

There are many ways to perform clustering. Effectiveness and precision (coincidence with human intuition or knowledge) are the requirements that may be contradictory, in particular when dealing with textual information. To increase the speed, one can try to reduce the number of document comparisons. This may be achieved by performing two stages of computations: initialisation of clusters and classification. The particular, conceptually simple method under evaluation is as follows [19]:

Initialization

1. Take a possibly representative sub-collection of documents $\boldsymbol{B}_{ig}$ — 'initial group'.
2. Determine the similarity between all the documents of the sub-collection — formula (2).

3. Choose c different documents — prototypes of cluster centres ν_j . Their mutual similarity should lie under some threshold th.

Of course, a large number of variations for initialisation may be created. For example, the use of keywords may be simpler and faster (but not necessarily better) choice.

Grouping

1. Take a new document d_{new} from $\boldsymbol{B}$ / $\boldsymbol{B}_{ig}$.
2. Compute the similarity of d_{new} to all actual centres ν_j.
3. Add document d_{new} to the cluster of the highest similarity of its centre — (2), and change this centre if necessary — (1).
4. Are all the documents examined ?
   If *yes*, then go to step 5.
   If *not*, then go to step 1.
5. Prepare all the c clusters to access, and *stop*.

Remark: Owing to computations done in step 3 the similarity of all clusters as well as the similarity of all documents to all cluster centres are known. This information may be useful if one decides to give access to less relevant documents

- independently of their cluster membership;
- only through those clusters whose centres are somewhat similar to the most relevant one.

3.1 Text similarity

In a natural way, the method applied for the full text comparison uses comparison of single words. Here, a fuzzy relation

$$RW = \{(< w_1, w_2 >, \mu_{RW}(w_1, w_2)) : w_1, w_2 \in \mathbf{W}\},$$

on $\mathbf{W}$ — the set of all words within the universe of discourse, for example a considered language or dictionary — is proposed as a useful instrument.

A possible form of the membership function $\mu_{RW} : \mathbf{W} \times \mathbf{W} \to [0, 1]$ may be as proposed in [12, 13]

$$\mu_{RW}(w_1, w_2) = \frac{2}{N^2 + N} \sum_{i=1}^{N(w_1)} \sum_{j=1}^{N(w_1)-i+1} h(i, j) \tag{1}$$

where:
$N(w_1)$, $N(w_2)$ — the number of letters in words w_1, w_2, respectively;
$N = max\{N(w_1), N(w_2)\}$ — the maximal length of the considered words;
$h(i, j)$ — the value of the binary function, i.e. $h(i, j) = 1$ if a subsequence, containing i letters of word w_1 and beginning from the j-th position in w_1,

appears at least once in word w_2; otherwise $h(i,j) = 0$; $h(i,j) = 0$ also if $i > N(w_2)$ or $i > N(w_1)$.
Note that $0,5(N^2 + N)$ is the number of possible subsequences to be considered.

The fuzzy relation RW is reflexive: $\mu_{RW}(w,w) = 1$ for any word w; but in general it is not symmetrical. This inconvenience can be easily avoided in the following way:

$$\mu_{RW}(w_1, w_2) = min\,\{\mu_{RW}(w_1, w_2),\ \mu_{RW}(w_2, w_1)\} \tag{2}$$

Moreover, RW reflects the following human intuition:

- the bigger the difference in length of two words is, the more different they are;
- the more common the letters contained in two words, the more similar the words are.

However, the value of the membership function contains no information on the sense or semantics of the arguments.

Example

Let us take the sample pair of words $w_1 = centre$ and $w_2 = center$.
Here $N(w_1) = 6$ and $N(w_2) = 6$, $N = max\,\{N(w_1), N(w_2)\} = 6$,

$$\mu_{RW}(w_1, w_2) = \mu_{RW}(w_2, w_1) = \tfrac{6+3+2+1}{21} \equiv 0.57.$$

It is easy to note that the method in its basic form (1) is time-consuming. The cost can be estimated by the formula

$$ces = \sum_{i=1}^{N} \{[N(w_1) - i + 1] \cdot [N(w_2) - i + 1] \cdot i\} \tag{3}$$

The key operation is comparison of two letters. To check whether a substring of word w_1 appears in word w_2, the whole word w_2 needs to be checked, and this procedure must be repeated for each searched substring. The need for reduction of needless operations and optimization of the comparison of words is evident. Experience shows that the use of strings of constant length, say k, is a proper solution. Then, the summation in (3) and the outer sum in (1) are reduced to only one component. Consequently, equation (1) takes the form [16]

$$\mu_{RW}(w_1, w_2) = \frac{1}{N - k + 1} \sum_{i=1}^{N(w_1)-k+1} h(k, j) \tag{4}$$

which is analogous to the n-gram method described in [2]. The improvement is remarkable. For instance, the similarity computation between an exemplary eleven- and four-letter strings via (4) under assumption that $k = 3$ is nine times less costly than the same comparison obtained via (1); the resulting loss of precision is less than 5%.

Basing on (1) or (4) it is possible to introduce the similarity measure for documents. Here, the document is considered to be a set (not a sequence) of words. The respective formula is of the form of fuzzy relation on the set of all documents $\boldsymbol{B}$, with the membership function μ_{RB} [12, 13, 18]

$$\mu_{RB}(d_1, d_2) = \frac{1}{N} \sum_{i=1}^{N(d_1)} max_{j \in \{1,2,\ldots,N(d_2)\}} \; \mu_{RW}(w_i, w_j) \tag{5}$$

where:
d_1, d_2 — the documents to be compared;
$N(d_1)$, $N(d_2)$ — the number of words in document d_1 and d_2 (the length of d_1 and d_2, respectively);
$N = max\,\{N(w_1), N(w_2)\}$;
μ_{RW} — the similarity measure for words defined by (1).

This relation models the similarity between two given documents, and is defined on the set of all words — $\mathbf{W}$. The value 1.0 of that function means that the documents considered are identical; on the other hand the value close to zero suggests that the documents are different.

3.2 Representation of documents, centres, and membership assignment

In the simplest case, a sequence of keywords/indexing terms can be used. The documents may also be represented by the most representative part (for example by abstract, summary, or both when considering scientific papers). The original full text version is also acceptable. Of course, preprocessing that involves elimination of stopwords and lexical analysis is a very advantageous procedure. For the similarity measure proposed in section 3.1, stemming of the remaining words (i.e. removing affixes) is not necessary.
Cluster centre is a document that is most similar to all the other documents in the sub-collection it represents. It is assumed that the centre is a (usually modified by preprocessing) document of collection $\boldsymbol{B}$ or its representation.

Assuming that the sub-collection of documents B_k $\{d_1, d_2, d_3, \ldots, d_k\}$, $k < N$, form a cluster due to the similarity of documents, the cluster center ν_i (reference document) is defined [19] as the document d_i, $i \leq k$, whose index i maximizes the sum

$$\sum_{j=1}^{k} \mu_{RB}(d_i, d_j) \text{ for all } i = 1, 2, \ldots, k \tag{6}$$

with μ_{RB} computed according to (2).

It is worth remarking, however, that groups likely to become clusters need to be predetermined.

For each document in collection $\boldsymbol{B}$, the degree of membership in a cluster is determined on the basis of the maximum of similarity of d_i to each cluster center

$$\mu_{i,j} = max_{i \leq N, j \leq c}\ \mu_{RB}(d_i, \nu_j) \tag{7}$$

where:
i, j — is the grade of membership of the i-th document in the cluster ν_j;
$\mu_{RB}(d_i, \nu_j)$ — is the fuzzy similarity of document d_i to the centre of cluster ν_j;
ν_j — denotes the centre of the j-th cluster.

Formula (7) means that the given document d_i belongs fuzzily to the cluster whose center ν_j is most similar to d_i over all cluster centres $\nu_1, \nu_2, \ldots, \nu_c$. Note that by this statement we preserve the fuzzy membership but on the other hand we actually perform a hard partitioning.

4 Practical evaluation

As mentioned, the evaluation was performed on two different textual databases. The first collection is the documentation of the Linux [21], the second one collects documents of a newsgroup [22]. The quality of grouping was evaluated using two indices, called precision p, and completeness (recall) r

$$p = \frac{n_{exp}}{n_{cl}} \qquad r = \frac{n_{exp}}{n_{ctg}} \tag{8}$$

where:
n_{exp} — the number of documents in the examined cluster which belong to the considered category defined by expert;
n_{cl} — the number of all documents in the cluster;
n_{ctg} — the number of all documents in the considered category defined by expert.

Several experiments by different categorizations

- reference (original) categorization,
- modified (more detailed) categorization,

and text bases for clustering, i.e.

- full text (after elimination of stopwords),
- keywords

have been performed. Different similarity thresholds th have also been applied.

Collection I — original

In the collection, five categories exist in the documentation [21]. It should be noted that these original categories are rather general and sometimes the topical relation is questionable. Nevertheless, we respect them as reference categories (Table 1).

Table 1. Number of categorized documents

LinuxOS	Hardware	Networking	Applications	Programming
81	81	146	77	36

The initial group chosen for experiments of $p(\boldsymbol{B}_{ig}) = 30$ is shown in Fig. 1. In Table 2 the list of eliminated words is shown. The results for full text document grouping that uses (1) and (8) with $th = 0.5$ are as shown in Tables 3 and 4.

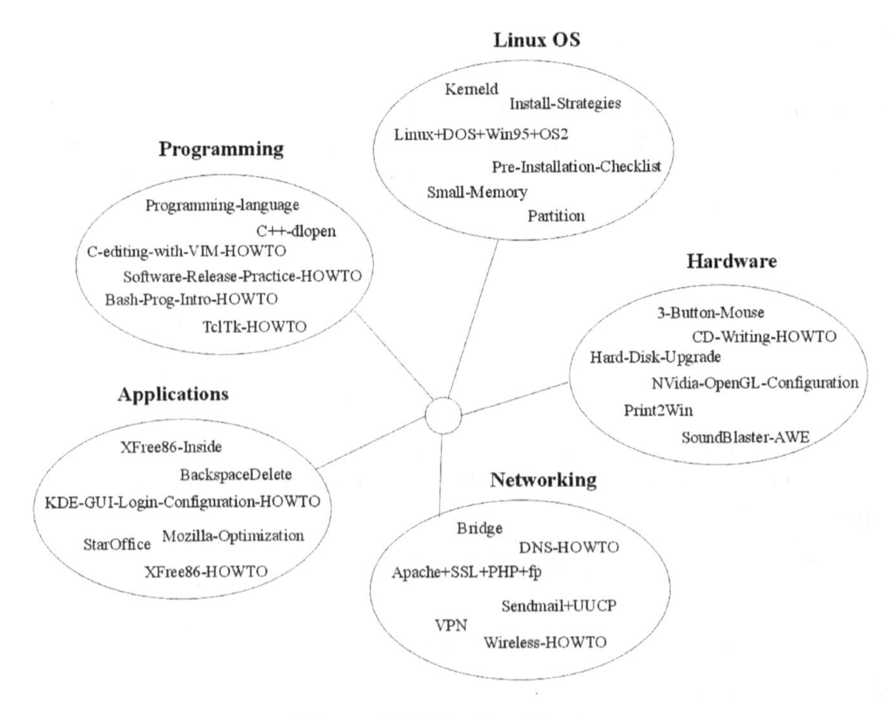

Fig. 1. Initial sub-collection

Because of possible modifications of cluster centers during the second stage of the algorithm the membership of some documents to particular clusters with respect to their distance to final centers may be not correct. The results of membership corrections are shown in square brackets — Table 4.

Table 2. List of eliminated stopwords

a an the to from by for at of in over but with as on off about and or that which wherever whatever whichever while if until unless how what where why do don't does doesn't did didn't have haven't has hasn't had hadn't can can't could couldn't may might shall should will would get gets got gotten take taken took takes be is isn't am amn't are aren't was wasn't were weren't been no not never some any very much many lot other another this that these those now moreover furthermore I you he she it me my his her your its their

Table 3. Full text grouping quality — initialization

cluster number	1	2	3	4	5	6
category of the center	Hardware	Programming	Applications	Linux OS	Linux OS	Applications
r	17%	33%	17%	33%	17%	33%
p	100%	17%	50%	20%	100%	50%
number of documents	1	12	2	10	1	4

Table 4. Full text grouping — result and its correction

claster number	LinuxOS	Hardware	Networking	Applications	Programming
1	0[0]	1[1]	0[0]	0[0]	0[0]
2	14[13]	12[15]	35[31]	24[18]	17[7]
3	0[0]	0[0]	1[0]	1[1]	0[0]
4	15[20]	13[11]	10[20]	11[17]	12[9]
5	2[3]	1[1]	3[2]	1[1]	2[0]
6	2[0]	2[1]	4[4]	2[3]	3[2]

Grouping have also been performed on the basis of keywords given in Table 5. Seven clusters have been identified for $th = 0.5$; the results are presented in Tables 6 and 7.

Table 5. List of chosen keywords

linux installation install hardware requirement boot root floppy partition dos windows drive administratration admin lilo bootloader zip win multiboot grub bsd freebsd solaris hard disk format partition file filesystem ext fat kernel io module hardware cpu network patch mod gui encrypt encryption raid driver cable dma udma scsi via ide bios modem connect connection ethernet card ip tcp router net network subnet internet address dhcp http ftp mail sendmail smtp pop imap ppp localhost link online offline gnu class download develop software bash programming program login shell serial paralell irq dial dialer pnp plug play pci isa odbc configure compile algorithm route routing tunnel masquerading masquerade proxy server arp broadcast lan apple packet vpn ssh firewall client virtual private xfree xwindows char font keyboard kde gnome manager database oracle Sybase mmbase base transaction apache www

Table 6. Keywords grouping quality — initialization

cluster number	1	2	3	4	5	6	7
category of the center	Hardware	Linux OS	Programming	Linux OS	Programming	Applications	Linux OS
r	17%	33%	50%	17%	33%	50%	50%
p	100%	33%	50%	50%	29%	50%	75%
number of documents	1	6	5	2	7	5	4

Table 7. Keywords grouping — result and its correction

claster number	LinuxOS	Hardware	Networking	Applications	Programming
1	0[0]	1[1]	0[0]	0[0]	0[0]
2	6[5]	9[9]	14[9]	8[7]	5[1]
3	5[3]	5[4]	22[15]	14[13]	12[7]
4	3[2]	1[1]	1[1]	2[2]	0[0]
5	4[14]	8[11]	18[35]	11[19]	10[10]
6	0[0]	0[0]	2[1]	3[2]	1[1]
7	15[12]	4[5]	1[1]	1[0]	2[0]

Collection I — modified

The collection has been obtained by more detailed determination of expert categories in the same documentation [21] — Table 8.

Table 8. Number of documents in ten categories determined by expert

Linux install	Kernel	Harddisks	Modems	Routing	VPN	GUI	Database (DB)	Mail	http
18	10	10	9	6	8	21	11	14	9

Sample results for keyword-based text document grouping using (1) and (8) with $th = 0.55$ are as shown in Table 9.

Table 9. Keyword-based grouping quality — initialization

cluster number	1	2	3	4	5	6	7	8	9
category of the center	DB	Kernel	VPN	DB	Mail	GUI	GUI	Linux install	GUI
r	25%	33%	20%	50%	10%	16%	33%	83%	66%
p	20%	50%	20%	50%	33%	100%	40%	50%	33%
number of documents	5	2	5	4	3	1	5	10	3

Collection II

Here, the following five newsgroups from [22] have been considered: sci.med, rec.motocycles, comp.graphics, alt.atheism, talk.politics.mideast. For the first stage of the heuristic algorithm 50 papers have been used, whereas for the second one — 500. Sample results are presented in Tables 10 and 11.

Table 10. Result of full text grouping; $th = 0.3$

cluster number	rec.motorcycles	comp.graphics	alt.atheism	sci.med	talk.politics.mideast
1	10	0	2	0	0
2	0	0	1	2	5
3	0	5	2	2	1
4	0	1	1	1	0
5	0	4	5	5	4

Table 11. Full text grouping quality — initialization

cluster number	1	2	3	4	5
r	100%	50%	50%	10%	50%
p	86%	63%	50%	30%	28%

5 Summary

In textual database searching or clustering, the methods used so far for text comparison have been based mainly on the classical identity relation, according to which two given texts are either identical or not (if we neglect some simple preprocessing operations). This work is a contribution to a group of methods that show that the *fuzziness* in the sense of Zadeh [20] can be applied to text similarity analysis. The solution described makes the consideration of language inflection easier; the method is also not sensitive to grammatical mistakes or other misshapen language constructions.

The clustering method identifies groups of documents which are similar to a given number of reference documents — cluster centres. The reference documents may be known a priori, or determined during the initialisation stage from the so-called 'initial group' of examined documents. The method is able to deal with full text documents, with the most informative document fragments, sentences, or more traditionally — with keywords/index terms.

The practical evaluation performed on two different types of documents allows one to expect that the method performs better if the topics of clusters are well reflected in the vocabulary and construction of sentences. In the first collection [21], the same specialized words are used for description of different technological issues, and the other way round, different words and names (like Oracle, Sysbase, etc.) are used in relation to the same topic (e.g. databases). Consequently, additional deep human knowledge of the topic is needed to classify the documents correctly. In the second collection [22], the similarity of contents is well reflected in the vocabulary which makes the task of automatic grouping easier.

Obviously, the method is not a panacea for every task of clustering or ranking but it may be useful for rough sorting and grouping of certain types of documents.

Acknowledgement. The author is grateful to Mr P. Rozycki for performing computations.

References

1. Baeza-Yates R, Ribeiro-Neto B (1999) Modern Information Retrieval. Addison Wesley, New York
2. Bandemer H, Gottwald S (1995) Fuzzy sets, Fuzzy Logic, Fuzzy Methods with Applications. John Wiley and Sons
3. Bezdek J C (1981) Pattern Recognition with Fuzzy Objective Function. Plenum Press, New York
4. Bezdek J C (1980) A Convergence Theorem for the Fuzzy ISODATA Clustering Algorithms. IEEE Trans. on Pattern Analysis and Machine Intelligence, 2:1-8
5. Bezdek J C, Hathaway R J, Sabin M J, Tucker W T (1987) Convergence Theory for Fuzzy c-Means: Counterexamples and Repairs. IEEE Trans. on Systems, Man, and Cybernetics, 17:873-877
6. Kraft D H, Chen J (2001) Integrating and Extending Fuzzy Clustering and Inferencing to Improve Text Retrieval Performance. In: Larsen H L, et al. (eds) Flexible Query Answering Systems. Physica-Verlag, A Springer-Verlag Company, Heidelberg, New York
7. Kraft D H, Chen J, Martin-Bautista M J, Amparo-Vila M (2003) Textual Information Retrieval with User Profiles using Fuzzy Clustering and Inferencing. In: Szczepaniak P S, Segovia J, Kacprzyk J, Zadeh L (eds) Intelligent Exploration of the Web. Physica-Verlag, A Springer-Verlag Company, Heidelberg, New York
8. Jain A K, Dubes R C (1988) Algorithms for Clustering Data. Englewood Cliffs, Prentice Hall
9. Larsen H L, Kacprzyk J, Zadrozny S, Andreasen T (2001) Flexible Query Answering Systems. Physica-Verlag, A Springer-Verlag Company, Heidelberg
10. Lebart L, Salem A, Berry L (1998) Exploring Textual Data, Kluwer Academic Publisher
11. Ho T B , Kawasaki S, Nguyen N B (2003) Documents Clustering using Tolerance Rough Set Model and Its Application to Information Retrieval. In: Szczepaniak P S, Segovia J, Kacprzyk J, Zadeh L (eds) Intelligent Exploration of the Web. Physica-Verlag, A Springer-Verlag Company, Heidelberg, New York

12. Niewiadomski A (2000) Appliance of fuzzy relations for text document comparing. Proceedings of the 5th Conference NNSC (Zakopane, Poland, June 6–10):347–352
13. Niewiadomski A, Szczepaniak P S (2001) Intutionistic Fuzzy Relations in Approximate Text Comparison. Published in Polish: Intuicjonistyczne relacje rozmyte w przyblizonym porownywaniu tekstow. In: Chojcan J, Leski J (eds) Zbiory rozmyte i ich zastosowania. Silesian Technical University Press, Gliwice, Poland:271–282; ISBN 83-88000-64-0
14. Niewiadomski A, Szczepaniak P S , (2002) Fuzzy Similarity in E-Commerce Domains. In: Segovia J, Szczepaniak P S, Niedzwiedzinski M (eds) E-Commerce and Intelligent Methods. Physica-Verlag, A Springer-Verlag Company, Heidelberg, New York
15. Niewiadomski A, Kryger P, Szczepaniak P S (2004) Fuzzyfication of Indescernibility Relation for Structurizing Lists of Synonyms and Stop-Lists for Search Engines. In: Rutkowski L, Siekmann J, Tadeusiewicz R, Zadeh L A (eds) Artificial Intelligence and Soft Computing ICAISC 2004. Proceedings of the Seventh International Conference on Neural Networks and Soft Computig. Zakopane, Poland, 2004. Series: Lecture Notes in Artificial Intelligence — LNAI 3070, Springer-Verlag, Berlin, Heidelberg, New York, 2004:504–509. ISBN 3-540-22123-9
16. Niewiadomski A, Kryger P, Szczepaniak P S (2004) Fuzzy Comparison of Strings in FAQ Answering. In: W.Abramowicz (eds) BIS'2004. Proceedings of 7th International Conference on Business Information Systems. Wydawnictwo Akademii Ekonomicznej, Poznan, Poland:355–362. ISBN 83-7417-019-0
17. Pal S K, Talwar V, Mitra P (2002) Web Mining in Soft Computing Framewarks: Relevance, State of the Art and FutureDirections. IEEE Trans. on Neural Networks, vol.13, no.5.
18. Szczepaniak P S, Niewiadomski A (2003) Internet Search Based on Text Intuitionistic Fuzzy Similarity. In: Szczepaniak P S, Segovia J, Kacprzyk J, Zadeh L (eds) Intelligent Exploration of the Web. Physica-Verlag, A Springer-Verlag Company, Heidelberg, New York
19. Szczepaniak P S, Niewiadomski A (2003) Clustering of Documents on the Basis of Text Fuzzy Similarity. In: Abramowicz W (eds) Knowledge-Based Retrieval and Filtering from the Web. Kluwer Academic Publishers, USA:219–230; ISBN 1-4020-7523-5
20. Zadeh L (1965) Fuzzy Sets. Information and Control, 8:338–353.
21. http://tldp.org/HOWTO/HOWTO-INDEX/categories.html
22. http://people.csail.mit.edu/u/j/jrennie/public_html/20Newsgroups

Oriented k-windows: A PCA driven clustering method

D.K. Tasoulis[1], D. Zeimpekis[2], E. Gallopoulos[2], and M.N. Vrahatis[1]

[1] Department of Mathematics, University of Patras, GR-26110 Patras, Greece {dtas,vrahatis}@math.upatras.gr
[2] Computer Engineering & Informatics Dept., University of Patras, GR–26500 Patras, Greece {dsz,stratis}@hpclab.ceid.upatras.gr

Summary. In this paper we present the application of Principal Component Analysis (PCA) on subsets of the dataset to better approximate clusters. We focus on a specific density-based clustering algorithm, k-Windows, that holds particular promise for problems of moderate dimensionality. We show that the resulting algorithm, we call Oriented k-Windows (OkW), is able to steer the clustering procedure by effectively capturing several coexisting clusters of different orientation. OkW combines techniques from computational geometry and numerical linear algebra and appears to be particularly effective when applied on difficult datasets of moderate dimensionality.

1 Introduction

Density based methods are an important category of clustering algorithms [4, 13, 23], especially for data of low attribute dimensionality [6, 17, 26]. In these methods, clusters are formed as regions of high density, in dataset objects, surrounded by regions of low density; proximity and density metrics need to be suitably defined to fully describe algorithms based on these techniques though there are difficulties for problems of high attribute dimensionality; cf. [1, 16]. One recent technique in this category is "Unsupervised k-Windows" (UkW for short) [28], that utilizes hyperrectangles to discover clusters. The algorithm makes use of techniques from computational geometry and encapsulates clusters using linear containers in the shape of d-dimensional hyperrectangles that are aligned with the standard Cartesian axes and are iteratively adjusted with movements and enlargements until a certain termination criterion is satisfied; cf. [24, 28]. An advantage of the algorithm is that it allows a reduction in the number of objects examined at each step, something especially useful when dealing with large collections. Furthermore, with proper tuning, the algorithm can detect clusters of arbitrary shapes. We show here a general approach that appears to improve the effectiveness of UkW. This approach relies on techniques from linear algebra to orient the hyperrectangles

Mark Last et al. (Eds.): Advances in Web Intelligence and Data Mining (SCI) **23**, 319-328 (2006)
www.springerlink.com

into directions that are not necessarily axes-parallel. The technique used in this process is PCA [18], implemented via singular value decomposition (SVD) [14]. The paper is organized as follows. The details of unsupervised k-Windows are described in Section 2. Section 3 describes OkW, while Section 4 presents experimental evidence of its efficiency. Finally, Section 5 contains concluding remarks. We would be assuming that datasets are collections of d-dimensional objects, and can be represented by means of an $n \times d$ object-attribute matrix; unless specified otherwise, "dimension" refers to attribute cardinality, d.

2 Unsupervised k-Windows clustering

UkW aims at capturing all objects that belong to one cluster within a d–dimensional window. Windows are defined to be hyperrectangles (orthogonal ranges) in d dimensions; cf. [28] for details. UkW employs two fundamental procedures: *movement* and *enlargement*. The movement procedure aims at positioning each window as close as possible to the center of a cluster. The enlargement process attempts to enlarge the window so that it includes as many objects from the current cluster as possible. The two steps are illustrated in Figure 1. UkW provides an estimate for the number of clusters that describe

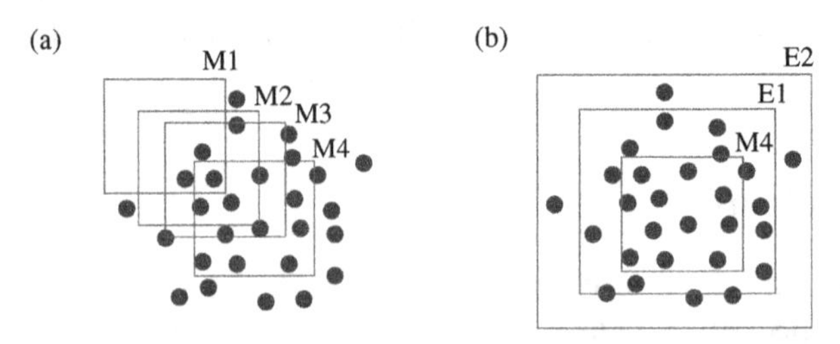

Fig. 1. (a) Sequential movements M2, M3, M4 of initial window M1. (b) Sequential enlargements E1, E2 of window M4.

a dataset. The key idea is to initialize a large number of windows. When the movement and enlargement of all windows terminate, all overlapping windows are considered for merging by considering their intersection. An example of this operation is exhibited in Figure 2.

The computationally intensive part of k-Windows is the determination of the points that lie in a hyperrectangle. This is the well known "orthogonal range search" problem [22]. The high level description of the algorithm is as follows:

algorithm UkW $(a, \theta_e, \theta_m, \theta_c, \theta_v, \theta_s, k)$ {
 execute W=DetermineInitialWindows(k,a)
  for each hyperrectangle w_j in W **do**
    repeat

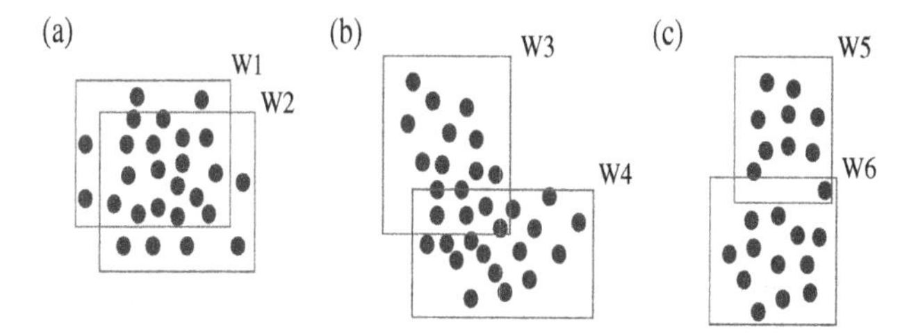

Fig. 2. (a) W_1 and W_2 satisfy the similarity condition and W_1 is deleted. (b) W_3 and W_4 satisfy the merge operation and are considered to belong to the same cluster. (c) W_5 and W_6 have a small overlap and capture two different clusters.

```
        execute movement(θ_v,w_j)
        execute enlargement(θ_e,θ_c,θ_v,w_j)
      until the center and size of w_j remain unchanged
  execute merging(θ_m,θ_s,W)
  Output {clusters c_{l_1}, c_{l_2}, ... so that: c_{l_i} = {i : i ∈ w_j, label(w_j) = l_i} }
}
function DetermineInitialWindows(k,a) {
  initialize k d-dimensional hyperrectangles w_{m_1}, ..., w_{m_k} with edge length a
  select k points from the dataset and
          center the hyperrectangles at these points
  return a set W of the k hyperrectangles
}
function movement(θ_v, hyperrectangle w) {
  repeat
    find the objects that lie in the hyperrectangle w
    calculate the mean m of these objects
    set the center of w equal to m
  until the Euclidean distance between m and
        the previous center of w is less than θ_v
}
function enlargement(θ_e,θ_c,θ_v, hyperrectangle w) {
  repeat
    foreach dimension i do
      repeat
        enlarge w along current dimension by θ_e%
        execute movement(θ_v,w)
      until increase in number of objects
            along current dimension is less than θ_c%
  until increase in number of objects
        is less than θ_c% along each dimension
}
function merging(θ_m,θ_s, set of hyperrectangles W) {
  for each hyperrectangle w_j in W not marked do
```

```
mark w_j with label w_j
if ∃ w_i ≠ w_j in W, that overlaps with w_j
  compute the number of points n that
          lie in the overlap area
  if (n/|w_i|) ≥ θ_s and |w_i| < |w_j|
    disregard w_i
  if 0.5 * (n/|w_j| + n/|w_i|) ≥ θ_m
    mark all w_i labeled hyperrectangles in W with label w_j
}
```

3 k-Windows with PCA based steering

Results with UkW in [28] showed that a procedure based on moving and enlarging hyperrectangles together with merging is effective in discovering clusters. All movements, however, were constrained to be parallel to any one of the standard cartesian axes. We next consider the possibility of allowing the hyperrectangles adapt both their orientation and size, as means to more effective cluster discovery. Assume, for example, that k d-dimensional hyperrectangles of the UkW algorithm have been initialized, as in function `DetermineInitialWindows` above. After executing the movement function, the center and contents of each hyperrectangle might change. It would then be appropriate to re-orient the hyperrectangle from its original position (axes-parallel in the first step, maybe different in later ones) so as to take into account the local variabilities present in the data. This is accomplished by means of PCA on a translation of the subset of points (objects) included in the specific hyperrectangle.

Let, for example, $A^{(i)} \in \mathbb{R}^{n_i \times d}$ be the object-attribute matrix corresponding to objects in a subset $\mathcal{P}$ of the original dataset. Matrix $B^{(i)} := A^{(i)} - \frac{1}{n_i} ee^{\top} A^{(i)}$, where $e \in \mathbb{R}^{n_i}$ is a vector of all 1's, contains the points $B^{(i)}$, that is the points of $\mathcal{P}$ written in terms of the cartesian axes centered at the centroid $g = \frac{1}{n_i} e^{\top} A^{(i)}$ of $\mathcal{P}$. Let the SVD of $B^{(i)}$ be $B^{(i)} = U^{(i)} \Sigma^{(i)} (V^{(i)})^{\top}$ and $\{u_j, \sigma_j, v_j\}, j = 1, ..., d$ the nontrivial singular triplets of $B^{(i)}$, in decreasing order, i.e. $\sigma_1 \geq \sigma_2 \geq \cdots \geq \sigma_d \geq 0$. For simplicity we omit the index i. One way to build a hyperrectangle bounding the elements of $\mathcal{P}$ is to use the fact that the columns of V_i are the principal directions of the elements of $\mathcal{P}$. The hyperellipsoid with semiaxes σ_j encloses all of $\mathcal{P}$. It can, however, cover a much larger volume than the volume enclosed by the span of $\mathcal{P}$ and even the hyperellipsoid of minimal volume containing $\mathcal{P}$. On the other hand, computing either of these minimal bodies and keeping track of geometrical operations with them is difficult; e.g. see [20]. Instead, since OkW was based on hyperrectangles, we prefer to preserve that choice here as well. We search, then for hyperrectangles that bound the points of $\mathcal{P}$, but no longer requiring that they are axes-parallel. A simple, yet effective choice, is to exploit the equality $B^{(i)}V^{(i)} = U^{(i)}\Sigma^{(i)}$ which shows that the projections of the n_i rows of $B^{(i)}$ on any of the principal axes, say v_j, is given by the elements of column j of $U^{(i)}$ multiplied by σ_j. Therefore, we can construct a bounding hyperrectangle for $\mathcal{P}$ by taking g as center and the values $2\sigma_j\|u_j\|_\infty$ as edge lengths. An alternative (sometimes better) approximation would be to project all points on j^{th} principal direction, and use as corresponding edge length parallel to axis j, the largest distance between the projected points. Note that adapting the edge lengths in this manner makes the algorithm less sensitive to the size of the original hyperrectangle edge length.

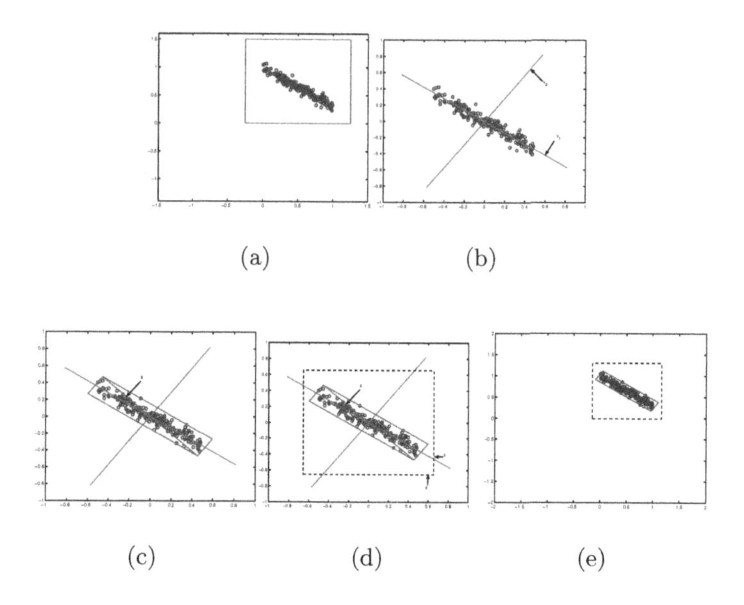

Fig. 3. The initial data and window (a) are shifted and the principal directions are computed (b). We compute the minimum bounding principal directions oriented hyperrectangle (c) and the corresponding hypercube (cf. next subsection 3.1) (d). Finally, we shift the data back (e).

An example of the above process is illustrated in Figure 3. Initially the data points are centered and the principal components of the corresponding matrix are computed. Then, we compute the minimum principal directions oriented hyperrectangle with edge lengths $2\sigma_j\|u_j\|_\infty, j = 1, ..., d$. Finally, we compute a suitable hypercube (we analyze this issue in the next subsection, 3.1) and shift the data back. Then, the enlargement process takes place along each dimension of the rotated window. We call the proposed algorithm *Oriented k-Windows (OkW)* and summarize it below.

algorithm OkW(u, θ_e, θ_m, θ_c, θ_v,k) {

execute W=DetermineInitialWindows(k,u)
for each d dimensional hyperrectangle w_j in W **do**
  execute movement(θ_v,w_j)
  execute SVD-enlargement(θ_e,θ_c,θ_v,w_j)
execute merge(θ_m,θ_s,W)
Output $\{$clusters $c_{l1}, c_{l2}, \ldots$ so that: $c_{l_i} = \{i : i \in w_j, \text{label}(w_j) = l_i\}\}$

function SVD-enlargement(θ_e,θ_c,θ_v, hyperrectangle w) $\{$
  repeat
  compute the principal components V of the points in w
  compute a hyperrectangle with edge length $2\sigma_1 \|u_1\|_\infty, \ldots, 2\|u_d\|_\infty$
  foreach dimension **do**
    repeat
      enlarge w along current dimension by θ_e%
      execute movement(θ_v,w)
      until increase in number of objects
        along current dimension is less than θ_c%
    until increase in number of objects
        is less than θ_c% along every dimension
$\}$

3.1 Computational issues

A challenge in the proposed approach is the efficient handling of non axes-parallel hyperrectangles. Unlike the axes-parallel case, where membership of any d-dimensional point can be tested via the satisfaction of $2d$ inequalities (one per interval range, one per dimension), it appears that oriented hyperrectangles necessitate storing and operating on 2^d vertices. To resolve this issue, we follow a trick from computational geometry designed to handle queries for non-axes parallel line segments, which amounts to embedding each segment in an axes-parallel rectangle [11]. Similarly, in OkW, we embed each hyperrectangle in an axes-parallel "bounding hypercube" that has the same center, say c, as the hyperrectangle, and edge length, s, equal to the maximal diagonal of the hyperrectangle. This edge length guarantees enclosure of the entire oriented hyperrectangle into the axes-parallel hypercube. Thus, the bounding hypercube can be encoded using a single pair (c, s). To perform range search and answer if any given point, x, in the bounding hypercube also lies in the oriented hyperrectangle, it is then enough to check the orthogonal projection of the point onto the d principal directions. If $l_j, j = 1, ..., d$ denote the edge lengths of the hyperrectangle, then x lies in the hyperrectangle if it satisfies all inequalities $|x^\top v_j| \le \frac{l_j}{2}$ for $j = 1, ..., d$. Therefore, we only need to store $O(d)$ elements, namely $(c, s, l_1, ..., l_d)$.

OkW necessitates the application of SVD on every centered data subset in the course of each enlargement step. It is thus critical to select a fast

SVD routine, such as LAPACK [3], for boxes of moderate size without any particular structure, or from [7], for sparse data. In any case, it is important to exploit the structure of the dataset to reduce the cost of these steps. The call to an SVD routine for problems without specific structure carries a cost of $O(n_i d^2)$, which can be non-negligible. In the applications of greater interest for OkW, however, the attribute dimensionality d is typically much smaller than n and even n_i, so that we can consider the asymptotic cost of the SVD to be linear in the number of objects. Furthermore, when d is much smaller than n_i, it is preferable to first compute the QR decomposition of $B^{(i)}$ and then the SVD of the resulting upper triangular factor [14].

4 Experimental Results

UkW and OkW were implemented in C++ under Linux using the gcc 3.4.2 compiler. The SVD was computed using a C++ wrapper[3] around SVD-PACK[4]. All experiments were performed on an AMD Athlon(tm) 64 Processor 3200+, with $1GB$ of RAM. We outline here some results and defer to [27] for more detailed experiments.

Two artificial and two real datasets were used. The former are considered to be difficult, even for density-based algorithms. The first, $Dset_1$, consists of four clusters (three convex and one non-convex) for a total of 2,761 objects, while the second one, $Dset_2$, consists of 299 objects, organized in two clusters forming concentric circles. The values of the parameters $\{\theta_e, \theta_m, \theta_c, \theta_v\}$ were set to $\{0.8, 0.1, 0.2, 0.02\}$ for both UkW and OkW. We used 64 initial windows for $Dset_1$ and 32 for $Dset_2$. Results are depicted in Figure 4. For both datasets, OkW was able to discover the correct clustering. For $Dset_1$, UkW was able to identify correctly only the three convex clusters, while it split the non-convex one into three clusters. Finally, for $Dset_2$, UkW split the outer circle in four clusters. UkW was able to identify correctly the outer cluster only when the number of initial windows was increased to over 200.

The real datasets were $Dset_{UCI}$ and $Dset_{DARPA}$. The former is the Iris dataset from the UCI machine learning repository [8], that consisted of 150 objects of 4 attributes each, organized in three classes, namely Setosa, Versicolour, Virginica. The final real dataset was from the KDD 1999 intrusion detection contest [19]. It contained 100,000 objects of 37 (numerical) attributes, $77,888$ objects corresponded to "normal connections" and $22,112$ corresponding to "Denial of Service" (DoS) attacks. The confusion matrices for $Dset_{UCI}$ were obtained from each one of UkW and OkW for 32 initial windows with the same parameter setting, are depicted in Table 1. Both algorithms successfully identified the number of clusters. However, the number of points that have different class and cluster labels, were 8 for UkW and 7 for OkW. Finally,

[3] http://www.cs.utexas.edu/users/suvrit/work/progs/ssvd.html.
[4] http://www.netlib.org/svdpack/

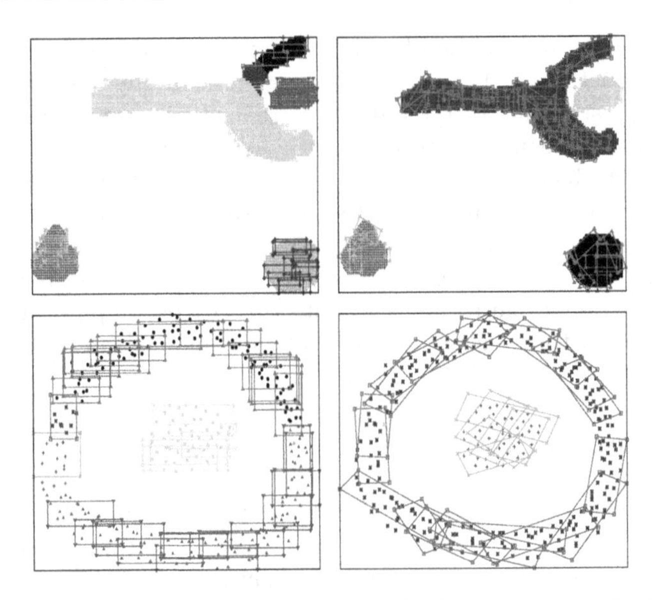

Fig. 4. $Dsets_{1,2}$ with the result of UkW (left) and OkW (right).

the application of UkW on $Dset_{\text{DARPA}}$ with 32 initial windows, estimated the presence of 6 clusters. One of them contained 22,087 DoS objects; the remaining clusters contained objects corresponding to normal connections, with the exception of one cluster that also contained 37 DoS objects. Therefore, UkW's performance was moderately adequate. The application of OkW on the above dataset estimated the presence of five clusters. All, except one cluster contained exclusively either normal or DoS objects. Only 2 DoS objects were assigned in a cluster of 747 normal objects. Therefore, OkW resulted in considerably fewer misclassifications, and thus greater accuracy, than UkW.

Table 1. Confusion data for $Dset_{\text{UCI}}$: The elements in each pair corresponds to the confusion matrices for UkW and OkW.

	Iris class		
Cluster id	Setosa	Versicolour	Virginica
1	50, 50,	0, 0	0, 0
2	0, 0	46, 48	4, 5
3	0, 0	4, 2	46, 45

5 Related work and discussion

The use of PCA in OkW offers a paradigm of synergy between computational linear algebra and geometry that appears to significantly enhance the ability of k-Windows to capture clusters having difficult shapes. OkW follows the recent trend in clustering algorithms to capitalize on any local structure of the data to produce better clustering of datasets of moderate dimensionality; see e.g. [9, 10, 12]. Preliminary results appear to indicate that OkW competes well with algorithms such as these and density based workhorses such as DB-SCAN [13]. We plan to report on these issues in the near future [27] as well as on the challenges posed by OkW, e.g. when it is applied on problems of high dimensionality and the associated costs for large numbers of objects. We close by noting that oriented hyperrectangles have been under study in other contexts, including fast interference detection amongst geometric models and in computing approximations to the set of reachable states in a continuous state space in hybrid systems research [5, 15, 21, 25].

Acknowledgments

We thank the organizers of the "3rd World Conference on Computational Statistics & Data Analysis" and E. Kontoghiorghes in particular, for giving us the opportunity to present early segments of our work during that meeting. We thank the authors of [9] for allowing us to access their implementation of the 4C algorithm. We also thank P. Drineas for a useful discussion early in the course of this work and the reviewers for their careful reading and suggestions. This research was supported in part by a University of Patras "Karatheodori" grant. The second author was also supported by a Bodossaki Foundation graduate fellowship.

References

1. C.C. Aggarwal, A. Hinneburg, and D.A. Keim. On the surprising behavior of distance metrics in high dimensional space. In *Proc. 8th Int'l Conf. Database Theory (ICDT)*, pages 420–434, London, 2001.
2. P. Alevizos, D.K. Tasoulis, and M.N. Vrahatis. Parallelizing the unsupervised k-windows clustering algorithm. In R. Wyrzykowski, editor, *Lecture Notes in Computer Science*, volume 3019, pages 225–232. Springer-Verlag, 2004.
3. E. Anderson et al. *LAPACK Users' Guide.* SIAM, Philadelphia, 1999. 3d ed.
4. M. Ankerst, M. M. Breunig, H.-P. Kriegel, and J. Sander. Optics: Ordering points to identify the clustering structure. In *Proceedings of ACM-SIGMOD International Conference on Management of Data*, 1999.
5. G. Barequet, B. Chazelle, L.J. Guibas, J.S.B. Mitchell, and A. Tal. BOXTREE: A hierarchical representation for surfaces in 3D. *Computer Graphics Forum (CGF)*, 15(3):C387–396, Aug. 1996.
6. P. Berkhin. A survey of clustering data mining techniques. In J. Kogan, C. Nicholas, and M. Teboulle, editors, *Grouping Multidimensional Data: Recent Advances in Clustering*, pages 25–72. Springer, Berlin, 2006.
7. M.W. Berry. Large scale singular value decomposition. *Int. J. Supercomp. Appl.*, 6:13–49, 1992.
8. C.L. Blake and C.J. Merz. UCI repository of machine learning databases, 1998.

9. C. Bohm, K. Kailing, P. Kroger, and A. Zimek. Computing clusters of correlation connected objects. In *ACM SIGMOD international conference on Management of data*, pages 455–466. ACM Press, 2004.
10. K. Chakrabarti and S. Mehrotra. Local dimensionality reduction: A new approach to indexing high dimensional spaces. In *VLDB*, pages 89–100. Morgan Kaufmann Publishers Inc., 2000.
11. M. de Berg, M. Van Kreveld, M. Overmars, and O. Schwarzkopf. *Computational Geometry: Algorithms and Applications.* Springer, 2000.
12. C. Domeniconi, D. Papadopoulos, D. Gunopulos, and S. Ma. Subspace clustering of high dimensional data. In *SIAM Data Mining 2004*, 2004.
13. M. Ester, H.P. Kriegel, J. Sander, and X. Xu. A density-based algorithm for discovering clusters in large spatial databases with noise. In *Proc. 2nd Int'l. Conf. on Knowledge Discovery and Data Mining*, pages 226–231, 1996.
14. G.H. Golub and C.F. Van Loan. *Matrix Computations.* The Johns Hopkins University Press, Baltimore, 3d edition, 1996.
15. S. Gottschalk, M. Lin, and D. Manocha. OBB-Tree: A hierarchical structure for rapid interference detection. In *Proc. of ACM SIGGRAPH*, pages 171–180, 1996.
16. A. Hinneburg, C. Aggarwal, and D. Keim. What is the nearest neighbor in high dimensional spaces? In *The VLDB Journal*, pages 506–515, 2000.
17. A. K. Jain, M. N. Murty, and P. J. Flynn. Data clustering: a review. *ACM Computing Surveys*, 31(3):264–323, 1999.
18. I. T. Jolliffe. *Principal Component Analysis.* New York: Spriger Verlag, 2002.
19. KDD Cup data. http://kdd.ics.uci.edu/databases/kddcup99/kd dcup99.html, 1999.
20. P. Kumar, J.S.B. Mitchell, and E.A. Yildirim. Approximate minimum enclosing balls in high dimensions using core-sets. *J. Exp. Algorithmics*, 8, 2003.
21. J. McNames. Rotated partial distance search for faster vector quantization encoding. *IEEE Signal Processing Letters*, 7(9):244–246, September 2000.
22. F. Preparata and M. Shamos. *Computational Geometry.* Springer Verlag, New York, Berlin, 1985.
23. C.M. Procopiuc, M. Jones, P.K. Agarwal, and T.M. Murali. A Monte Carlo algorithm for fast projective clustering. In *Proc. 2002 ACM SIGMOD*, pages 418–427, New York, NY, USA, 2002. ACM Press.
24. M. Rigou, S. Sirmakessis, and A. Tsakalidis. A computational geometry approach to web personalization. In *IEEE International Conference on E-Commerce Technology (CEC'04)*, pages 377–380, San Diego, California, July 2004.
25. O. Stursberg and B.H. Krogh. Efficient representation and computation of reachable sets for hybrid systems. In O. Maler and A. Pnueli, editors, *HSCC*, volume 2623 of *Lecture Notes in Computer Science*, pages 482–497. Springer, 2003.
26. P.-N. Tan, M. Steinbach, and V. Kumar. *Introduction to Data Mining.* Pearson Addison-Wesley, Boston, 2005.
27. D.K. Tasoulis, D. Zeimpekis, E. Gallopoulos, and M.N. Vrahatis. Manuscript (In preparation), 2006.
28. M. N. Vrahatis, B. Boutsinas, P. Alevizos, and G. Pavlides. The new k-windows algorithm for improving the k-means clustering algorithm. *Journal of Complexity*, 18:375–391, 2002.

A cluster stability criteria based on the two-sample test concept

Z. Volkovich[1], Z. Barzily[2], and L.Morozensky[3]

[1] Software Engineering Department, ORT Braude Academic College, Karmiel 21982, Israel.
Affiliate Professor. Department of Mathematics and Statistics. The University of Maryland, Baltimore County, USA. vlvolkov@ort.org.il

[2] Software Engineering Department, ORT Braude Academic College, Karmiel 21982, Israel. zbarzily@ort.org.il

[3] Software Engineering Department, ORT Braude Academic College, Karmiel 21982, Israel. leonatm@bezeqint.net

1 Abstract

A method for assessing cluster stability is presented in this paper. We hypothesize that if one uses a "consistent" clustering algorithm to partition several independent samples then the clustered samples should be identically distributed. We use the two sample energy test approach for analyzing this hypothesis. Such a test is not very efficient in the clustering problems because outliers in the samples and limitations of the clustering algorithms heavily contribute to the noise level. Thus, we repeat calculating the value of the test statistic many times and an empirical distribution of this statistic is obtained. We choose the value of the "true" number of clusters as the one which yields the most concentrated distribution. Results of the numerical experiments are reported.

2 Introduction

Clustering methods are widely used in many different areas as a practical tool to understand structure in complex data. A crucial issue in clustering is to determine how many clusters are presented in the data. This problem has been recognized as one of the most difficult problems in cluster analysis.

Most of the iterative clustering methods consist of two procedures: the first procedure determines, for a given number of clusters, an optimal partition. The second procedure tests the validity of the partition This testing procedure typically employs an intuitive rule for the goodness of the partition. In majority of the cases it is based on one of the following tools :

Mark Last et al. (Eds.): Advances in Web Intelligence and Data Mining (SCI) **23**, 329-338 (2006)
www.springerlink.com

- Indexes, based on multivariate statistics, which compare dispersions within and between the clusters (see, [3], [14], [18], [24]).
- A stability (similarity, merit) function that measures the consistency of labels assignments to samples elements (see, [21], [2], [19]).
- A measure assessing the likelihood of a solution (see, [23], [13]).

We discuss in this paper a statistical approach to the cluster stability problem based on multivariate two-sample test concept. We consider as a model a Euclidian space $\mathcal{X}$ with a partition $\mathcal{C} = \{c_j\}$, $j = 1, ..., k$ and a set of associations

$$v(x, c_j) = \begin{cases} 1 \text{ if } x \in c_j \\ 0 \text{ otherwise} \end{cases} .$$

I.e. the underlying distribution $\mu_{\mathcal{X}}$ of $\mathcal{X}$ is

$$\mu_{\mathcal{X}} = \sum_{j=1}^{k} p_{c_j} \mu_{c_j}, \tag{1.1}$$

where p_j, $j = 1, .., k$ are the cluster probabilities and μ_{C_j}, $j = 1, .., k$ are the inner clusters distributions. This mixture model leads to the following joint distribution of $x \in \mathcal{X}$ and its cluster label

$$P(x, c_j) = v(x, c_j) p_{c_j} \mu_{c_j} . \tag{1.2}$$

Our approach assumes the existence of such a distribution $P(x, c_j)$ in the case where the number of clusters k is chosen correctly and the absence of such a distribution on $\mathcal{X} \times \mathcal{C}_k$ otherwise. Let us suppose we sequentially draw N samples $\mathcal{S}_1, \mathcal{S}_2, ..., \mathcal{S}_N$ from $\mathcal{X}$ for each value of $k = 2, 3, ..., k^*$. The parameter k^* is some predefined number offering a maximal considered number of clusters. Let a clustering algorithm Cl be available. In the ideal case the algorithm delivers the true partition if the value of k is correct. I.e. the clustered samples $(\mathcal{S}_1, \mathcal{C}_1), (\mathcal{S}_2, \mathcal{C}_2), ..., (\mathcal{S}_N, \mathcal{C}_N)$, generated by the algorithm, are selected from a population distributed according to P.

The substance we are dealing with, actually belongs to the subject of the hypothesis testing. Here, no prior knowledge of the distribution is available thus; this test has to be a distribution-free two sample test. Outliers in the sample and limitations of clustering algorithms contribute heavily to the noise level of the test. To overcome this difficulty we have to get our conclusion based on a sufficiently large amount of the sampled information. This is acquired by bootstrapping and clustering of the samples. Under the null hypothesis: the existance of the measure P, these sets are drawn from the same distribution. Furthermore, we would expect to obtain the least "random" distribution under the "true" number of clusters. Due to the noise level it is unlikely to get a rejection of the null hypothesis in a direct test. However, it is natural to assume that the most concentrated distribution is achieved under the true number of the clusters. From this point of view, our procedure can be roughly described as the generation of the empirical distribution with a consequent concentration test. Note that the outliers detection is most essential in this purpose.

3 Mixture clustering model

In this section we state several facts about the mixture clustering model from the information theory the point of view (see, [20], [22]). The main principle of the model is that each item in the space $\mathcal{X}$ belongs with certain probability to each cluster. Here a clustering solution is given by the set of probability distributions for associating points to clusters. This association is termed "fuzzy membership in clusters". A partial case, where each point belongs to one cluster, corresponds to the hard clustering situation. Determining an optimal association distribution is the goal of probabilistic clustering approach. From the information theory point of view, clustering is the basic strategy for the data lost compression. According to this approach the data is divided to groups which are described, in the most efficient way, in the terms of the bit rate employing a representative to each group. A clustering procedure is intended to compress the initial data throwing away the insignificant information. A distinction between the relevant and unsuitable information in the data is provided with the help of a distortion function, which usually measures a similarity among the data items. Let us suppose that a distortion function in each cluster is $d_j(x, y)$, $j = 1, ..., k$. A lossy compression can be made by assigning the data to clusters so that the mutual information

$$I(\mathcal{C}, \mathcal{X}) = \sum_{x,j} P(c_j|x)P(x)\log_2 \frac{P(c_j|x)}{P(c_j)}$$

is minimized. The minimization is constrained by fixing the expected distortion

$$\overline{d}(\mathcal{X}, \mathcal{C}) = \sum_{x,j} P(c_j|x)P(x)d_j(x, x_{c_j})$$

where x_{c_j} is the representative of the cluster c_j. The formal solution of this task is provided by a Boltzmann distribution

$$P(c_j|x) = \frac{P(c_j)}{Z(x,T)} \exp\left(-\frac{d_j(x, x_{c_j})}{T}\right),$$

where

$$Z(x,T) = \sum_{j} P(c_j) \exp\left(-\frac{d_j(x, x_{c_j})}{T}\right)$$

is the normalization constant and T is the Lagrange multiplier. If we suppose that in the case of the hard clustering each cluster is specified by the parameter vectors $Y = \{y_j\}$, $j = 1, ..., k$ then the marginal distribution of Y becomes

$$P(Y) = \frac{e^{-\frac{F}{T}}}{\sum_{Y} e^{-\frac{F}{T}}}$$

where

$$F = -T \sum_{x} \ln \left(\sum_{j} \exp \left(-\frac{d_j(x, y_j)}{T} \right) \right)$$

is the so called "the free energy" of the partition. It is evident from this equation that the most probable values of Y minimize F. Therefore, optimal estimations of the cluster parameters can be indeed achieved by means of minimizing the free energy. An important example of this construction is a partition corresponding to the distortions

$$d_j(x, y_j) = ((x - y_j)\Sigma_j^{-1}(x - y_j)),$$

where y_j is the centroid of the cluster c_j and Σ_j is the covariance matrix of the cluster. In this case the expression for F can be rewritten as

$$F = -T \sum_{x} \ln \left(\sum_{j} \exp \left(-\frac{(x - y_j)\Sigma_j^{-1}(x - y_j)}{T} \right) \right).$$

If we fix the matrices Σ_j then we obtain from the equations

$$\frac{\partial F}{\partial y_j} = 0, \; j = 1, ..., k,$$

the result

$$y_j = \frac{\sum_{x} x P(c_j|x)}{\sum_{x} P(c_j|x)}, \; j = 1, ..., k.$$

We would like to point out the similarity between this approach and the maximum-likelihood estimation of normal mixtures parameters. This model arises in clustering problems in which the underlying distribution is of the form

$$f(x) = \sum_{j=1}^{k} p_j G(x|y_j, \Sigma_j),$$

where $G(x|y, \Sigma)$ is the Gaussian density with the mean y and the covariance matrix Σ. Estimates of the model parameters are provided by means of the well known EM algorithm. In the special spherical case of $\Sigma_j = \sigma^2 I, \; j = 1, ..., k$, where I is the identity matrix and σ^2 is the unknown dispersion of each cluster, the standard k-means algorithm can be shown to be a version of the EM algorithm (see, for example [10], [4]). The essential distinction between the mentioned approaches is that in the free energy optimization approach no prior information on the data distribution is assumed. The distributions are directly derived from the corresponding Bolzman and Gibbs distributions and the suitable optimization task.

4 A two sample test

Let $\mathcal{X}$: $X_1, X_2, ..., X_n$ and $\mathcal{Y}$: $Y_1, Y_2, ..., Y_m$ be two samples of independent random vectors having the probability laws F and G, respectively. The classical two-sample problem consists of testing the hypothesis

$$H_0 : F(x) = G(x)$$

against the general alternative

$$H_1 : F(x) \neq G(x),$$

when the distributions F and G are unknown. Such a problem occurs in many areas of research. The Kolmogorov-Smirnov test, the Cram‘er-von Mises test, the Friedman's nonparametric ANOVA test and the Wald-Wolfowitz test must be reminded as the classical univariate procedures for this purpose. Many multivariate tests can be found in the literature (see, for example, [9], [5], [12]).

We use, in the current paper, the two-sample energy test approach which is described in [26]. This test can be relatively easily implemented. Note, that this approach is very similar to the one proposed in [27] and [15] resting upon the conditional positive definite kernels. Applications of this approach are discussed in [16] and [1]. The statistic Φ_{nm} of the test contains three parts correspond to the energies of the samples $\mathcal{X}$, $\mathcal{Y}$ and the their interaction energy.

$$\Phi_{nm} = \frac{1}{n^2}\sum_{i<j}^{n} R\left(|x_i - x_j|\right) + \frac{1}{m^2}\sum_{i<j}^{m} R\left(|y_i - y_j|\right) - \tag{1.3}$$

$$-\frac{1}{nm}\sum_{i=1}^{n}\sum_{j=1}^{m} R\left(|x_i - y_j|\right)$$

where $R(r)$ is a continuous, monotonic decreasing function of the Euclidean distance r. The function $R(r) = -\ln(r)$, proposed by the authors in [26], provides a scale invariant test having a good rejection power against many alternatives. In order to avoid situations in which a single projection dominates the value of the distance the sample values must be standardized by $z_{ik} = (z_{ik} - \mu_k)/\sigma_k$, where μ_k, σ_k are the mean value and the standard deviation of the projections.

We are applying this methodology to distinguish between distributions presented in (1.2). Let us introduce the function

$$\widetilde{R}\left(x_1, x_2, c_1, c_2\right) = R\left(|x_1 - x_2|\right)\chi(c_1 = c_2)$$

where $\chi(c_1 = c_2)$ is an indicator function of the event $c_1 = c_2$:

$$\chi(c_1 = c_2) = \begin{cases} 1 \text{ if } c_1 = c_2 \text{ is true} \\ 0 \text{ otherwise} \end{cases}.$$

For two measures μ and ν satisfying (1.2) we define:

$$L(\mu,\nu) = \sum_{j_1,j_2} \iint R(|x_1 - x_2|)\,\chi(c_1 = c_2) p_{j_1}(\mu)\mu_{c_{j_1}}(dx_1) p_{j_2}(\nu)\nu_{c_{j_2}}(dx_2) =$$
$$= \sum_{j=1}^{k} p_j(\mu)p_j(\nu) \iint R(|x_1 - x_2|)\,\mu_{c_j}(dx_1)\nu_{c_j}(dx_2)$$

and

$$Dis(\mu,\nu) = L(\mu,\mu) + L(\nu,\nu) - 2L(\mu,\nu).$$

Proposition 1. *Let μ and ν be two measures satisfying (1.2) such that $P_\mu(c|x) = P_\nu(c|x)$, then*

- $Dis(\mu,\nu) \geq 0$;
- $Dis(\mu,\nu) = 0$ *if and only if* $\mu = \nu$.

Proof. Let us define α and β to be two distributions and define

$$W(\alpha,\beta) = \iint R(|x_1 - x_2|)\,\alpha(dx_1)\beta(dx_2).$$

It was proved in the appendix of [26], for every α and β, that

$$W^*(\alpha,\beta) = (R(\alpha,\alpha) + R(\beta,\beta) - 2R(\alpha,\beta)) \geq 0,$$

and $W^*(\alpha,\beta) = 0$ if and only if $\alpha = \beta$. It is easy to see that

$$Dis(\mu,\nu) = \sum_{j=1}^{k} p_j^2(\mu) W^*(\mu_{c_j},\nu_{c_j}),$$

thus, the proposition is proved.

The distance $Dis(\mu,\nu)$ can be considered as a direct extension of the distance W^* between measures on $\mathcal{X}$ to multi-measures μ and ν defined on $\mathcal{X} \times \mathcal{C}$. Indeed, we are going to compare the marginal distributions μ_{c_j} and ν_{c_j}.

5 Methodology description and experiments

5.1 Methodology description

In this chapter we present our approach to the cluster stability problem. Recall, we consider a subset $\mathcal{X}$ of a metric space and suppose that a hard clustering algorithm Cl is available. The algorithm has the input parameters: a clustered sample $\mathcal{S}$ and a predefined number of clusters k. The output is a clustered sample $(\mathcal{S}, \mathcal{C}_k)$, where $\mathcal{C}_k$ is the vector of the cluster labels of the sample $\mathcal{S}$. We introduce following parameters:

- N_S - number of the samples;
- M - size of sample;
- k^*- maximal number of clusters to be tested;

For two given samples $\mathcal{S}_1$ and $\mathcal{S}_2$ we consider a clustered sample $(\mathcal{S}_1 \cup \mathcal{S}_2, \mathcal{C}_k)$ .We denote by $c(x)$ the mapping, induced by the clustering, from this sample to $\mathcal{C}_k$. Let us introduce

$$L(\mathcal{S}_1, \mathcal{S}_2) = \sum_{j=1}^{k} \frac{1}{|C_j|} \sum_{x_{j_1} \in \mathcal{S}_1} \sum_{x_{j_2} \in \mathcal{S}_2} \widetilde{R}(x_{j_1}, x_{j_2}, c(x_{j_1}) = j, c(x_{j_2}) = j) \chi(x_{j_1} \neq x_{j_2}),$$

where

$$|C_j| = \sum_{x_{j_1} \in \mathcal{S}_1} \sum_{x_{j_2} \in \mathcal{S}_2} \chi(c(x_{j_1}) = j) \chi(c(x_{j_2}) = j) \chi(x_{j_1} \neq x_{j_2}).$$

The algorithm consists of the following steps

1. for $k = 1$ to k^*
2. for $n = 1$ to N_S
3. $\mathcal{S}_n^{(1)} = sample(\mathcal{X}, M)$
4. $\mathcal{S}_n^{(2)} = sample(\mathcal{X} \setminus \mathcal{S}_n^{(1)}, M)$
5. Calculate

$$\mathcal{Q}_n = L(\mathcal{S}_n^{(1)}, \mathcal{S}_n^{(1)}) + L(\mathcal{S}_n^{(2)}, \mathcal{S}_n^{(2)}) - 2L(\mathcal{S}_n^{(1)}, \mathcal{S}_n^{(2)})$$

6. Standardize of the values of $\{\mathcal{Q}_n\}$, $n = 1, ..., N_S$

$$\widetilde{\mathcal{Q}}_n = (\mathcal{Q}_n - mean(\mathcal{Q}_n))/std(\mathcal{Q}_n).$$

7. Calculate index I_k of the sample $\{\widetilde{\mathcal{Q}}_n\}$, $n = 1, ..., N_S$.
8. Optimal value of k is chosen according to the appropriate extremum point of the index.

The procedure $sample(\mathcal{X}, M)$ means a selection of a sample without replacement of size M from the set, where each observation is equally likely to be chosen. $mean(\mathcal{N}_n)$ and $std(\mathcal{N}_n)$ are the routines intended to calculate the arithmetic mean and standard deviation of $\mathcal{N}_n$. The quality of an obtained partitions is evaluated (step 7 of the algorithm) by three concentration statistics: the Friedman's Index (see [11]), the famous Kullback-Leibler distance (see, for example [6]) from the normal distribution and the Kurtosis. The parameter selection for the clustering sample procedures is remained as an open problem. Usually, it is recommended to choose the parameters as large as possible, for instance, to split the data (see, for example [21]). We try for our purpose several parameters' sets.

5.2 Numerical experiments

In order to evaluate the performance of the described methodology we performed several numerical experiments on a public dataset. The samples obtained (steps 3 and 4 of the algorithm) are clustered by applying the k-means algorithm such that initial centroids are chosen randomly for each of the clustered samples and the initial centroids of the united sample are calculated as the weighted mean value of the final sample centroids. We ignore samples having empty clusters. The obtained clustering results are compared to the "true" partition. This dataset is available in http://www.dcs.gla.ac.uk/idom/ir_resources/test_collections/) and it consists of

- DC0–Medlars Collection (1033 medical abstracts).
- DC1–CISI Collection (1460 information science abstracts).
- DC2–Cranfield Collection (1400 aerodynamics abstracts).

This data set was analyzed in [8] by means of the spherical k–means algorithm and has been also considered in several ensuing works (see, for example, [17] and [25]). Following to the well known approach " bag of words" 300 and 600 "best" terms (see [7] for term selection details) were selected. This dataset is known to be well separated by means of the two leading components. In our experiments we choose, in (1.3), $R(r) = -ln(r)$ and $k^* = 7$. Obviously, a better partition is predictable in the case of the 600 essential terms. This fact is demonstrated in table 1, where the situation of three clusters is clearly designated by all indexes. In the worse separated representation by means of 300 terms, under a similar set of the procedure parameters, only the kurtosis index detects the true number of clusters (see, table 2). However, an appropriate parameters selection yields the correct value (see, table 3). Meanwhile, the parameters' choice problem remains unsolved here and is an issue for a future research. Moreover, we would like to note that the Friedman index appears to be in our situation the most responsive.

	2	3	4	5	6	7
Friedman index	0.0294	0.0081	0.0221	0.0351	0.0414	0.0467
KL-distance	0.0813	0.0409	0.0575	0.0707	0.0962	0.0904
Kurtosis	3.0139	2.5497	3.0097	3.4822	3.7487	4.0102

Table 1. N_S =1000, M=1000, the number of terms is 600.

	2	3	4	5	6	7
Friedman index	0.0130	0.0234	0.0190	0.0104	0.0097	0.0078
KL-distance	0.0380	0.0626	0.0513	0.0453	0.0404	0.0582
Kurtosis	3.1374	2.9766	3.6760	3.6368	3.5848	4.0647

Table 2. N_S =1500, M=1000, the number of terms is 300.

	2	3	4	5	6	7
Friedman index	0.0116	0.0076	0.0199	0.0214	0.0189	0.0198
KL-distance	0.0343	0.0282	0.0515	0.0524	0.1579	0.0534
Kurtosis	2.6355	2.5544	2.8118	3.3735	3.4828	3.9053

Table 3. N_S =1500, M=1500, the number of terms is 300.

6 Summary

In this paper we have determined a method for assessing cluster stability and demonstrated its performance. From the results of the numerical experiments one can deduce that the algorithm performs fairly well. In future work we intend to examine the performance of the method under other criteria and kernels so that a better performing method should be obtained.

References

1. Ya. Belopolskaya, L. Klebanov, and V. Volkovich. Characterization of elliptic distributions. *Journal of Mathematical Sciences*, 127(1):1682–1686, 2005.
2. A. Ben-Hur, A. Elisseeff, and I. Guyon. A stability based method for discovering structure in clustered data. In *Pacific Symposium on Biocomputing*, pages 6–17, 2002.
3. R. Calinski and J. Harabasz. A dendrite method for cluster analysis. *Commun Statistics*, 3:1–27, 1974.
4. G. Celeux and G. Govaert. A classification EM algorithm for clustering and two stochastic versions. *Computational Statistics and Data Analysis, 14:315*, 1992.
5. W. J. Conover, M. E. Johnson, and M. M. Johnson. Comparative study of tests of homogeneity of variances, with applications to the outer continental shelf bidding data. *Technometrics*, 23:351–361, 1981.
6. T. M. Cover and J.A. Thomas. *Elements of Information Theory.* New York: Wiley, 1991.
7. I. Dhillon, J. Kogan, and Ch. Nicholas. Feature selection and document clustering. In M. Berry, editor, *A Comprehensive Survey of Text Mining*, pages 73–100. Springer, Berlin Heildelberg New York, 2003.
8. I. S. Dhillon and D. S. Modha. Concept decompositions for large sparse text data using clustering. *Machine Learning*, 42(1):143–175, January 2001. Also appears as IBM Research Report RJ 10147, July 1999.
9. B. S. Duran. A survey of nonparametric tests for scale. *Communications in statistics - Theory and Methods*, 5:1287–1312, 1976.

10. C. Fraley and A.E. Raftery. How many clusters? Which clustering method? Answers via model-based cluster analysis. *The Computer Journal*, 41(8):578–588, 1998.
11. J. H. Friedman. Exploratory projection pursuit. *J. of the American Statistical Association*, 82(397):249–266, 1987.
12. J. H. Friedman and L. C. Rafsky. Multivariate generalizations of the Wolfowitz and Smirnov two-sample tests. *Annals of Statistics*, 7:697–717, 1979.
13. A. K. Jain and J. V. Moreau. Bootstrap technique in cluster analysis. *Pattern Recognition*, 20(5):547– 568, 1987.
14. J.Hartigan. Statistical theory in clustering. *J Classification*, 2:6376, 1985.
15. L. Klebanov. One class of distribution free multivariate tests. *SPb. Math. Society, Preprint*, 2003(03), 2003.
16. L. Klebanov, T. Kozubowskii, S.Rachev, and V.Volkovich. Characterization of distributions symmetric with respect to a group of transformations and testing of corresponding statistical hypothesis. *Statistics and Probability Letters*, 53:241–247, 2001.
17. J. Kogan, C. Nicholas, and V. Volkovich. Text mining with information–theoretical clustering. *Computing in Science and Engineering*, pages 52–59, November/December 2003.
18. W. Krzanowski and Y. Lai. A criterion for determining the number of groups in a dataset using sum of squares clustering. *Biometrics*, 44:2334, 1985.
19. E. Levine and E. Domany. Resampling method for unsupervised estimation of cluster validity. *Neural Computation*, 13:2573–2593, 2001.
20. K. Rose, E. Gurewitz, and G. Fox. Statistical mechanics and phase transitions in clustering. *Physical Review Letters*, 65(8):945–848, 1990.
21. V. Roth, V. Lange, M. Braun, and Buhmann J. Stability-based validation of clustering solutions. *Neural Computation*, 16(6):1299 – 1323, 2004.
22. S. Still and W. Bialek. How many clusters? An information-theoretic perspective. *Neural computation*, 16(12):2483 – 2506, December 2004.
23. C. Sugar and G. James. Finding the number of clusters in a data set : An information theoretic approach. *J of the American Statistical Association*, 98:750–763, 2003.
24. R. Tibshirani, G. Walther, and T. Hastie. Estimating the number of clusters via the gap statistic. *J. Royal Statist. Soc. B*, 63(2):411423, 2001.
25. V. Volkovich, J. Kogan, and C. Nicholas. k–means initialization by sampling large datasets. In I. Dhillon and J. Kogan, editors, *Proceedings of the Workshop on Clustering High Dimensional Data and its Applications (held in conjunction with SDM 2004)*, pages 17–22, 2004.
26. G. Zech and B. Aslan. New test for the multivariate two-sample problem based on the concept of minimum energy. *The Journal of Statistical Computation and Simulation*, 75(2):109 – 119, february 2005.
27. A.A Zinger, A.V. Kakosyan, and L.B Klebanov. Characterization of distributions by means of the mean values of statistics in connection with some probability metrics. In *Stability Problems for Stochastic Models, VNIISI*, pages 47–55, 1989.